Waves and Ocean Structures

Waves and Ocean Structures

Editors

Jaw-Fang Lee
Ray-Yeng Yang

MDPI • Basel • Beijing • Wuhan • Barcelona • Belgrade • Manchester • Tokyo • Cluj • Tianjin

Editors
Jaw-Fang Lee
National Cheng Kung University
Taiwan

Ray-Yeng Yang
National Cheng Kung University
Taiwan

Editorial Office
MDPI
St. Alban-Anlage 66
4052 Basel, Switzerland

This is a reprint of articles from the Special Issue published online in the open access journal *Journal of Marine Science and Engineering* (ISSN 2077-1312) (available at: https://www.mdpi.com/journal/jmse/special_issues/zara_waves_ocean_structures).

For citation purposes, cite each article independently as indicated on the article page online and as indicated below:

LastName, A.A.; LastName, B.B.; LastName, C.C. Article Title. *Journal Name* **Year**, *Volume Number*, Page Range.

ISBN 978-3-0365-2361-3 (Hbk)
ISBN 978-3-0365-2362-0 (PDF)

Contents

About the Editors

Jaw-Fang Lee is currently a research professor at Tainan Hydraulics Laboratory, National Cheng Kung University. He retired as a Professor from the Department of Hydraulic and Ocean Engineering, NCKU, 2020. He published four books (in Chinese): Fundamental Coastal Hydraulics, Finite Element Method, Boundary Element Method, and Wave Hydrodynamics of Ocean Structures. He has devoted to promoting the academic field of Coastal and Ocean Engineering in Taiwan.

Ray-Yeng Yang got his Ph.D. degree from Institute of Naval Architecture and Ocean Engineering, National Taiwan University, Taiwan, in 1995. Right now he is an Associate Professor of Department of Hydraulic and Ocean Engineering, National Cheng Kung University (NCKU), Taiwan. He was research fellow and the former deputy director of Tainan Hydraulics Laboratory, NCKU (2015–2017). Now, he is also the executive director of Taiwan Society of Ocean Engineering and deputy secretary-general of Chinese Ocean & Underwater Technology Association. He has his expertise in ocean (coastal) engineering, ocean energy (wave & ocean current energy), ocean environment science (internal wave & Langmuir circulation), hydrodynamic stability (double diffusion, salt-finger convection), floating offshore structure (offshore wind turbine & marine aquaculture cage net) and physical modeling simulation. Until now, more than 250 (55 SCI, 40 EI papers & 160 conference) papers, two book chapters and 92 technical research reports have been published by Dr. Yang. For the latest years, his academic work has been published in international renowned, scholarly journals including "Journal of Fluid Mechanics", "Journal of Mechanics", "Geophysical Research Letter", "IJOPE", "European Journal of Fluid Mechanics-B/Fluid", "Acta Astronautica", "Ocean and Coastal Management", "The Open Ocean Engineering Journal", "Journal of Ocean and Coastal Engineering", "Physics of Wave Phenomena" , "Pure and Applied Geophysics" "Pure and Applied Geophysics", "IEEE-Ocean", "Journal Marine Science and Engineering", and "Ocean Engineering" etc.

Preface to "Waves and Ocean Structures"

It is interesting to know that the cease of the Journal of Coastal and Ocean Engineering in Taiwan has caused the inauguration of this special issue. The guest editor thought this special issue might be a good alternative for interested authors to submit their masterpieces. Nevertheless, the guest-editors have their particular major research fields, and the rise of the topic "Ocean Waves and Structures." This is the first attempt of the guest-editors, and they also welcome related subjects to be submitted to this special issue. The guest-editors appreciate very much all contributors to this special issue. This special issue contains articles related to the main subjects, namely, ocean deployments that subjected to actions of ocean waves, and require research investigations studying mechanisms of wave actions. We hope the related research results will also continue to submit to this special issue. Thanks are also given to all assistant editors, without their help it would be impossible for the special issue to run in order and in effective ways to be published. Further, because the number of articles in this special issue has reached the milestone, and the special issue can be published as a book. This is also a prize for all authors and the guest-editors: Jaw-Fang Lee and Ray-Yeng Yang. The guest-editors also like to mention that this special issue will be extended to the next year with "Ocean and Waves and Structures II" as the topic. We hope all of us could work hard consistently to have a second continuous success.

Jaw-Fang Lee, Ray-Yeng Yang
Editors

Journal of
Marine Science and Engineering

MDPI

Editorial

Waves and Ocean Structures

Jaw-Fang Lee * and Ray-Yeng Yang

Department of Hydraulics and Ocean Engineering, National Cheng Kung University, Tainan City 701, Taiwan; ryyang@mail.ncku.edu.tw
* Correspondence: jflee@mail.ncku.edu.tw

check for updates

Citation: Lee, J.-F.; Yang, R.-Y. Waves and Ocean Structures. *J. Mar. Sci. Eng.* **2021**, *9*, 305. https://doi.org/10.3390/jmse9030305

Received: 26 February 2021
Accepted: 3 March 2021
Published: 9 March 2021

This Special Issue concentrates on the problems of interactions between water waves and ocean structures. The methodologies used to solve the problems are either theoretical, numerical, or experimental approaches. New findings are particularly welcome. Coupling effects caused from either side of water waves or ocean structures are the main issue of the problems. Various types of ocean structures can be considered, such as fixed, movable, porous, flexible, free-floating, or with moorings.

The start of this Special Issue was right at the time when the Journal of Coastal and Ocean Engineering belonging to the Taiwan Society of Ocean Engineering ended. The guest co-editors Professor R-Y Yang and I believed that promising articles could find this Special Issue to submit for publishing. After all the paper collecting days and giving efforts, we are very happy to accept the results. The guest editors are particularly grateful for administration help and contribution from the assistant editor Ms. Zara Liu and others. With the conclusion of this year, we very much appreciate contributions from all authors. We will continue this Special Issue with extension II for the year 2021. Further, Professor Yang has the intention to promote another Special Issue; therefore, the co-guest editor will be replaced by Professor Ruey-Shan Shih. Welcome, Professor Shih. Together, we expect to accomplish another issue this year.

We encourage university scholars or industrial personnel with good research results to continue support and contribute to this Special Issue.

Various ocean structures are included in this issue, and coupling effects between waves and structures are emphasized. Lee et al. [1] presented a concept of a water column type wave power converter. Li et al. [2] considered submerged breakwaters. Lin et al. [3] studied an ocean current turbine system. Thiagarajan and Moreno [4] investigated oscillating heave plates in wind turbines. Chiang et al. [5] proposed an actuator disk model. Tseng et al. [6] investigated Bragg reflections of periodic surface-piercing submerged breakwaters. Lee et al. [7] analyzed caisson structures with a wave power conversion system installed. Yeh et al. [8] reported motion reduction in offshore wind turbines. Wu and Hsiao [9] considered submerged slotted barriers. Tang et al. [10] studied floating platforms with fishnets. Chen et al. [11] calculated mooring drags of underwater floating structures with moorings. Jeong et al. [12] estimated the motion performance of light buoys using ecofriendly and lightweight materials. Zhang et al. [13] considered vibrations of deep-sea risers. On the other hand, Shugan et al. [14] studied the effects of plastic coating on sea surfaces.

Author Contributions: Both authors contributed equally to this manuscript. All authors have read and agreed to the published version of the manuscript.

Funding: This research received no external funding.

Acknowledgments: We want to express our sincere thanks to all the authors and the reviewers.

Conflicts of Interest: The authors declare no conflict of interest.

References

1. Lee, H.H.; Chen, G.-F.; Hsieh, H.-Y. Study on an Oscillating Water Column Wave Power Converter Installed in an Offshore Jacket Foundation for Wind-Turbine System Part I: Open Sea Wave Energy Converting Efficiency. *J. Mar. Sci. Eng.* **2021**, *9*, 133. [CrossRef]
2. Li, C.-Y.; Shih, R.-S.; Weng, W.-K. Visualization Investigation of Energy Dissipation Induced by Eddy Currents for a Solitary-Like Wave Passing over Submerged Breakwater Sets. *J. Mar. Sci. Eng.* **2020**, *8*, 834. [CrossRef]
3. Lin, S.-M.; Chen, Y.-Y.; Hsu, H.-C.; Li, M.-S. Dynamic Stability of an Ocean Current Turbine System. *J. Mar. Sci. Eng.* **2020**, *8*, 687. [CrossRef]
4. Thiagarajan, K.; Moreno, J. Wave Induced Effects on the Hydrodynamic Coefficients of an Oscillating Heave Plate in Offshore Wind Turbines. *J. Mar. Sci. Eng.* **2020**, *8*, 622. [CrossRef]
5. Chiang, Y.-C.; Hsu, Y.-C.; Chau, S.-W. Power Prediction of Wind Farms via a Simplified Actuator Disk Model. *J. Mar. Sci. Eng.* **2020**, *8*, 610. [CrossRef]
6. Tseng, I.-F.; You, C.-S.; Tsai, C.-C. Bragg Reflections of Oblique Water Waves by Periodic Surface-Piercing and Submerged Breakwaters. *J. Mar. Sci. Eng.* **2020**, *8*, 522. [CrossRef]
7. Lee, H.H.; Wu, T.-Y.; Lin, C.-Y.; Chiu, Y.-F. Structural Safety Analysis for an Oscillating Water Column Wave Power Conversion System Installed in Caisson Structure. *J. Mar. Sci. Eng.* **2020**, *8*, 506. [CrossRef]
8. Yeh, P.-H.; Chung, S.-H.; Chen, B.-F. Multiple TLDs on Motion Reduction Control of the Offshore Wind Turbines. *J. Mar. Sci. Eng.* **2020**, *8*, 470. [CrossRef]
9. Wu, Y.-T.; Hsiao, S.-C. Propagation of Solitary Waves over a Submerged Slotted Barrier. *J. Mar. Sci. Eng.* **2020**, *8*, 419. [CrossRef]
10. Tang, H.-J.; Huang, C.-C.; Yang, R.-Y. Numerical Study of the Influence of Fishnet Mesh Size on a Floating Platform. *J. Mar. Sci. Eng.* **2020**, *8*, 343. [CrossRef]
11. Chen, C.-T.; Lee, J.-F.; Lo, C.-H. Mooring Drag Effects in Interaction Problems of Waves and Moored Underwater Floating Structures. *J. Mar. Sci. Eng.* **2020**, *8*, 146. [CrossRef]
12. Jeong, S.-M.; Son, B.-H.; Lee, C.-Y. Estimation of the Motion Performance of a Light Buoy Adopting Ecofriendly and Lightweight Materials in Waves. *J. Mar. Sci. Eng.* **2020**, *8*, 139. [CrossRef]
13. Zhang, J.; Guo, H.; Tang, Y.; Li, Y. Effect of Top Tension on Vortex-Induced Vibration of Deep-Sea Risers. *J. Mar. Sci. Eng.* **2020**, *8*, 121. [CrossRef]
14. Shugan, I.; Yang, R.-Y.; Chen, Y.-Y. An Experimental and Theoretical Study of Wave Damping due to the Elastic Coating of the Sea Surface. *J. Mar. Sci. Eng.* **2020**, *8*, 571. [CrossRef]

Journal of
Marine Science and Engineering

MDPI

Article

Study on an Oscillating Water Column Wave Power Converter Installed in an Offshore Jacket Foundation for Wind-Turbine System Part I: Open Sea Wave Energy Converting Efficiency

Hsien Hua Lee *, Guan-Fu Chen and Hsiang-Yu Hsieh

Department of Marine Environment and Engineering, National Sun Yat-sen University, Kaohsiung 804, Taiwan; good821210@gmail.com (G.-F.C.); un7221429@gmail.com (H.-Y.H.)
* Correspondence: hhlee@mail.nsysu.edu.tw

Abstract: This study is focused on the wave energy converter of an oscillating water column (OWC) system that is integrated with a jacket type infrastructure applied for an offshore wind turbine system. In this way, electricity generation by both wind power and wave power can be conducted simultaneously to maximize the utilization of sustainable energy. A numerical analysis was performed in this research to model and simulate the airflow response and evaluate the converting efficiency of wave energy from an OWC system integrated with an offshore template structural system. The performance of the system including the generating airflow velocity, air-pressure in the chamber, generating power and then the converting efficiency of power from waves are all analyzed and discussed in terms of the variations of the OWC system's geometrical parameters. The parameters under consideration include the exhale orifice-area of airflow, gate-openings of inflow water and the submerged chamber depth. It is found that from the analytical results the performance of the OWC wave energy converter is influenced by the dimensional parameters along with the design conditions of the local environment. After a careful design based on the in-situ conditions including water depth and wave parameters, an open OWC system can be successfully applied to the template structure of offshore wind power infrastructure as a secondary generating system for the multi-purpose utilization of the structure.

Keywords: offshore wind power; template structure system; oscillating water column; wave power converting system

check for updates

Citation: Lee, H.H.; Chen, G.-F.; Hsieh, H.-Y. Study on an Oscillating Water Column Wave Power Converter Installed in an Offshore Jacket Foundation for Wind-Turbine System Part I: Open Sea Wave Energy Converting Efficiency. *J. Mar. Sci. Eng.* **2021**, *9*, 133. https://doi.org/10.3390/jmse9020133

Received: 3 January 2021
Accepted: 25 January 2021
Published: 28 January 2021

1. Introduction

The intensive development in the high technology industry and the over- consumption of products such as clothes, shoes and electronic devices demands more power supply for industries than ever. Besides the requirements from the industries, the supply of power required for a more comfortable civil life has also increased massively, mainly in newly developed countries. Unfortunately, most electricity power is generated from combustion power plants and the large amounts of carbon dioxide discharged into the air inevitably cause many environmental problems. Those environmental problems not only pollute people's normal life and affect their health but moreover, eventually lead to dramatic climate change, which has been evidenced and discussed by many studies [1,2]. Therefore, alternative power production from natural resources such as solar energy, wind energy, ocean energy or other forms of non-fossil combustion energy that do not cause environmental impact must be considered.

Ocean energy has been studied as a massive power source for a long time. Among all kinds of ocean energy, wave energy is the one being studied most thoroughly because it is well distributed in the oceans around the world and the energy is abundant. Devices for the exploitation of wave energy have been extensively studied, among which oscillating water column (OWC) wave energy converting systems are ones with relatively higher efficiency.

To date studies on OWC type wave energy converting systems are still in a very active status. Some focus on the improvement of the mechanism of energy harvesting such how as to change the shape of traditional OWC system into a U-OWC [3,4] or a so called backward bent duct buoy OWC (BBDB-OWC) by using a backward-bent duct buoy [5] or by applying a double-chamber to improve the energy harvest ability in the deep water [6] or how to bend the front wall of the chamber to study its influence on the energy conversion [7]. Research focused on the efficiency of turbine performance for outflow and inflow motions is also performed [8] while a similar study released recently [9] is also for a turbine system, where an axial-flow impulse turbine was installed on an OWC model to replace the traditional one and a model was built and placed in a wave flume for the experimental tests under regular wave conditions. Investigations on the wave-height and power taking-off damping effect were carried out experimentally and numerically [10]. Some are focused on the performance of the air-chamber [11], where the effect of neglecting the air compressibility has been studied that an experimental test model scaled down to 1/50 may result in an overestimation up to 15% for the air pressure in the OWC chamber. It points out that for a scaled-down model tested in experimental laboratory, attention must be paid while a full scaled model test will be more realistic.

Mostly of the mentioned studies on OWC wave energy converting systems are focused on systems located close to a shoreline with acting waves, where incident waves will propagate right into the air-chamber of the OWC wave energy converter and react inside the chamber. In this way the wave energy is easier to capture and larger converting efficiencies may be achieved. Only very rare studies have examined OWC wave energy converters situated in an open sea environment. When an OWC wave energy converter is installed in open sea conditions, because the waves will mostly bypass the structure, the energy that can be captured from the incident waves will usually, be much less than a with a system situated by the shoreline. On the other hand, trying to improve the energy capture capability by expanding the size of the facility or installing additional devices may induce severe strains on the main structure where the OWC device is situated. Therefore, there is a dilemma for an open sea OWC wave energy converting system: building an OWC wave energy converting system in open sea that may have high energy converting efficiency comparable to that of a system situated on the shoreline will cost a fortune and yet the durability is still a major concern.

As published lately [12], an OWC wave energy converting system integrated with the column of a jacket platform was studied. In that study the OWC was installed in one of the legs of the jacket structure and a theoretical analysis using an ANSYS finite element model was applied to the proposed OWC energy converter to determine the parameters. It was found that the integrated device had certain effectiveness and feasibility. Another study is an application of an OWC wave energy converting system to buoys for sensors [13]. The analysis revealed that these self-powered sensor buoys are able to provide the required power output for the considered wave climate. It also concluded that the power absorption capacity of these small diameter (less than 25 cm) buoys is rather low when compared with the expected electrical power output of a larger diameter buoy, which is designed for electrical grid supply. The case of OWC installed on a buoy [13] is on a floating structure as some other similar applications on the floating structures were also presented [14,15], of which the behavior will not be able to apply to the OWC installed on a fixed offshore structure. Therefore, some interesting issues are: how will the OWC perform for a device similarly situating in open sea but with much larger diameter and fixed stiff legs on the seabed? How would the main structure respond when it is attached with an additional OWC wave energy converting device in an open sea conditions? Some new studies on so-called "hybrid wind-wave energy converters" (or HWWECs) applied to jacket structures are the ones that led to the innovative idea to integrate the wind-turbine offshore template infrastructure with an OWC wave energy converter [16,17]. Both a mathematical model analysis and a scaled-down experimental study were carried out for the study of the

HWWEC system. Since these studies are pioneers in the study of an OWC wave energy converter situated in open sea, the data obtained in the research are valuable.

Similarly, in this study an OWC wave energy converting system which is integrated with an offshore jacket platform is studied. The purposes of this study will focus not only on the effectiveness of the OWC system in open sea, which will be treated as an extra bonus for the cost to build the OWC device can be limited but also on the responses of the main structure after wave energy has been captured by the added OWC device. The OWC wave energy converting system studied here will be built right in the space among the legs of an offshore jacket structure instead of on the legs. In this way the air-chamber can be enlarged so that a small ratio between the outlet orifice and the water surface contained in the chamber, an essential parameter for the OWC device, can be achieved. Avoiding installing an OWC device around the legs of the template structure [12] may have other advantages such as reducing the turbulence and additional vibration caused by the OWC on the legs, which are the main structural members. As the first part of the research, this study will focus on the conversion efficiency of the OWC device installed on the offshore template structure.

As presented in a recent study [18–20], the parameters for the application of a conventional OWC wave energy converting system to the caisson breakwater through upgrading on the safety and function of the caisson-based system were examined. Parameters examined in that study include the dimensions of the OWC chamber such as the orifice of the air-chamber allowing airflow in/output, the chamber length along the direction of incident wave, the size of the opening gate for incident waves and the submerged depth of the air-chamber. All of those parameters are presented in dimensionless form, while wave-height and period of waves are considered in a range of variations. Similarly, parameters to be examined in the study for the performing efficiency of the OWC wave converter integrated in the offshore template structure will include the variations of the orifice, the size of the air-chamber in terms of its height and the openings of the gate facing the incident waves. In addition to those parameters the shape of the air-chamber will also be investigated.

The analysis in this study will focus on the conversion efficiency of the OWC wave energy converting system installed on an offshore template structure that includes the airflow velocity from the air-chamber, the pneumatic power and the converting efficiency in terms of a ratio between the pneumatic power and the energy of incident waves. Parameters to be examined in this study include the dimensions of the OWC chamber such as the orifice of the air-chamber allowing airflow in/output, the size of the opening gate for incident waves and the submerged depth of the air-chamber in terms of the height of the chamber. All of these parameters are presented in dimensionless form, while wave-height and period of waves in annual bases for certain site of the installation are considered.

2. Analytical Model and Environmental Forces

2.1. Analytical Model of the Structure

In this study, a typical template or jacket type structure for the offshore wind-turbine foundation located in water deeper than 30.0 m is designed with the installation of accommodated OWC system. As shown in Figure 1, it is a schematic 3-D view of the jacket type structure associated with an OWC wave energy converting system, where the front side of the air-chamber will allow a square gate if necessary. This offshore jacket type of infrastructure for the wind-power station is adapted from a preliminary design of Taipower from Taiwan, which is designed to locate a wind-turbine of 4–6 MW electricity power capacity for annual average wind-speed while the structure is based on a design-wave condition for a storm of 50 years return period at the proposed location.

The feasibility assessment of the generator system has taken into considerations of geographical characteristics such as seabed situation, wind, wave, current and other environmental conditions. Therefore, this study will focus on the associated OWC wave energy converting effect. The dimension of the prototype OWC chamber will basically follow the internal space of the frame of the template structure so that the side walls of the chamber are 12 m wide while the height of the wall is a variable. Mainly, the position of top ceiling of the OWC chamber is decided first, which is at a level 7.0 m above the water and then the height of the wall is varied from 15 m to 21 m (8–14 m submerged depth), a ratio of 0.5 to 0.7 to the water depth.

Figure 2 shows the side-view of the jacket type structure. Dimensions for the structure and the size of structural members are also shown in the figure, where the height of the offshore template infrastructure is 43.0 m above the seabed and width of a square platform is 14.0 m. The diameters of the cylinder members are: 2.0 m for the vertical members and the top frames, 0.9 m for the inclined members and 0.8 m for the horizontal members.

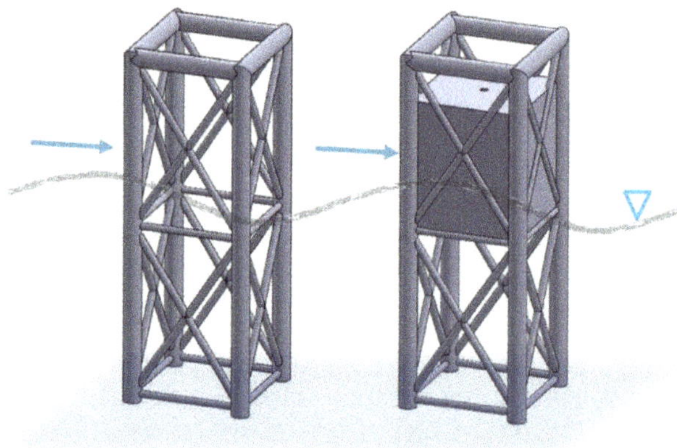

Figure 1. Schematic drawing of a jacket type structure installed with OWC converter.

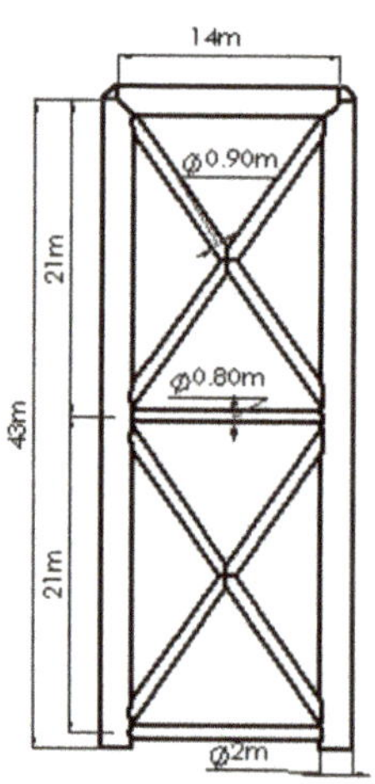

Figure 2. Side view and the dimension of the jacket type structure.

2.2. Analytical Methods and Environmental Conditions

The theorem applied in this study includes two parts. The first part is for the calculation of the fluid and wave motions and their influence to the airflow in the OWC chamber such as the velocity of the airflow through the opening orifice. The second part is about the estimation of the efficiency of the power converted from wave-energy of an OWC system to pneumatic power that can drive a turbine generator installed on the OWC system, of which the related theorem has been presented previously [18].

2.2.1. Basic Theorems for Fluid

In this study a theorem of unsteady Navier-Stokes Equations [21,22] in conservation form consisting of continuity equations, equation of momentums and equation of turbulence dynamics for fluid with density ρ and velocity U are applied and shown as follows.

Continuity equation:

$$\frac{\partial \rho}{\partial t} + \nabla \cdot (\rho U) = 0 \tag{1}$$

Equation of momentum:

$$\frac{\partial \rho U}{\partial t} + \nabla \cdot (\rho U \times U) - \nabla \cdot \left(\mu_{eff} \nabla U\right) = \nabla \cdot p' + \nabla \cdot \left(\mu_{eff} \nabla U\right)^T + B \tag{2}$$

where B is the sum of body force, μ_{eff} is the effective viscosity, p' is the revised pressure. The effective viscosity and the revised pressure can be presented as:

$$\mu_{eff} = \mu_t + \mu \tag{3}$$

$$p' = p + \frac{2}{3}\rho k \tag{4}$$

It is also noticed that μ_t is the viscosity of the turbulence, which according to the assumption of k-ε model, is related to the dynamic energy and the dissipation of the dynamic energy as presented as:

$$\mu_t = C_\mu \rho \frac{k^2}{\varepsilon} \tag{5}$$

where k, are obtained directly from the equation of dynamic energy and equation of energy dissipation presented as follows:

$$\frac{\partial (\rho k)}{\partial t} + \nabla \cdot (\rho U k) = \nabla \cdot [\left(\mu + \frac{\mu_t}{\sigma_k}\right) \nabla k] + P_k - \rho \varepsilon \tag{6}$$

$$\frac{\partial (\rho \varepsilon)}{\partial t} + \nabla \cdot (\rho U \varepsilon) = \nabla \cdot \left[\left(\mu + \frac{\mu_t}{\sigma_\varepsilon}\right) \nabla \varepsilon\right] + \frac{\varepsilon}{k}(C_{\varepsilon 1} P_k - C_{\varepsilon 2} \rho \varepsilon) \tag{7}$$

where $C_{\varepsilon 1}$, $C_{\varepsilon 2}$, σ_k, σ_ε are constant parameters while P_k is related to viscosity and floating force and can be presented as:

$$P_k = \mu_t \nabla U \cdot \left(\nabla U + \nabla U^T\right) - \frac{2}{3} \nabla \cdot U (3\mu_t \nabla \cdot U + \rho k) + P_{kb} \tag{8}$$

The numerical simulation method of this study uses ANSYS CFX to simulate the performance of the air flow field in the air chamber, and the turbulence model uses the k-ε model. It is assumed that the structure itself does not move by other forces, and only consider the movement of the fluid and the air flow field, especially for the air motions influenced by the waves in the air-chamber of the OWC system.

2.2.2. Calculation for Pneumatic Powers

The pneumatic power for the air flow inward and outward the chamber is obtained from the kinetic work shown in Equation (9). The pneumatic power presented as Ea from the airflow with density ρ_a and velocity v_a through a cross-section area A_a with a wave-period T can be presented as:

$$E_a = \frac{1}{2} m {v_a}^2 = \frac{1}{2} \rho_a A_a {v_a}^3 \cdot T \tag{9}$$

2.3. Numerical Process Applied to the Analysis

The numerical process applied in this study as indicated is based on the unsteady Navier-Stokes equations theorem in conservation form as presented in the previous section. The verification of this numerical process has been carried out and compared with an experimental data [18,19] from an on-field study, a full size OWC wave energy converter situated at Sakeda (Japan) led by Goda, a pioneering researcher in the field [23,24]. In the study for the verification of the numerical method, the comparisons were made for the variations of both water elevation and pressures in the chamber of a full size OWC wave energy converter. It is noticed from the comparison that the analytical results are in very good match with experimental data in general. Therefore, it is confident that this numerical tool may appropriately describe the behavior of an OWC wave converting device integrated into an offshore template structural system.

Presented as following steps are the numerical analysis process along with the boundary conditions:

- Step 1: Establishing the model geometry according to the analytical object.

For the proposed OWC wave converting device integrated into an offshore template structural system, the model geometry was constructed for the water-body with open surface-field implemented along with the structure.

- Step 2: Meshing and re-meshing the model if needed after a pre-analysis.

To implement meshes of the analytical model, both the fluid body and the offshore template structure under study are segregated into small elements that can appropriately describe the properties and behavior of the body.

- Step 3: Setting the boundary conditions based on the analytical environment.

Figure 3 shows a schematic view of the boundary conditions. On the left hand side is the inlet boundary allowing the input of incident waves. In the numerical analysis, a piston-type wave maker method was applied to produce incident waves [25]. The seabed is impermeable condition for fluid like a wall while a free surface between the water and air was applied at the interlayer. The open area is on the top boundary for the air except for the air in the chamber that is confined depending on the variation of the chamber-geometry. At the right hand side it is an outlet boundary allowing the dissipation of the waves propagating away. For the side-condition of analytical domain a symmetrical condition is presumed. The symmetry conditions is not like a wall boundary condition, which is a "non-slip" condition in which it assumes the velocity at the wall is zero no matter what wall roughness value is applied. While the "slip" boundary (slip/symmetry boundary) would allow the flow to move along the boundary, it means that the velocity at the slip/symmetry boundary is not zero.

- Step 4: Performing the analysis.

The analysis will include several pre-analyses in order to adjust the solution quality. A re-mesh will also be performed if the solution does not converge based on the convergence condition. In order to match the convergence requirement in this study that the RMS (root mean square) of residuals for all solutions must be less than 1.0×10^{-4}, the numbers of nodes and elements applied in this study based on the shape of the air-chamber are varied from 1,286,110–1,244,131 and 1,608,690–1,675,950. Therefore, the running time of computation for each case will be around 72 h in a computer with CPU of eight cores/16 threads.

The incident wave is assumed to propagate in the direction normal to the front face of the jacket structure. The exerting pressure on the jacket structure then can be calculated based on the properties of the wave conditions for the normal operation of the wind-turbine system. The reference site for this study was selected at 10 km off Changhua, west coast of Taiwan Strait, with a water depth of 30 m. The applied wave conditions, unlike the ones for the structural safety design purposes, are based on the local annual average wave statistics. The wave height (H) is 1.5 m, the wave period (T) is 7.5 s, and the corresponding wave length (L) is 84.51 m. The on-field data are also from the preliminary design data of Tai-power.

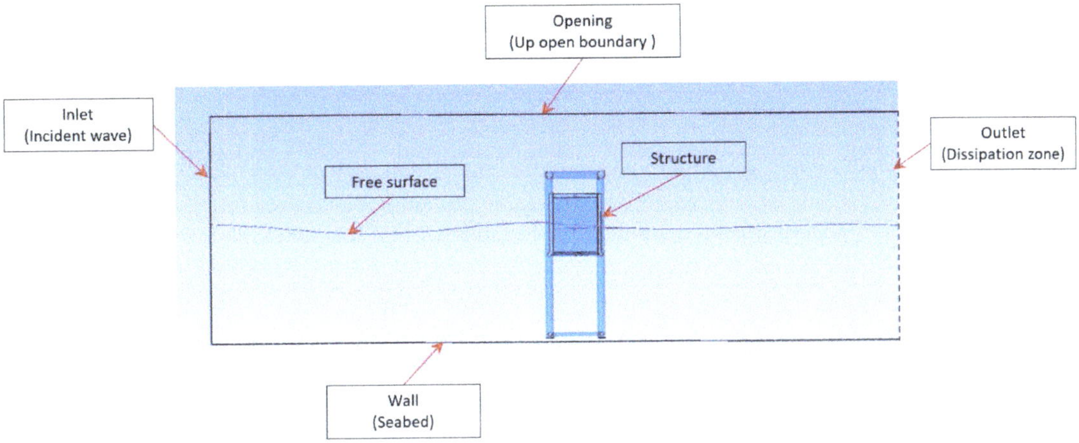

Figure 3. Schematic drawing of the boundary conditions.

3. Dimensional Parameters Applied to the Study

3.1. Various Roof Shapes of the Air-Chamber

Before the study for dimensional parameters applied to the OWC wave energy converter, the shape of the air-chamber was examined first. Under conditions of the space allowed in the template type offshore structure for the installation of an OWC wave energy converter, a cubic box with various shapes of roof was taken into consideration. Since the roof is the location for the in/outlet of the compressed air, the efficiency of the energy converted from the captured wave energy in the chamber may be influenced by the shape of the air-chamber. As presented in Figure 4, three air-chambers located in the template structure with various shapes of roof were considered. The first one has a quadrilateral triangle shape cap direct from the bottom of the chamber and the second one also has a quadrilateral triangle cap but from the middle part of the chamber-box while the third one maintains as a cubic box. All of them have the same size of in/outlet orifice on the apex of the roof.

Figure 4. Consideration for various shapes of air-chamber of the OWC wave energy converter installed in an offshore template structure.

3.2. Dimensional Parameters to Be Examine

The parameters chosen in the analysis basically are related to the dimension of the air-chamber for the OWC converter situating on an offshore template structure. Parameters considered include area ratio between the opening of orifice A and the water surface A_w in the chamber presented as $R_A = A/A_w$, the ratio of the open-gate O in the front side facing the incident wave to the chamber height Z presented as $R_O = O/Z$ and the ratio of chamber height of OWC converter to the water depth h presented as $R_Z = Z/h$ under a condition that during the operation even in a low water level the whole device is submerged in the water.

Figure 5a–c present schematic drawings of the side view of the OWC converter installed in an offshore template structure, where parameters such as the ratio of orifice area R_A, ratio of chamber height Rz and ratio of the front gate opening Ro are correspondingly presented in Figure 5a–c. The dimensions of the device related to parameters to be examined are all marked in the figure and shown in variables as A, O and Z. A table for the variation of these parameters is presented in Table 1. It is noted that because too many parameters are to be analyzed when one parameter is a variable the other parameters will be set as constant. Such a referenced model is set as: $R_A = 0.4\%$ ($A = 144$ m^2), $R_O = 0\%$ and $R_Z = 50\%$, where they are shown in bold letters in the table.

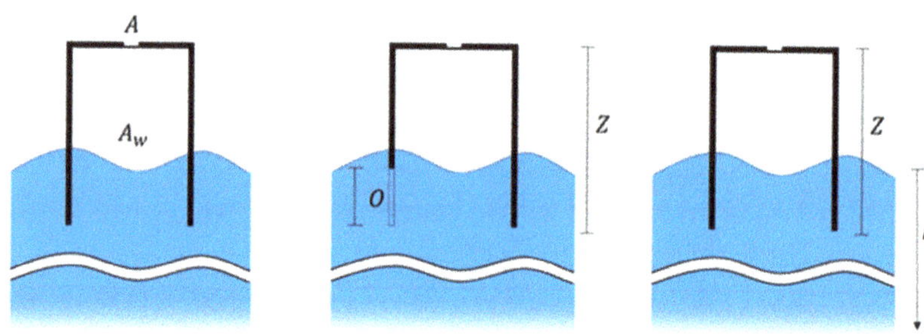

Figure 5. Parameters to be examined in the study: (**a**) Ratio of orifice area $R_A = A/A_w$; (**b**) Ratio of front gate opening $R_O = O/Z$; (**c**) Ratio of chamber height $R_Z = Z/h$.

Table 1. Parameters utilized in the analysis.

Parameter	Code	Variables
Area-ratio of orifice	R_A	0.1%, 0.2%, **0.4%**, 0.8%, 1.6%
Opening-ratio of gate	R_O	0%, 25%, 31.25%, 37.5%, 43.75%, 45.31%, 50%
Height-ratio of chamber	R_Z	**0.50**, 0.55, 0.60, 0.65, 0.70

During the analysis, responses of the OWC system for the converting efficiency to be discussed include the velocity of the airflow through the orifice, the air pressure, the power produced by the airflow and the converting efficiency in terms of the ratio of pneumatic power and power of incident waves. Comparisons of these responses corresponding to the given wave conditions are presented and discussed in the following sections in terms of the parameters of area-ratio of orifice opening R_A, opening ratio of OWC gate to the height of the chamber Ro and the ratio of the height of the air chamber to water depth Rz.

4. Numerical Results and Discussion

During the analysis the variation of the air-flow in the chamber can be obtained. Presented in Figure 6 is the water-flow around and through the template structure. It shows that the turbulence is small but, if the dimensions of the structure are larger than 1/5 wavelength, the diffraction effect must be considered as the analysis performed for the large-size tension-leg platform [26,27]. The other factor is that the applied waves based on local environmental conditions are quite small in terms of the wave-height and length.

Presented in Figure 7 is the typical variation of air-flow in the chamber, where the flow presented in a vector form can be observed clearly. As shown in the figure, the strongest velocity of the air-flow concentrates in the central space of the chamber along with the vertical line of the orifice while near the orifice in the upper part of the air-chamber a significant vortex response can be observed. Along with the time the variations of the air-flow can also be observed. As is shown in Figure 8, an obvious alternately inward and outward flow of the air are presented along with the time variations. By following the time history of the flow-motions shown in the figures, the inward/outward motions of air-flow are in a period pattern along with the wave heaving motion. It also shows that the outward velocity is larger than the inward velocity. This is due to the effect that when the air flows inward it may encounter the water surface inside the chamber and affect the velocity.

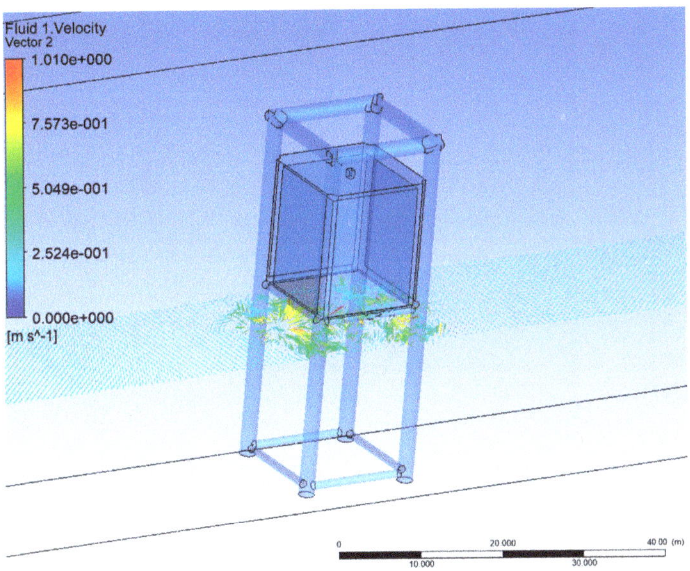

Figure 6. Fluid motions around the template structure with an air-chamber of the OWC.

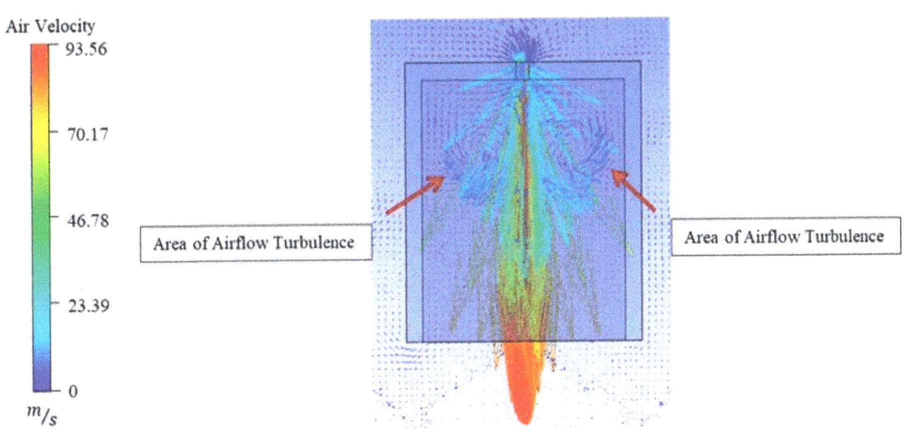

Figure 7. Typical air-flow and vortex variations in the air-chamber of the OWC.

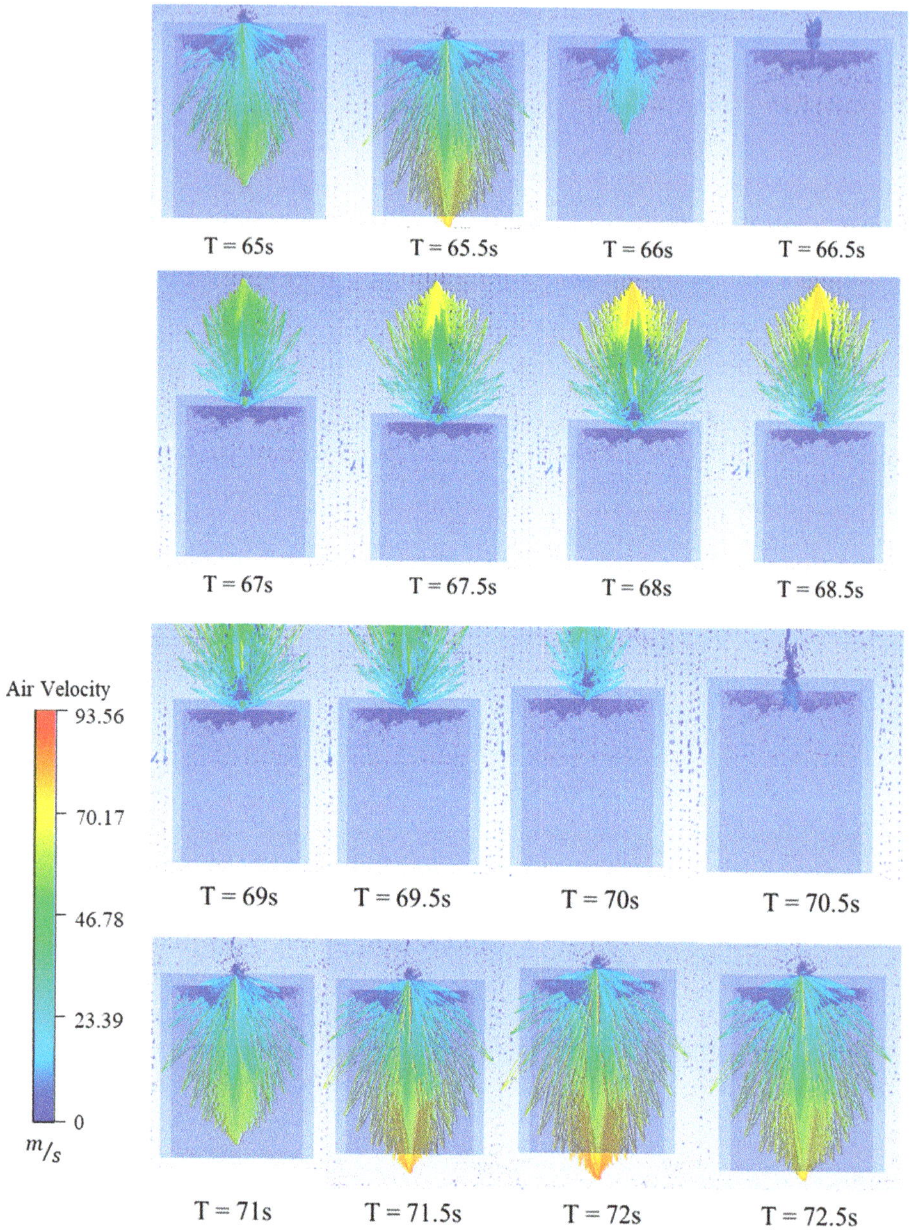

Figure 8. Variations of air-flow in the air-chamber of the OWC corresponding to time.

4.1. Responses Corresponding to Various Shapes of the Air-Chamber

When the conditions such as the opening ratio of the in/outlet orifice, the ratio of opening gate and the ratio of submerged depth are set as a constant as R_A = 0.4%, R_O = 0.5%, and Rz = 0, the speeds of the air-flow from the orifice for three types of shape of the air-chamber are presented in Figure 9a–c corresponding to various shapes of the chamber. It is noticed that the velocity of outflow and inflow are both presented positively, which are taken as a spatial average across the orifice area. The maximum velocity is

51.4 m/s, 52.1 m/s and 71.8 m/s corresponding to three types of chamber roof, namely, quadrilateral triangle, semi-quadrilateral triangle and cubic shape. The average velocities corresponding to these three types of chamber-shape are 39.2, 46.1 and 59.1 m/s. It shows that for maximum velocity of the air-flow from the orifice, there is not much difference between the first two types of chamber-shape but for the last type the cubic one, the maximum velocity can reach a value higher than the other two by 38%. The average velocity from the orifice of the cubic type of air-chamber is 28% higher than that from the semi-quadrilateral triangle type and 50% higher than that from the quadrilateral triangle type of air-chamber. Therefore, it was recommended that the chamber shape for the OWC wave energy converter installed on the offshore template structure under study will be a typical cubic box with in-outlet orifice on the apex of the chamber. Except for the shape, the dimensions for the orifice area and height of the chamber and additional opening of front gate will be discussed in the following sections.

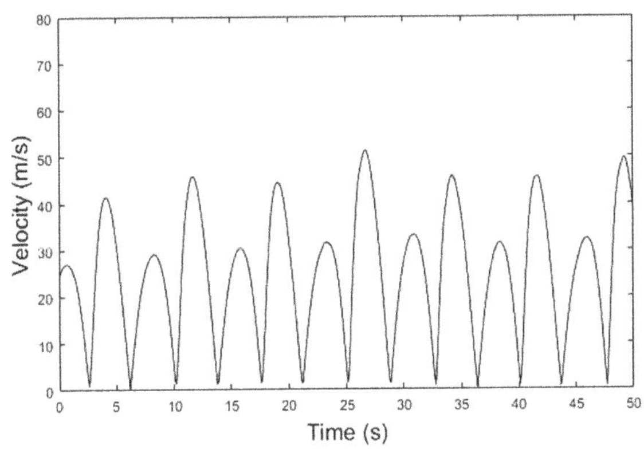

(**a**) Quadrilateral triangle shape

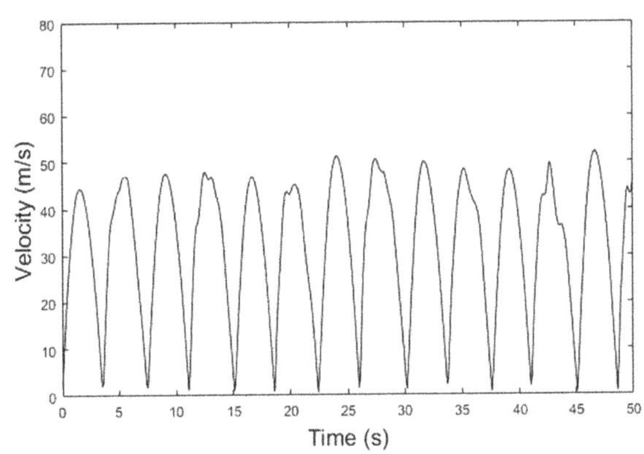

(**b**) Semi-quadrilateral triangle shape

Figure 9. *Cont.*

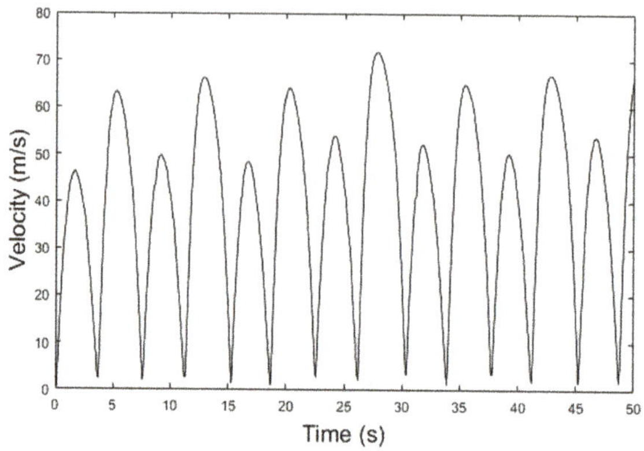

(**c**) Cubic shape

Figure 9. Velocity of air-flow corresponding to various shapes of OWC chamber.

4.2. Responses Corresponding to Area-Ratio R_A of Orifice

The area-ratio of the orifice cross-section of OWC will have influence on the performance of an OWC converting system. Therefore, taking into consideration in this subsection is a dimensionless ratio of area of the OWC orifice cross-section to the area of water surface in the chamber as was indicated as R_A and listed in Table 1. Since the area of water surface confined in the chamber is a constant while the cross-section of the orifice is variable, the ratio R_A is ranged from 0.1% to 1.6%.

4.2.1. Velocity of Airflow from the OWC

The velocity presented in Figure 10 is the average of first 1/3 maximum (significant) velocity of each time-history analysis corresponding to the area ratio of orifice cross-section R_A. It is observed that corresponding to the increase of the opening ratio R_A, the airflow velocity decreases nonlinearly. The airflow velocity drops slowly at the early stage corresponding to the area-increment of orifice opening and then decreases fast after the opening ratio of the orifice area R_A is larger than 0.8%. When the opening-ratio R_A increases to 1.6%, the airflow velocity is reduced to 60 m/s.

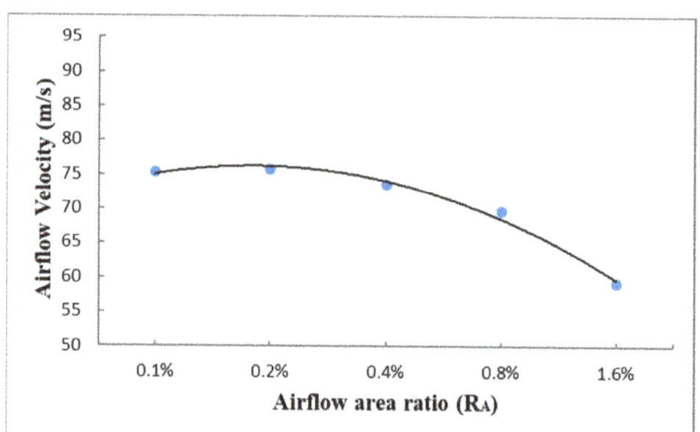

Figure 10. Airflow velocity corresponding to area ratio of orifice cross-section R_A.

4.2.2. Air-Pressure from the OWC

Presented in Figure 11 is the average of 1/3 maximum (significant) air-pressure corresponding to various area-ratios of orifice of the OWC converter. It is observed that corresponding to the increase of ratio of the orifice opening, the air-pressure will remain at a high level till it reaches certain value and then drops rapidly. When the cross sectional area of the orifice to the area of water surface in OWC chamber is smaller than 0.4%, the air-pressure remain in a stable value larger than 4566 Pa. Similar to the velocity response, the air-pressure will decrease in a higher rate corresponding to the area-increment of orifice opening when the opening-ratio is larger than 0.8%. Moreover, the air-pressure can reach a value over 4500 Pa for most cases.

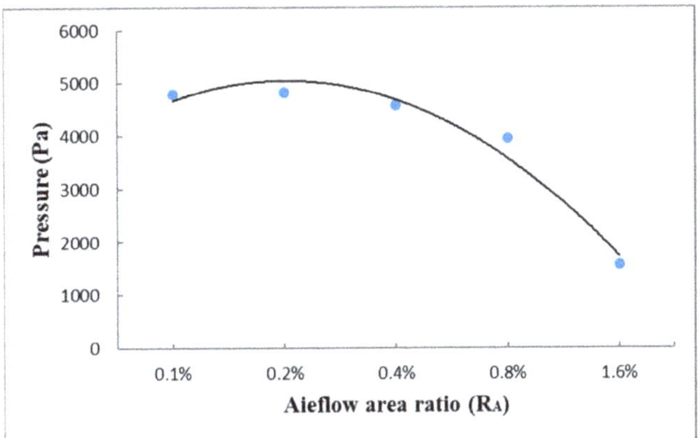

Figure 11. Air-pressure corresponding to area ratio of orifice cross-section.

4.2.3. Pneumatic Power from the OWC

Presented in Figure 12 is the average of 1/3 maximum (significant) pneumatic power corresponding to various area-ratios of orifice of the OWC converter. It is observed that corresponding to the increase of ratio of the orifice opening the pneumatic power will slightly increases till it reaches a peak value in the lower ratio of the orifice area and then decreases nonlinearly. The ratio corresponding to the peak of pneumatic power is 0.2% for the cross-sectional area of the orifice to the area of water surface in OWC chamber. Similar to the velocity response, the pneumatic power will decrease in a higher rate corresponding to the area-increment of orifice opening when the opening-ratio is larger than 0.8%. The pneumatic power of the OWC wave energy converter can reach a value larger than 250 kW for most cases.

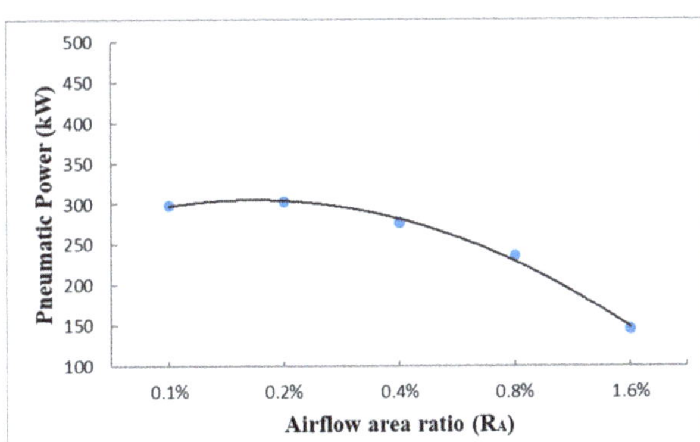

Figure 12. Pneumatic power corresponding to area ratio of orifice cross-section.

4.2.4. Efficiency of Power Converted from the OWC

For the examination of efficiency of power converted from wave-energy of an OWC system to pneumatic power that can drive a turbine generator, an estimation for the ratio between energy induced by airflow and produced from incident waves is applied [18]. As presented in Figure 13 is the average of 1/3 maximum (significant) pneumatic power corresponding to various area-ratios of orifice of the OWC converter. It is observed that corresponding to the increase of the opening ratio, the converting-efficiency of OWC decreases. The converting-efficiency stays at a high level first and then drops fast at the area-ratio approaches 0.8%. Basically after the opening-ratio increases to 0.8%, the efficiency of power converted is reduced to 10%. It is also found that the converting efficiency from the OWC system installed in an offshore template structure is only slightly over 10% for most cases without an additional opening of the front gate.

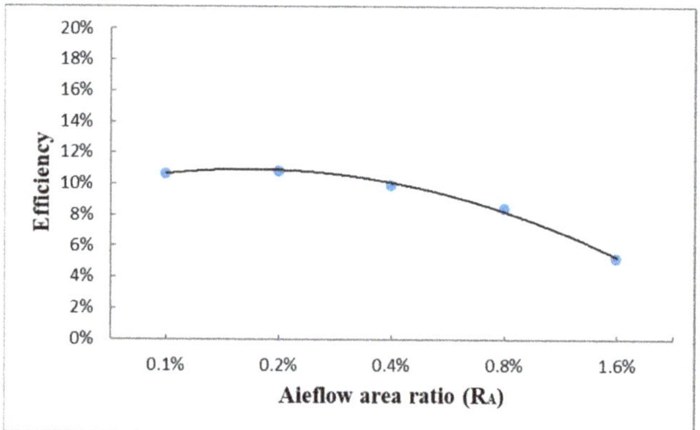

Figure 13. Efficiency corresponding to area ratio of orifice cross-section.

4.3. Responses Corresponding to Opening-Ratio of OWC Gate R_O

Considering the prevailing direction of waves, adding an opening gate to the original chamber is expected to capture more energy from the incident waves and upgrade its energy conversion efficiency. However, if the wave direction changed and is not aligned with the opening gate, it will be an issue that can be studied in a future study. The opening of the OWC gate is set as a ratio of the gate-opening to the height of the chamber for the OWC converter as indicated as $O/Z = R_O$ in Figure 6. The height of the chamber of the OWC converter is a constant of 15-m while the opening-ratio of the gate is a variable as presented in Table 1 ranged from 0% to 50%.

4.3.1. Velocity of Airflow from the OWC

Presented in Figure 14 is the average of first 1/3 maximum velocity of airflow for various ratios of gate-opening of the OWC converter to the chamber height. It is observed that corresponding to the increase of the opening-ratio of the air chamber gate, the airflow velocity increases until it reaches the ultimate value. The maximum average velocity of the airflow found from the orifice is 88.24 m/s when the opening ratio is 45.31% as set in this study. It is also noticed that for the largest opening ratio of the gate when it is 50%, corresponding to which the air-velocity drops suddenly to less than 60 m/s. For the gate opened by 50% of the height of chamber, it means that the submerged part of the chamber is 0.5 m. During the heaving motion of the waves the gate may be over part of the water surface. Once the gate opening is over the lowest water level, the velocity drops suddenly because the air chamber is wild opened then and the air is not confined in the chamber to be compressed through the orifice. Therefore, the determination of the largest value that must be corresponding to the highest water level is important and essential to the efficiency of the conversion effect.

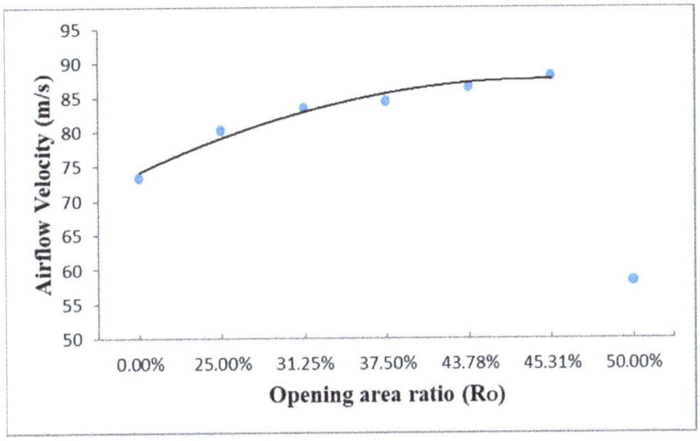

Figure 14. Airflow velocity corresponding to opening ratio of OWC gate.

4.3.2. Air-Pressure in the Air Chamber of the OWC

Presented in Figure 15 is the average of first 1/3 maximum air-pressure for various ratios of gate-opening of the OWC converter to the height of the air-chamber of the OWC. It is observed that corresponding to the increase of the opening-ratio of the air chamber gate, similar to the velocity responses, the air-pressure increases till it reaches an ultimate value. Only a slightly flatter slope was found for the variation of the air-pressure increment when compared to the velocity variation corresponding the increase of the opening-ratio of the gate. The maximum air-pressure found in the air-chamber 5800 Pa, but when the opening-gate is over the lowest water level the air-pressure also drops suddenly to 3783 Pa, which is reduced by about 35%.

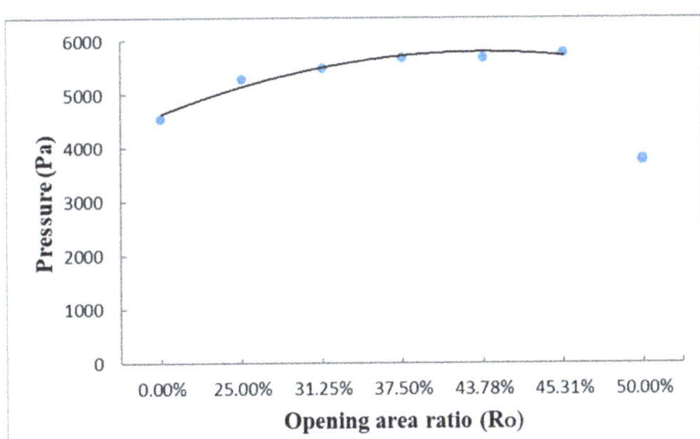

Figure 15. Air-pressure corresponding to opening ratio of OWC gate.

4.3.3. Pneumatic Power from the OWC

Presented in Figure 16 is the average power of airflow in 1/3 maximum pneumatic power for various opening-ratios of OWC gate to the height of the chamber of OWC. It is observed that corresponding to the increase of opening-ratios of OWC gate the pneumatic power also increases till the ultimate peak value. The increment of the pneumatic power is in a nonlinear but stiffer slope compared to the air-pressure increment trend. The maximum power obtained corresponding to the gate-opening in this case is 480 kW when the gate-opening ratio is 45.31% to the height of the air-chamber of the OWC. However, when the opening-gate is above the lowest level of the water, the pneumatic power drops to 139 Pa, which is only 29%, less than 1/3 of the highest power that can be obtained.

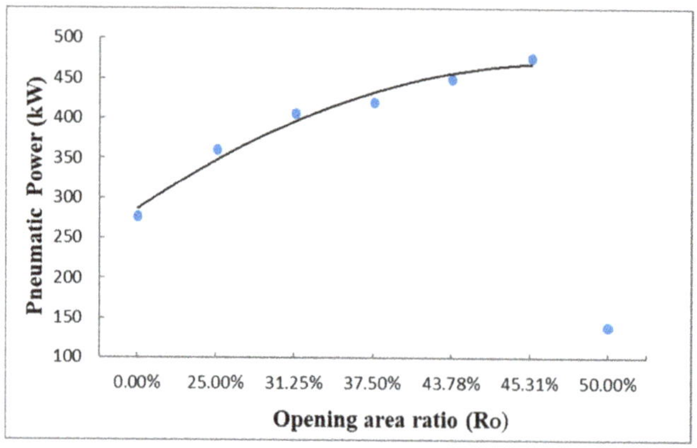

Figure 16. Pneumatic power corresponding to opening ratio of OWC gate.

4.3.4. Efficiency of Power Converted from the OWC

Presented in Figure 17 is the efficiency of power of airflow converted from the OWC for various opening-ratios of OWC gate. It is observed that corresponding to the increase of opening-ratios of OWC gate the converting efficiency of the power also increases till the ultimate peak value. The increment of the pneumatic power is also in a stiffer slope compared to the air-pressure increment trend. The maximum converting efficiency of the power obtained corresponding to the gate-opening in this case is 17.2% when the gate-opening ratio is 45.31% to the height of the air-chamber of the OWC. Compared to the case without a gate-opening, of which the converting efficiency is 9.9%, the converting efficiency obtained from a larger gate opening is much higher. As indicated in the numbers the increase of the converting efficiency can be raised by 74% from the largest opening of the gate. It is also noticed that the converting efficiency will drop dramatically if the gate opened above the lowest water level. As is shown in Figure 17, the efficiency is 5.0% for an over opened gate (R_O = 50%). That is reduced by 70% compared to a case, where the gate is opened in a ratio of 45.31%, slightly over the lowest water level.

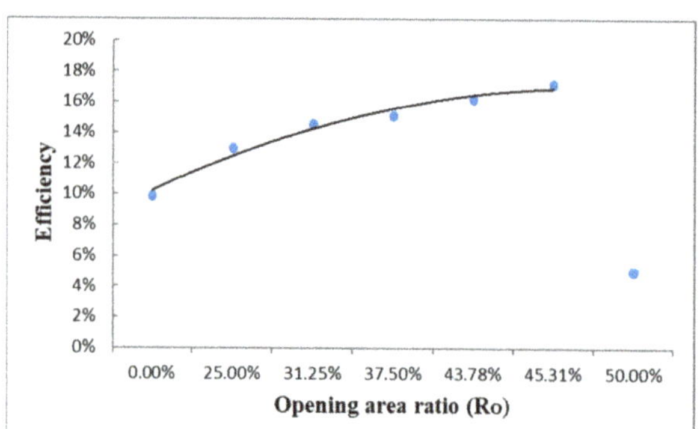

Figure 17. Efficiency corresponding to opening ratio of OWC gate.

The results show that increasing the opening area can effectively increase the efficiency of wave energy conversion. As the opening area increases, the wave conversion power is also higher. When the gate-opening ratio is 0.4531, the highest average speed is 88.24 (m/s), the generated power is 480 kW, and the conversion efficiency can be calculated as 17.2%. However when the gate opened is higher than the water surface as the case that the opening ratio is 50%, the internal air will escape and fail to flow through the airflow outlet, resulting in a significant reduction in wave energy conversion efficiency. Therefore, it is recommended that the opening area ratio is 37.5–45.31% based on this study.

4.4. *Responses Corresponding to the Ratio of Chamber-Height R_Z*

The submerged depth of the OWC may have influence to the performance of an OWC converting system. Therefore, taking into consideration in this section is a dimensionless ratio of chamber-height of the OWC converter to the water depth as indicated as R_Z and listed in Table 1. Since the apex of the OWC chamber located at the frame of the template structure is fixed based on the sea-water level, the main variation factor is the dimension of the chamber below the water level, or the height of the chamber.

4.4.1. Velocity of Airflow from the OWC

Presented in Figure 18 is the average of 1/3 maximum (significant) velocity of airflow for various ratios R_Z of the chamber-height of the OWC converter to the water depth. It is found that corresponding to the increase of the chamber-height of the OWC converter, the airflow velocity would not be influenced for the first two variations of ratio but after that it decreases almost linearly.

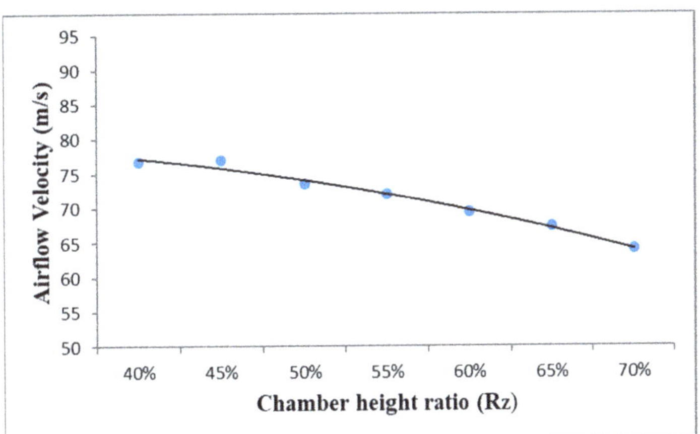

Figure 18. Airflow velocity corresponding to height-ratio R_Z of OWC chamber.

The maximum average velocity of the airflow is 76.97 m/s, corresponding to the ratio of chamber-height $Rz = 0.45$, while the second largest average velocity is 76.71 m/s, corresponding to the ratio $Rz = 0.40$ the smallest height of the chamber designed in the OWC converter.

Compared to the gate opened in the front wall, the reduction of the height of the whole chamber seems less effective for the energy conversion. As indicated in the second case for gate-opening that a 25% height of front wall is opened to reduce the front wall to 11.25 m high of which the velocity of air-flow is 80.38 m/s. When compared to the first case of the height-variation of the chamber that the chamber-height is 40% ($Rz = 0.40$) of water depth meaning that the air-chamber is 12 m high, the velocity from the gate-opening seems compatibly higher.

4.4.2. Air-Pressure in the air Chamber of the OWC

Presented in Figure 19 is the average of 1/3 maximum (significant) air-pressure for various ratios R_Z of the chamber-height of the OWC converter to the water depth. It is found that corresponding to the increase of the chamber-height of the OWC converter, the air-pressure would not be significantly influenced for the first several variations of ratio and then decreases almost suddenly when the ratio Rz is larger than 0.65. As was observed in ratios of $Rz = 0.40$ to $Rz = 0.60$, the air-pressure varies from 4932 Pa to 4062 Pa, which decreases in a slow rate till $Rz = 0.65$. When $Rz = 0.7$ the air-pressure drops to 1764 Pa almost in a sudden. This suggests that for the chamber-height in a lower ratio range the influence of the ratio on the air-pressure is minor, however, if the ratio is larger than 0.65 the influence on the air-pressure can be very significant.

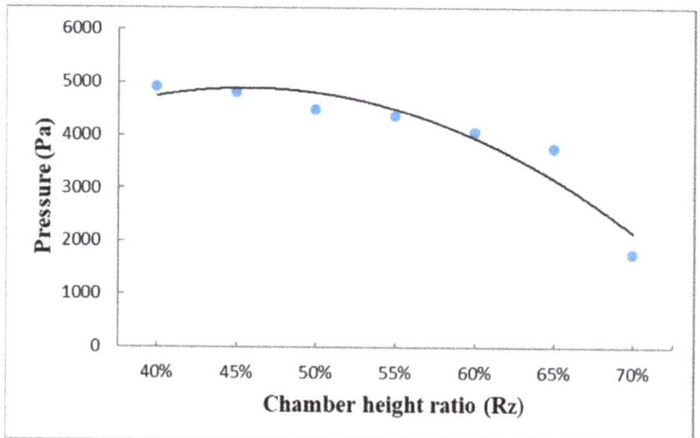

Figure 19. Air-pressure corresponding to height ratio R_Z of OWC chamber.

4.4.3. Power of Airflow from the OWC

Presented in Figure 20 is the pneumatic power corresponding to various ratios of OWC chamber height to the water depth. It is observed that in the first two ratios the pneumatic power slightly increases from 314 kW to 317 kW and then decreases corresponding to the increase of height-ratios of OWC chamber to the water depth. The pneumatic power decreases in an almost linear trend corresponding to the ratio increase of OWC height to the water depth.

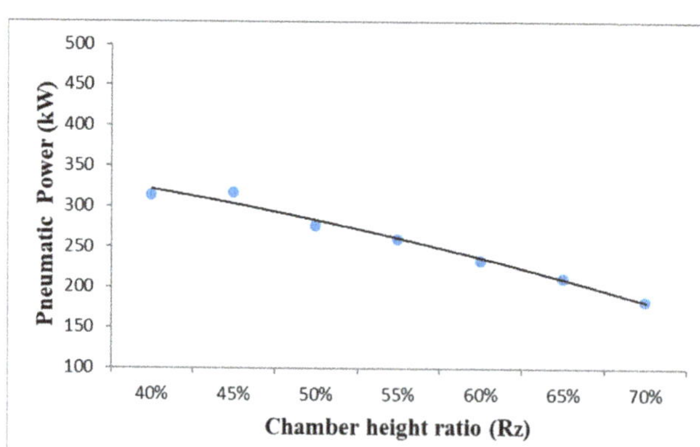

Figure 20. Pneumatic power corresponding to height-ratio R_Z of OWC chamber.

4.4.4. Efficiency of Power Converted from the OWC

Presented in Figure 21 is the efficiency of power of airflow converted from the OWC for various ratios of OWC chamber height to the water-depth. It is found that a pattern similar to the pneumatic power is presented for the first two ratios of the OWC height to the water depth, where the converting efficiency is slightly increase from 11.3 to 11.4 when the ratios are increased from 0.4 to 0.45. After that the converting efficiency of the wave energy to pneumatic power decreases corresponding to the increase of the height-ratio of the OWC height to the water depth.

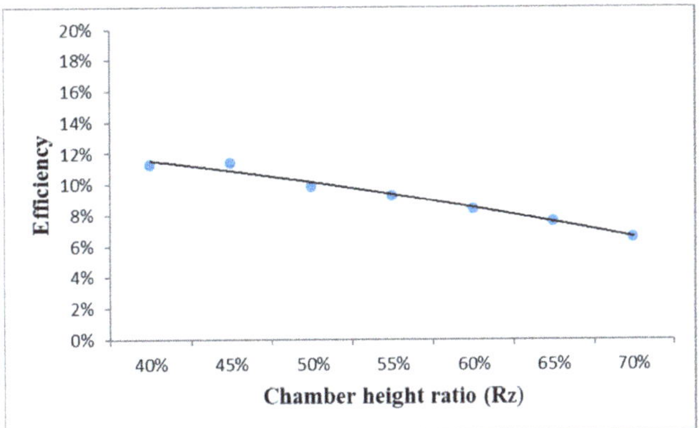

Figure 21. Efficiency corresponding to height-ratio of OWC chamber.

4.5. General Discussions on the Power Converting Efficiency

A general discussion for the power converting efficiency examined in this study will be compared to available experimental data and discussed here. Since only very few experimental studies have been performed and even for the rare one, the model type for analysis, shape, corresponding size from scaled-down to prototype and loading conditions are merely comparable, this subsection still tries to make a comparison to the very rare experimental data for the OWC wave converter integrated in an offshore template structure. The experimental data were presented in reference [16], where a scaled-down model of 1:50 for an OWC wave converter integrated in an offshore template structure was tested under regular and irregular waves to find the hydrodynamic response. As indicated, since the model and the analytical parameters are different from this study, we'll choose the converting efficiency that based on nearly similar analytical conditions to compare and discuss.

As presented in the referred article [16], the capture width ratio (CWR) is the parameter used to evaluate the performance of the OWC sub-system. Corresponding to the wave-steepness, which is 0.017 quite small as applied to this study but was a representative in-situ condition, the performance parameters are located in a range lower than 10 except for the case that the period of the applied wave is 7 s. As comparison, the wave-period applied in this study is 7.5 s and the corresponding dimensionless damping coefficient B*, (please refer to [16]) is about 37 in terms of a similar scaled-down model so that the performance parameters obtained from the experimental data are located in the range between 5 and 10. It is found that compared to the performance parameters obtained from the experiment, the converting efficiencies of this study are in a range between 5.2 and 10.1 as indicated in Figure 13 (efficiency corresponding to area ratio of orifice cross-section). The results obtained from this study are compactible to an experimental result and quite encouraging.

5. Conclusions

In this research an OWC wave energy converting system was installed and combined to an offshore template structure for a wind turbine system as a subsystem for power generation. The theoretical results showed that though the converting efficiency in open sea conditions is not very high, it is workable and the efficiency can also be raised by 70% through addition of a front gate. Some substantial conclusions from the study are listed as follows:

- For the variation of chamber shapes, three kinds of shape were examined for the performance of the power converting, namely, quadrilateral triangle, semi-quadrilateral triangle and cubic shape. Among these three shapes, the chamber of cubic shape has the best performance in terms of the velocity of air-flow through the out-let orifice. The average velocity from the orifice of the cubic type of air-chamber is 28% higher than that from the semi-quadrilateral triangle type and 50% higher than that from the quadrilateral triangle type of air-chamber.
- Corresponding to the increase of the area-ratio R_A of orifice-opening, the conversion-efficiency decreases. For the early stage the conversion efficiency remains 10% till the area-ratio approaches 0.8%. It is also found that the conversion efficiency from the OWC system installed in an offshore template structure is only slightly over 10% for most cases if gate opening is not allowed.
- When gate-opening is allowed in the direction of incident waves the conversion efficiency can be greatly increased. For the case where the gate-opening ratio is 0.4531 to the height of the chamber, the generated power can reach as high as 480 kW, and the conversion efficiency is 17.2%, which represents an increase by 74% compared to the chamber without an opening in the front wall. The results show that the opening area can effectively increase the efficiency of wave conversion, but when the height of the opening area is higher than the water surface, internal air will escape, which leads to a reduction in the wave energy conversion efficiency. Therefore, it is recommended that the opening area ratio be 0.375~0.4531, based on the size of this experiment.
- For the ratio of chamber-height R_Z, since the height of chamber apex remains unchanged, the main changing factor is the underwater depth. The results show that an increase in the depth will directly reduce the outlet airflow velocity and power generation. It is recommended that the air chamber height ratio to the water depth be 0.4–0.5.

Therefore, according to this study for an OWC wave converter installed in an open sea situation, an OWC device with smaller outlet orifice and smaller height of the chamber relative to the water depth will have better performance in terms of the air-velocity and the pneumatic power converted from the wave power. Furthermore, if a gate-opening is allowed at the front side of the chamber (usually in the prevailing direction of the wave propagation), a ratio of gate-opening reaching 45% may give a great upgrade for the conversion efficiency.

It is noticed that the highest average power obtained in the system is 0.48 MW. When compared to the main structure, established for a wind turbine system of 5 MW power generation, the associated wave energy converting system may produce about 10% of the power obtained from the main system. For a sub-system, the results are quiet encouraging. However, in order to effectively increase the efficiency of ocean energy extraction and avoid negative impacts on the original structure and the environment, there are still many issues to be considered, hopefully to increase the efficiency of energy use and enhance the feasibility of sustainable development.

Author Contributions: Conceptualization, H.H.L.; methodology, H.H.L. and G.-F.C.; software, G.-F.C.; validation, G.-F.C., H.-Y.H. and H.H.L.; formal analysis, G.-F.C.; resources, H.H.L.; writing—original draft preparation, G.-F.C. and H.H.L.; writing—review and editing, H.H.L.; supervision, H.H.L.; project administration H.H.L.; funding acquisition, H.H.L. All authors have read and agreed to the published version of the manuscript.

Funding: This study has been financially supported in part by the Ministry of Science and Technology (MOST), Taiwan as indicated in the acknowledgements.

Institutional Review Board Statement: Not applicable for studies not involving humans or animals.

Informed Consent Statement: Not applicable.

Data Availability Statement: Not applicable.

Acknowledgments: This research has been financially supported in part by the Ministry of Science and Technology (MOST), Taiwan under grants "MOST 108-2221-E-110-025-MY2".

Conflicts of Interest: The authors declare no conflict of interest.

References

1. Rohli, R.V.; Vega, A.J. *Climatology*, 4th ed.; Jones & Bartlett Learning: Burlington, MA, USA, 2018; ISBN 9781284126563.
2. Ruddiman, W.F. *Earth's Climate: Past and Future*; W. H. Freeman and Company: New York, NY, USA, 2008; ISBN 9780716784906.
3. Spanos, P.D.; Strati, F.M.; Malara, G.; Arena, F. An approach for non-linear stochastic analysis of U-shaped OWC wave energy converters. *Probabilistic Eng. Mech.* **2018**, *54*, 44–52. [CrossRef]
4. Boccotti, P. Comparison between a U-OWC and a conventional OWC. *Ocean Eng.* **2005**, *34*, 799–805. [CrossRef]
5. Sheng, W. Power performance of BBDB OWC wave energy converter. *Renew. Energy* **2019**, *132*, 709–722. [CrossRef]
6. Ning, D.; Zhou, Y.; Zhang, C. Hydrodynamic modeling of a novel dual-chamber OWC wave energy converter. *Appl. Ocean Res.* **2018**, *78*, 180–191. [CrossRef]
7. Bouali, B.; Larbi, S. Contribution to the geometry optimization of an oscillating water Column wave energy converter. *Energy Procedia* **2013**, *36*, 565–573. [CrossRef]
8. Ansarifard, N.; Fleming, A.; Henderson, A.; Kianejad, S.; Orphin, J. Comparison of inflow and outflow radial air turbines in vented and bidirectional OWC wave energy converters. *Energy* **2019**, *1821*, 159–176. [CrossRef]
9. Liu, Z.; Xu, C.; Qu, N.; Cui, Y.; Kim, K. Overall performance evaluation of a model-scale OWC wave energy converter. *Renew. Energy* **2020**, *149*, 1325–1338. [CrossRef]
10. Elhanafi, A.; Kim, C.J. Experimental and numerical investigation on wave height and power take–off damping effects on the hydrodynamic performance of an offshore–stationary OWC wave energy converter. *Renew. Energy* **2018**, *125*, 518–528. [CrossRef]
11. Simonetti, L.C.; Elsafti, H.; Oumeraci, H. Evaluation of air compressibility effects on the performance of fixed OWC wave energy converters using CFD modelling. *Renew. Energy* **2018**, *119*, 741–753. [CrossRef]
12. Qiao, D.; Feng, C.; Ning, D.; Wang, C.; Liang, H.; Li, B. Dynamic response analysis of jacket platform integrated with oscillating water column device. *Front. Energy Res.* **2020**, *8*, 42. [CrossRef]
13. Henriques, J.C.C.; Portillo, J.C.C.; Gato, L.M.C.; Gomes, R.P.F.; Ferreira, D.N.; Falcao, A.F.O. Design of oscillating-water-column wave energy converters with an application to self-powered sensor buoys. *Energy* **2016**, *112*, 852–867. [CrossRef]
14. Singh, U.; Abdussamie, N.; Hore, J. Hydrodynamic performance of a floating offshore OWC wave energy converter: An experimental study. *Renew. Sustain. Energy Rev.* **2020**, *117*, 109501. [CrossRef]
15. Elhanafi, A.; Fleming, A.; Macfarlane, G.; Leong, Z. Numerical hydrodynamic analysis of an offshore stationary floating oscillating water column wave energy converter using CFD. *Int. J. Nav. Archit. Ocean Eng.* **2017**, *9*, 77–99. [CrossRef]
16. Perez-Collazo, C.; Greaves, D.; Iglesias, G. A novel hybrid wind-wave energy converter for jacket-frame substructures. *Energies* **2018**, *11*, 637. [CrossRef]
17. Michele, S.; Renzi, E.; Perez-Collazo, C.; Greaves, D.; Iglesias, G. Power extraction in regular and random waves from an OWC in hybrid wind-wave energy systems. *Ocean Eng.* **2019**, *191*, 106519. [CrossRef]
18. Lee, H.H.; Chen, C.-H. Parametric study for an oscillating water column wave energy conversion system installed on a breakwater. *Energies* **2020**, *13*, 1926. [CrossRef]
19. Lee, H.H.; Wu, T.-Y.; Lin, C.-Y.; Chiu, Y.-F. Structural Safety Analysis for an Oscillating Water Column Wave Power Conversion System Installed in Caisson Structure. *J. Mar. Sci. Eng.* **2020**, *8*, 506. [CrossRef]
20. Lee, H.H.; Chiu, Y.-F.; Lin, C.-Y.; Chen, C.-H.; Huang, M.-H. Parametric study on a caisson based OWC wave energy converting system. *World J. Eng. Technol.* **2016**, *4*, 213–219. [CrossRef]
21. Sarpkaya, T.; Isaacson, M. *Mechanics of Wave Forces on Offshore Structures*; Van Nostrand Reinhold Company: New York, NY, USA, 1981; ISBN 0-442-25402-2.
22. Milne-Thomson, L.M. *Theoretical Hydrodynamics*; The MacMillan Co.: New York, NY, USA, 1960.
23. Goda, Y.; Nakada, H.; Ohneda, H.; Suzuki, M.; Takahashi, S.; Shikamori, M. Results of field experiment of a wave power extracting caisson breakwater. *Proc. Ocean Dev.* **1991**, *7*, 143–148. [CrossRef]
24. Goda, Y.; Shinda, T.; Chiyama, S.; Ohneda, H.; Suzuki, M.; Takahashi, S.; Shikamori, M.; Takaki, Y. Experiment of a wave power extracting caisson breakwater. *Proc. Ocean Dev.* **1989**, *5*, 1–6.
25. Lal, A.; Elangovan, M. CFD simulation and validation of flap type wave maker. *World Acad. Sci. Eng. Technol.* **2008**, *22*, 76–82.
26. Lee, H.H.; Wang, W.-S. Analytical solution on the dragged surge vibration of TLPs with wave large body and small body multi-interactions. *J. Sound Vib.* **2001**, *248*, 533–556. [CrossRef]
27. Lee, H.H.; Wang, W.-S. On the dragged surge vibration of a twin TLP system with multi-interactions of wave and structures. *J. Sound Vib.* **2003**, *263*, 743–774. [CrossRef]

Journal of
Marine Science and Engineering

MDPI

Article

Visualization Investigation of Energy Dissipation Induced by Eddy Currents for a Solitary-Like Wave Passing over Submerged Breakwater Sets

Chi-Yu Li [1,2], Ruey-Syan Shih [3,*] and Wen-Kai Weng [3]

1 Bachelor Degree Program in Ocean Engineering and Technology, National Taiwan Ocean University, Keelung 20224, Taiwan; chiyuli@ntou.edu.tw
2 Center of Excellence for Ocean Engineering, National Taiwan Ocean University, Keelung 20224, Taiwan
3 Department of Harbor and River Engineering, National Taiwan Ocean University, Keelung 20224, Taiwan; wkweng@mail.ntou.edu.tw
* Correspondence: rsshih@mail.ntou.edu.tw; Tel.: +886-2-24622192 (ext. 6107)

Received: 18 September 2020; Accepted: 21 October 2020; Published: 22 October 2020

Abstract: Wave attenuation for the purpose of coastal protection has been an important topic in coastal engineering. Wave attenuation in relation to the vortices induced by a solitary-like wave propagating over submerged breakwaters (BWs) is discussed in this paper. A series of hydraulic model experiments was conducted to investigate the occurrence of eddies, the types and combinations of submerged BWs, and related phenomena of the range expansion of vortices. The microscopic changes in the flow field, the variation of eddies, and the distributions of streamlines were analyzed using the particle image velocimetry (PIV) technique. The measured transmission and reflection coefficients, along with the concept of energy conservation, were also examined to support the results. The results indicate that the attenuated wave energy is related to the induced vortices, and show that the total relative vortex energy for rectangular submerged BWs is larger than that for undulating submerged BWs in both the single and composite sets. The magnitude of the maximum vorticity of the undulating BW sets is larger and more concentrated than that of the rectangular BW sets; however, the total vortex energy is slightly smaller owing to the narrower vortex area range.

Keywords: particle image velocimetry; solitary wave; submerged obstacle; undulating breakwater; rectangular breakwater; vortex energy

1. Introduction

Waves are regarded as a form of energy transmission. Offshore waves can cause damage to coastal structures, which often comprise the infrastructure protecting lives and property. Coastal protection has therefore been an important topic in the field of coastal engineering. For this reason, various types of structures, including submerged artificial levees, breakwaters (BWs), and floating obstacles, have been studied to reduce the damage caused by waves. One approach to address these issues is to evaluate wave–structure interactions, and investigate how wave energy is attenuated. The physical characteristics of various types of undulating and sinusoidal topographies interacting with waves have been studied. Davis and Heathershaw [1] investigated the attenuation of surface-wave propagation over sinusoidally varying topography. Dalrymple and Kirby [2] investigated the interactions of small amplitude water waves with a patch of bottom ripples using the boundary integral equation method (BIEM). O'Hare and Davis [3] presented a method for modeling surface wave propagation over undulating topography. Kar et al. [4] analyzed the gravity wave transformation with a finite floating dock in the presence of bottom undulation trenches. Undulating topography has been applied and designed by Shih and Weng [5,6] and Shih et al. [7] as BWs, and the effectiveness of regular waves, irregular waves, and long wave attenuation was discussed in relation to their physical experiments. However, the interaction of the wave structure and the relationship between wave energy attenuation and vortices, especially the vortices induced by those undulating BWs, have not been widely investigated through experimental visualization.

In early research on the visualization of obstacle-induced vortices, Hino and Yamazaki [8] discussed the vortex formation and energy dissipation of periodic waves passing through vertical plates in a tank with dimensions 20 m long and 0.4 m wide. Viewed from the side, a 16 mm camera was used to shoot continuously at 12 fps. The hydrogen bubbles used to track the particles were generated by electrolysis with the application of a high DC voltage to the anode of a copper plate. The results showed that the energy dissipation rate can be estimated by the resistance coefficient. Ikeda et al. [9] investigated the vortex structure and energy dissipation near both sides of the edge of a vertical plate under wave action, and discussed the discrete vortices created by a submerged dike and the estimation of transmitted waves. It was found that the vortex can be numerically simulated with Lamb–Oseen vortices, confirming that the wave energy is dissipated as a result of the formation of the vortices. Hokamura et al. [10] studied the energy dissipation of eddies around the tip of a double-barrier floating breakwater (DBFB) using the dye flow visualization technique. The results showed that the reduction in wave height was caused by the loss of wave energy due to the vortex motion separated by the vertical plate tip. The DBFB caused vortices at the tips of two vertical plates, and these were widely distributed on the width scale of the floating body. The characteristics of velocity field variations when a solitary wave propagates over a reverse flow field were studied by Umeyama [11] using the super-resolution particle image velocimetry (PIV) technique. The velocity fields and flow patterns near a vertical and an inclined thin plate obstacle, induced by the propagation of a solitary wave, were also investigated by Zaghian et al. [12] using the PIV technique. Jiao et al. [13] investigated the flow field around three reefs using the PIV technique, which was used to validate the numerical simulation. Their simulation results showed good agreement with the results of the PIV experiments. Chang et al. [14] studied the interaction between a solitary wave and a vertical thin plate installed at the bottom, and discussed the fluid kinematics and vortex modes caused by the wave. Lin et al. [15] used the high-speed particle image velocimetry (HSPIV) technique to explore the dissipation process of solitary wave energy on very steep beaches, including its acceleration and pressure-gradient characteristics.

In order to determine the relationship between wave energy attenuation and vortices induced by submerged BWs, a series of hydraulic model experiments was conducted to investigate the eddy formation and flow field distributions generated by the interactions between waves and submerged structures in this study. The wave is a solitary-like wave generated by a landslide wave generator, and the submerged structures consist of a single set and a composite set of rectangular and undulating BWs. All experiments were filmed with a high-speed digital camera, and were analyzed using the digital particle image velocimetry tool PIVlab [16,17]. The distribution of the eddy velocity field and the vorticity strength were analyzed, and the results were used to calculate the proportion of eddy energy in the entire wave energy attenuation process. In addition, the measured transmission and reflection coefficients when the wave propagated over the submerged BWs, along with the concept of energy conservation, were adopted to confirm the calculations. Furthermore, the different properties of vortices and the distributions of streamlines induced by different submerged BWs are also discussed.

2. Experimental Setup and Methods

In this study, the flow field and formation of the vortices induced by the interaction between the generated solitary-like wave and different sets of submerged BWs were investigated experimentally. The model experiments were conducted in a two-dimensional open channel, which was about 5.0 m long, 0.08 m wide, and 0.2 m deep. The flume layout and experimental setup are schematically shown in Figure 1. Long waves, such as tsunamis, or waves resulting from large displacements of water, for example landslides and earthquakes, sometimes behave approximately like solitary waves. It is difficult to form a truly solitary wave in nature because trailing small dispersive waves are usually formed. Different generators of solitary or impulse waves have been widely adopted in many physical experiments to study the effects of long waves, including piston-type wave generators [18,19], Scott Russell wave generators [20,21], and landslide tsunamis [22–24]. In this study, a solitary-like wave generated by a sliding body was adopted. As can be seen in Figure 1, a landsliding system was adopted to serve as a solitary wave generator at the left end of the flume, and a solitary-like wave was generated by a concrete block sliding down a α = 67.5° slope. The generated solitary-like wave propagated from left to right, and a fixed vertical baffle with the same height as the water depth was installed to reduce reflected waves from the end wall. This is because the waveform and volume of the solitary waves are mainly above the water surface, and the waves can pass directly over the baffle and into a container when they reach the end of the flume. The water depth h was maintained at a constant 0.1 m, and sets of submerged BWs were placed 2.5 m away from the wave generator. The visualization area was set to focus on the environment surrounding the submerged BWs, as shown in Figures 1 and 2, which show a photograph of the experimental setup and the wave vortex measurement.

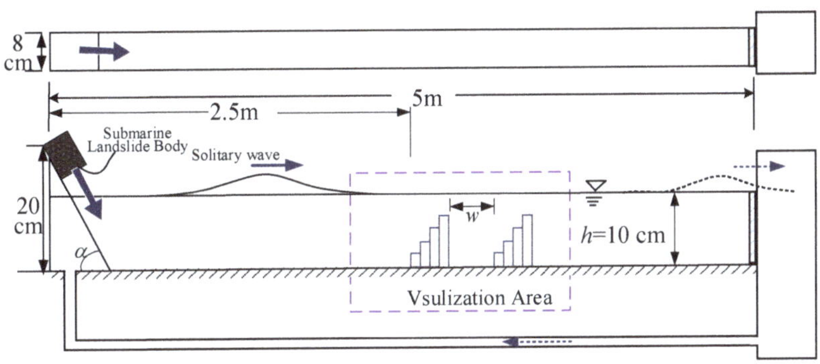

Figure 1. Schematic layout of wave flume and experimental setup.

Figure 2. Experimental setup and wave vortex measurement.

A set of submerged BWs was composed of four blocks of different heights and the same width, of 0.04 m, arranged in an ascending step formation from upstream to downstream. Their heights as a proportion of the water depth were 0.2, 0.4, 0.6 and 0.8, respectively. Figure 3 shows photographs of rectangular and undulating BW sets, resembling stair-step structures and underwater sand dunes, respectively, made of concrete material, as used in the experiments. A combination of two submerged BW sets, either both single or both composite sets, with the separation distance relative to the water depth w/h = 0.4, was also discussed in this study, as listed in Table 1.

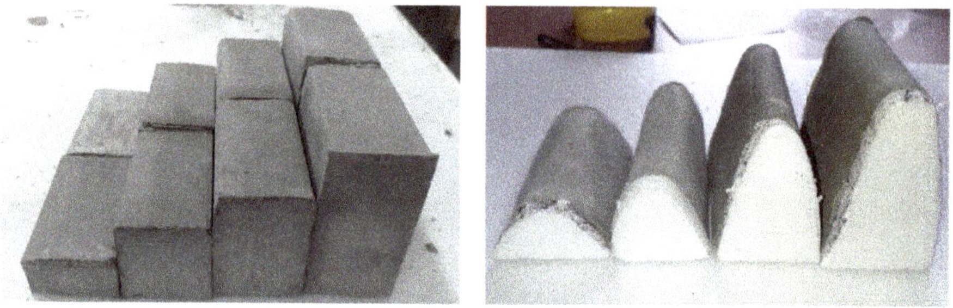

(**a**) (**b**)

Figure 3. Single (**a**) rectangular and (**b**) undulating breakwater set.

Table 1. Combination of breakwaters sets.

Breakwater Type	Schematic Diagram	Breakwater Height d/h	Separation Distance w/h
Single Set	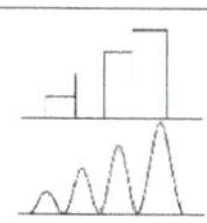	0.2, 0.4, 0.6, 0.8	—
Composite Set	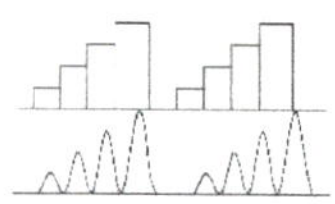	0.2, 0.4, 0.6, 0.8	0.4

A high-speed digital camera with a maximum frame rate of 960 fps at specific resolutions recorded the changes in the waveform to confirm the results measured by the wave gauges. At the highest resolutions, the camera supported images and/or videos with a resolution of up to 1920 × 1080 pixels. The video images were captured at different time intervals in the highest resolution for analysis. PVC-based glitter powder with a particle size of approximately 0.05 mm and a measured settling speed of 7.592 mm/s on average was chosen for use as the seeding particles. In this arrangement, the shooting observation range was relatively wide, and the tracking particles became comparatively blurred when shooting at 960 fps. To ensure the best results, a shooting speed of 480 fps to conduct the experiment was used in this study.

The digital particle image velocimetry tool for MATLAB, named PIVlab [16,17], was chosen as the PIV tool to analyze the captured images. With proper settings, including image preprocessing using contrast-limited adaptive histogram equalization (CLAHE), the cross-correlation of continuous image data could be derived. From this, the physical properties of the flow field, for example, horizontal velocity, u, vertical velocity, v, vorticity, ω, velocity intensity magnitude, and velocity vector angle, θ, which describe the wave–structure interactions for the generated solitary-like waves propagating over submerged BWs, could be derived.

According to Hino and Yamazaki [25], eddy capacity can be estimated by the kinetic energy per unit width of a vortex, that is, the calculated eddy energy. This can be expressed as:

$$E_v = \frac{\rho}{2}\int_0^a 2\pi r q^2 dr = \frac{1}{16}\pi\rho\omega^2 r_0^4\left(1 + 4\ln\frac{a}{r_0}\right) \tag{1}$$

where ω is the vorticity, ρ is the water density, r is the distance to the vortex center, r_0 is the radius of the boundary between the vortex area and the vortex-free area, and a is the maximum radius of the vortex representing the outer limit of the vortex, that is, $r_0 < a$, according to Hino and Yamazaki [25].

In addition, the speed q in the circumferential direction at radius r, and the vortex strength Γ (circulation), can be determined by

$$q = \frac{\omega}{2} r \tag{2}$$

$$\Gamma = \omega \pi r^2 \tag{3}$$

These physical properties of the flow field, including the horizontal velocity, u, vertical velocity, v, and speed, can be derived from each interrogation window during the PIV analyses using PIVlab. Therefore, the vorticity can be determined by

$$\omega = \frac{\partial v}{\partial x} - \frac{\partial u}{\partial y} \tag{4}$$

According to the Coastal Engineering Research Center [26], the total energy of a solitary wave can be expressed as

$$E_0 = \frac{8}{3} \rho g h^2 H \sqrt{H/3h} \tag{5}$$

where H is the solitary wave height and h is the water depth.

3. Experimental Validation

To confirm the appropriateness of applying the above-mentioned methods to the generated solitary-like waves and measured flow field by using PIV analyses, the measured waveform was compared with the theoretical results, and the cross-sectional distribution of the flow velocity in the flume was also verified using measurements made by an acoustic Doppler velocimeter (ADV).

3.1. Validation of the Solitary Wave Generation

As mentioned in the previous section, a solitary-like wave was generated by a landsliding system with a sliding concrete block. In order to confirm whether the analysis methods mentioned above were appropriate, the experimental wave elevation measured by two capacitance wave gauges was compared with the theoretically calculated elevation. Figure 4 shows a comparison of the experimental distribution of water elevation (waveform) and the theoretical distribution derived by the Boussinesq equation [27]:

$$\eta = H sech^2 \left(\sqrt{\frac{3H}{4h^3}} (x - ct) \right) \tag{6}$$

where η represents the water level and c is the wave speed. The comparison shows a satisfactory agreement between the experimental and theoretical results.

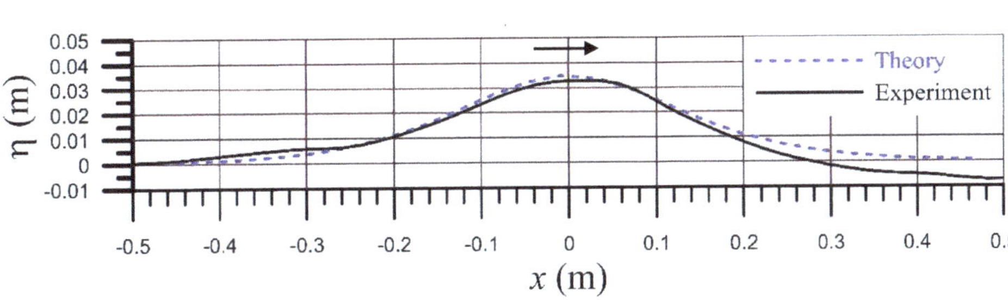

Figure 4. Comparison of measured waveform with the theoretical one.

3.2. Validation of the PIV-Analyzed Velocity Field

In order to confirm the applicability of the velocity field computed by the PIV analyses using PIVlab, the vertical water velocity profiles as measured by the ADV in a water tank with a controlled flow rate were compared to the PIV analyses. This approach was also conducted by Song and Chiew [28] and Ruonan et al. [29]. The ADV used in our study was a Vectrino (Nortek AS), as shown in Figure 5. A comparison of the vertical water velocity profiles derived by ADV and PIVlab with a controlled flow rate is shown in Figure 6. The cross-sectional average flow velocity was approximately 0.049 m/s in the water tank, serving as a reference for confirmation. According to the outlet velocity distribution along the cross-sectional direction of a current field without waves, the difference between the results from the PIV analyses using PIVlab and ADV is approximately 5.276%. More detailed information regarding the accuracy tests of the PIVlab can be found in a previously reported study [17].

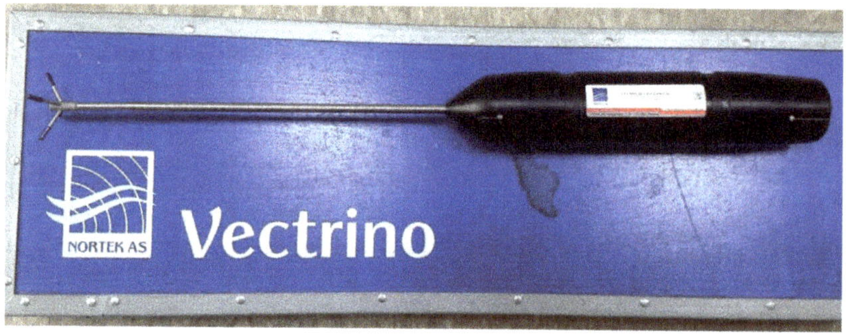

Figure 5. The acoustic Doppler velocimeter for the validation of the PIV-analyzed velocity field.

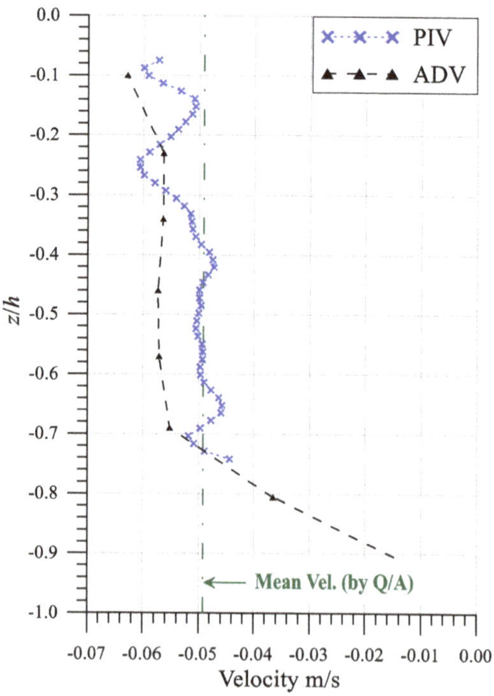

Figure 6. Comparison of the vertical velocity distribution derived by the acoustic Doppler velocimetry (ADV) and the particle image velocimetry (PIV) analyses.

4. Results and Discussion

4.1. Relative Vortex Energies Induced by a Single Set of Rectangular and Undulating BWs

The temporal variation of the flow field and the induced vortex when the generated solitary-like wave propagated over a single set of submerged rectangular and undulating BWs is presented in Figure 7. The data presented were derived from the PIV analyses, and the arrows represent the incident wave direction. From the circumferential velocity and vorticity, together with the current radius of each point, the effective eddy capacity can be calculated. The value of r_0 is determined by the maximum value of the positive and negative velocities, which usually occurs on the periphery of the vortex area. Based on the relative position of the vortex, the center of the entire vortex field is determined, and the energy distributions of each particle around the center point are obtained.

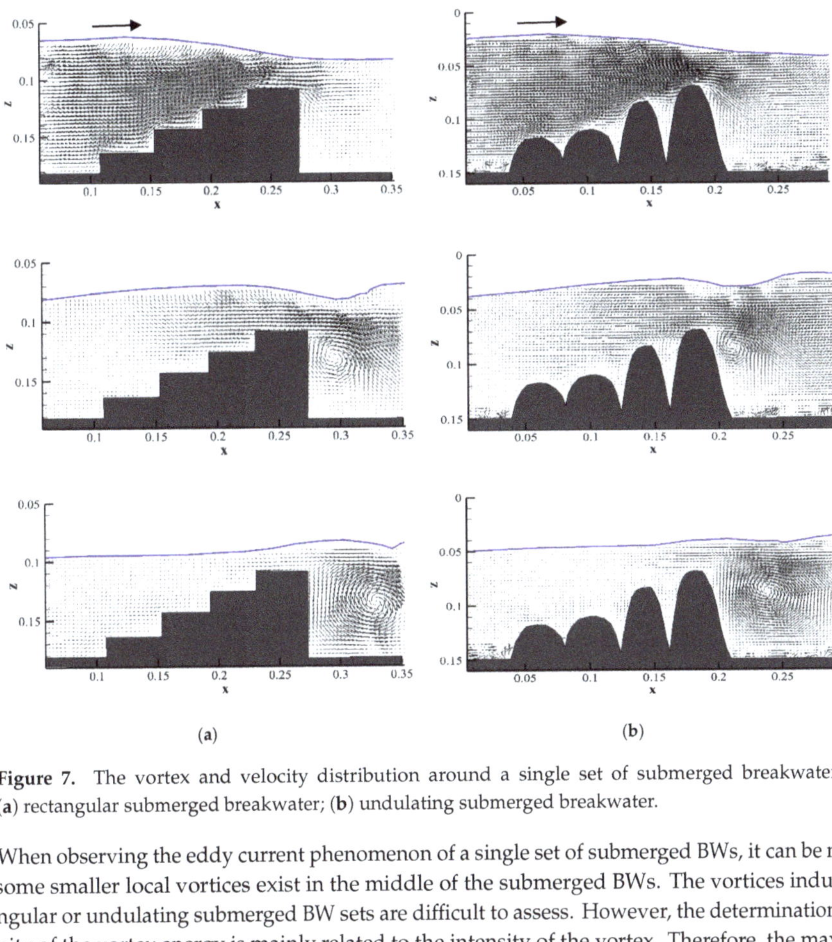

(**a**) (**b**)

Figure 7. The vortex and velocity distribution around a single set of submerged breakwaters: (**a**) rectangular submerged breakwater; (**b**) undulating submerged breakwater.

When observing the eddy current phenomenon of a single set of submerged BWs, it can be noticed that some smaller local vortices exist in the middle of the submerged BWs. The vortices induced by rectangular or undulating submerged BW sets are difficult to assess. However, the determination of the intensity of the vortex energy is mainly related to the intensity of the vortex. Therefore, the maximum vortices appearing behind the submerged BWs were derived from the subsequent results derived from the PIV analyses. From analysis of the vortex intensity, the maximum vorticities around the center of the vortex were 55.03 1/s and 107.81 1/s for a single set of rectangular and undulating submerged BWs, respectively, as shown in Figures 8a and 9a. Accordingly, the total relative vortex energy for a single set of rectangular submerged BWs was approximately 65.09%, with a = 0.051 m and r_0 = 0.025 m, and the total relative vortex energy behind a set of undulating submerged BWs was approximately 50.63%, with $a = 0.051$ m and $r_0 = 0.022$ m, as shown in Figures 8b and 9b. In addition, the transmission coefficient K_t and reflection coefficient K_r could be estimated via the ratio of the transmitted wave height to the incident wave height, and the ratio of the reflected wave height to the incident wave height, according to Goda and Suzuki [30] and Mansard and Funke [31]. The estimated transmission coefficient K_t and reflection coefficient K_r were, respectively, approximately 0.509 and 0.273 for a single set of rectangular submerged BWs, and approximately 0.811 and 0.178 for the single set of undulating submerged BWs, respectively. Moreover, from the concept of energy conservation, the sum of energy attenuation could be calculated from the equation $K_L = \sqrt{1 - K_r^2 - K_t^2}$, and this was found to be approximately 0.816 for the rectangular submerged BWs and 0.558 for the undulating ones.

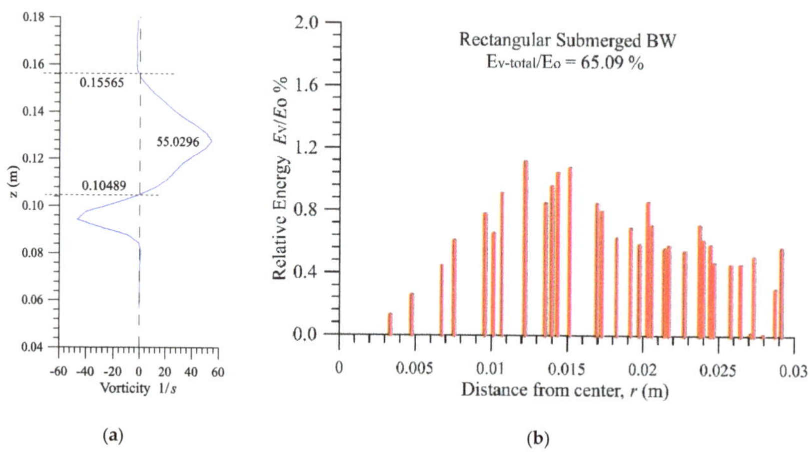

Figure 8. The vorticity and relative energy of the maximum vortex induced by a solitary-like wave propagating over a set of rectangular submerged breakwaters: (**a**) the vorticity distribution along the vertical line of the vortex center; (**b**) the relative vortex energy distribution.

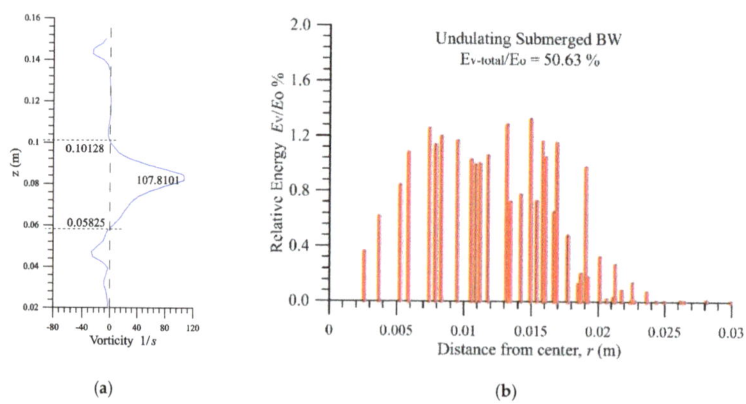

Figure 9. The vorticity and relative energy of the maximum vortex induced by a solitary-like wave propagating over a set of undulating submerged breakwaters: (**a**) the vorticity distribution along the vertical line of the vortex center; (**b**) the relative vortex energy distribution.

The results of the above-mentioned analyses show that the total vortex energy induced by a set of rectangular submerged BWs appears to be slightly larger because the vortex area range is wider than that of the undulating BW-induced vortex, although the maximum vorticity of the former is relatively smaller. In addition, the relative vortex energy percentage of the submerged BW-induced vortices shows that the main concentration range of vortex energy lies between 0 m and 0.03 m for the single set of rectangular submerged BWs, as shown in Figure 8b, and between 0 m and 0.025 m for the single set of undulating submerged BWs, as shown in Figure 9b. Moreover, the maximum relative energy, E_v/E_0, is approximately 1.12% for the single set of rectangular submerged BWs and approximately 1.36% for the single set of undulating submerged BWs. This shows that the rectangular submerged BW-induced vortex is relatively divergent, whereas the undulating submerged BW-induced vortex is relatively concentrated. E_v and E_0 denote the energy of the induced vortex and incident wave, as determined by Equations (1) and (5), respectively, and the variation of vorticity distribution through the vertical line of the vortex center can be obtained from the PIV analysis process using PIVlab.

Figures 10 and 11 show the temporal variation and distribution of streamlines when a solitary-like wave propagates over a set of rectangular or undulating submerged BWs, respectively. It can be clearly seen that some local small eddies occurred between the steps of the undulating submerged BWs, which are less noticeable in the case of the rectangular submerged BWs. Furthermore, when observing the speed gradient of the solitary wave crest passing through the set of submerged BWs, which can be traced through the streamline variation, it can be seen that the constant velocity gradient lines, which exhibit an upper and lower equal gradient pattern, occur at the front edges of the set of rectangular submerged BWs. This indicates that the water particles no longer move quickly in the forward direction after being blocked by the set of rectangular submerged BWs. By contrast, the blockages of water particles and the reduction of the propagation speed in the forward direction are less notable for the case of the solitary wave passing over the set of undulating submerged BWs.

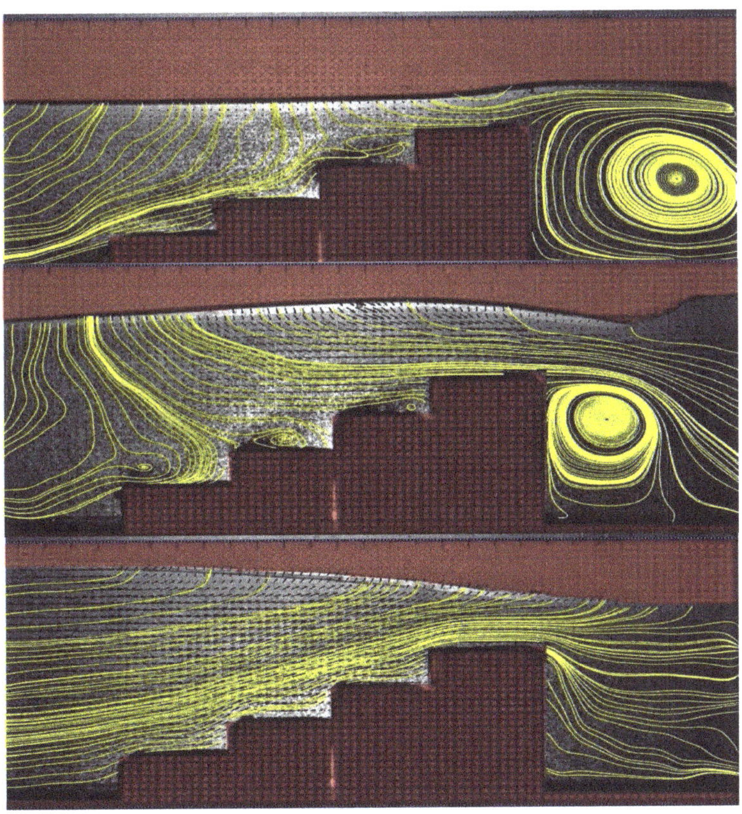

Figure 10. The streamline variation of a solitary-like wave propagating over a single set of the rectangular submerged breakwaters.

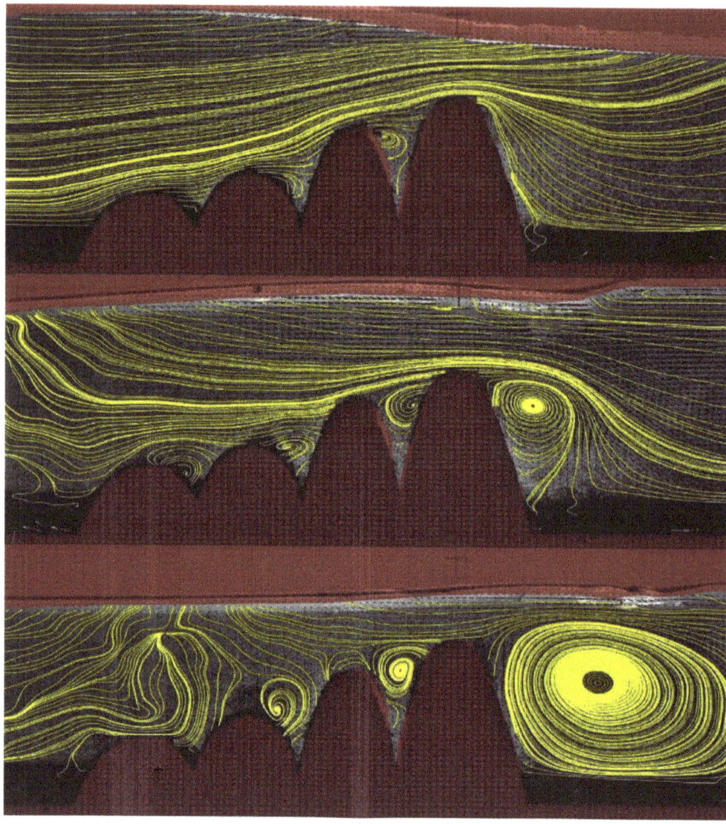

Figure 11. The streamline variation of a solitary-like wave propagating over a single set of the undulating submerged breakwaters.

4.2. Relative Vortex Energies Induced by a Composite Set of Rectangular and Undulating BWs

Similar to the previous section, the temporal variation of the flow field and the induced vortex when a generated solitary-like wave propagates over a composite set of submerged rectangular and undulating BWs is presented in Figure 12, with the arrows representing the incident wave direction. The omposite set of the submerged BWs is composed of two BWs, either rectangular or undulating, with a separation distance relative to the water depth w/h = 0.4, as listed in Table 1. The results acquired from a composite set of rectangular submerged BWs (Figure 12a) give an estimated transmission coefficient, K_t, and reflection coefficient, K_r, of approximately 0.454 and 0.256, respectively. Moreover, from the concept of energy conservation, the sum of energy attenuation, K_L, is approximately 0.854. From the analysis of the vortex intensity, the analyzed maximum vorticities around the center of the vortex for the first and second sets of the composite rectangular submerged BWs were approximately 55.03 1/s (Figure 13a) and 38.29 1/s (Figure 13b), respectively. The total relative vortex energy was obtained, and the relative percentage is shown in Figure 13. The total relative vortex energy of the first set of composite rectangular submerged BWs is approximately 28.29%, and approximately 36.99% for the second set. Therefore, the total vortex energy accounts for 65.28% of the entire incident wave energy.

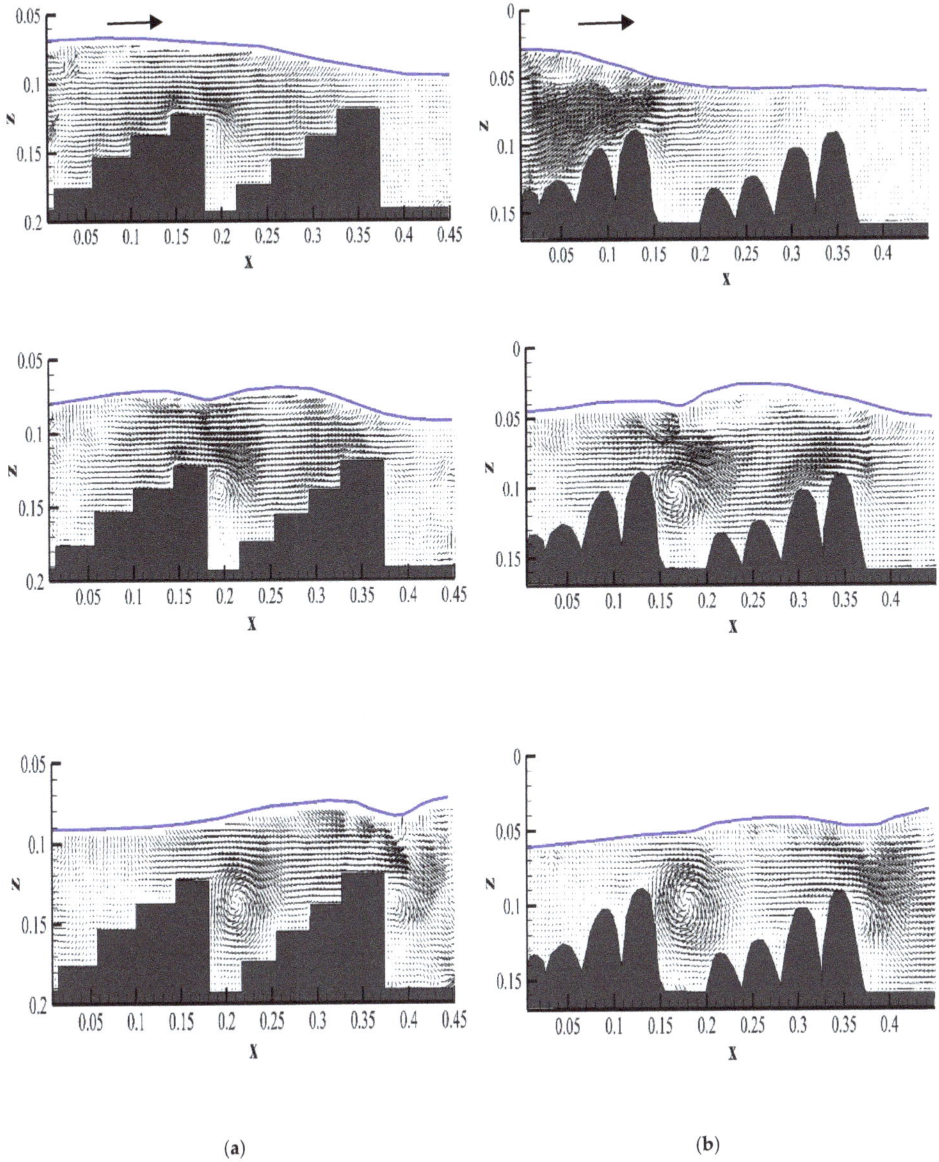

Figure 12. The vortex and velocity distribution around a composite set of submerged breakwaters with a separate distance of w/h = 0.4: (**a**) rectangular submerged breakwater; (**b**) undulating submerged breakwater.

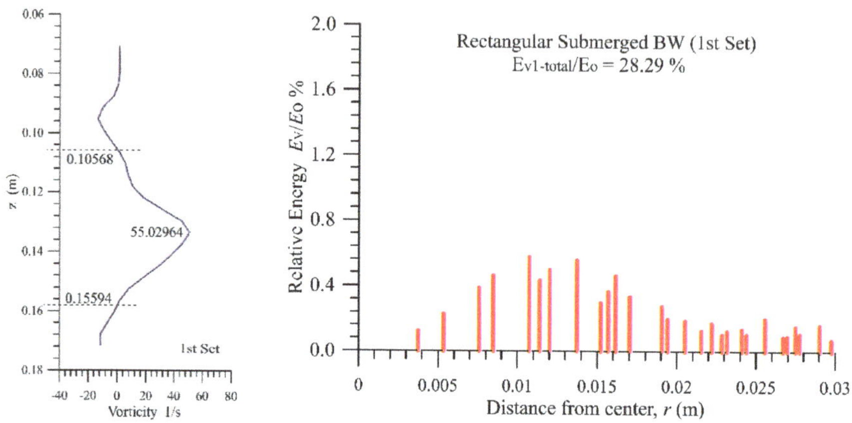

(**a**)

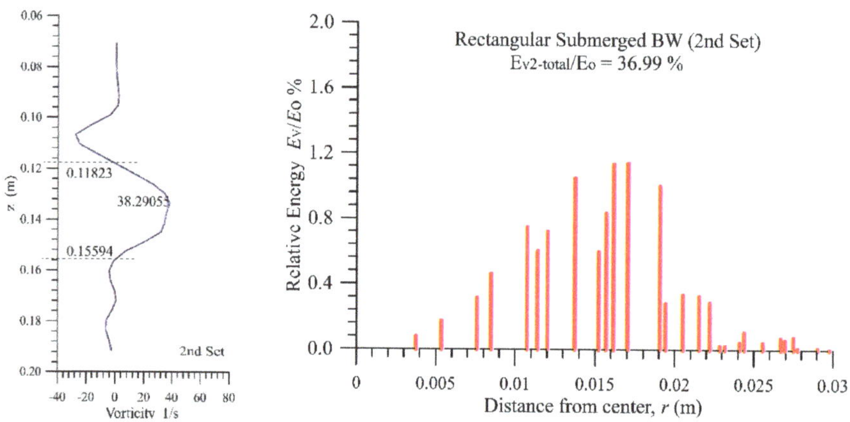

(**b**)

Figure 13. The vorticity distribution along the vertical line of the vortex center, and the relative vortex energy distribution of the maximum vortex induced by a solitary-like wave propagating over (**a**) the first set and (**b**) the second set of a composite set of rectangular submerged breakwaters.

Similarly, the vortex and the velocity field distribution induced by the landslide-generated, solitary-like wave propagating through the composite set of undulating submerged BWs are shown in Figure 12b. The vorticity distribution along the vertical line of the vortex center, and the distribution of relative vortex energy relative to the maximum vortex induced by the solitary-like wave propagating over the first and second sets of composite undulating submerged BWs, are shown in Figure 14. The maximum vorticities around the center of the first and second vortices are 86.11 1/s and 59.69 1/s, respectively, as shown in Figure 14a,b. The transmission coefficient, K_t, is estimated to be 0.736, and the reflection coefficient, K_r, is approximately 0.189. Thus, the sum of energy attenuation, K_L, can be calculated as 0.649 from the concept of energy conservation. The total relative vortex energies of the first and second sets of the composite undulating BWs were estimated to be 26.44% and 25.08%, respectively, as shown in Figure 14. Therefore, the total vortex energy accounts for approximately 51.52% of the entire incident wave energy. When comparing the total vortex energy of the single set case with that of the composite set case, it can be seen that the composite submerged BW sets cause more wave energy attenuation.

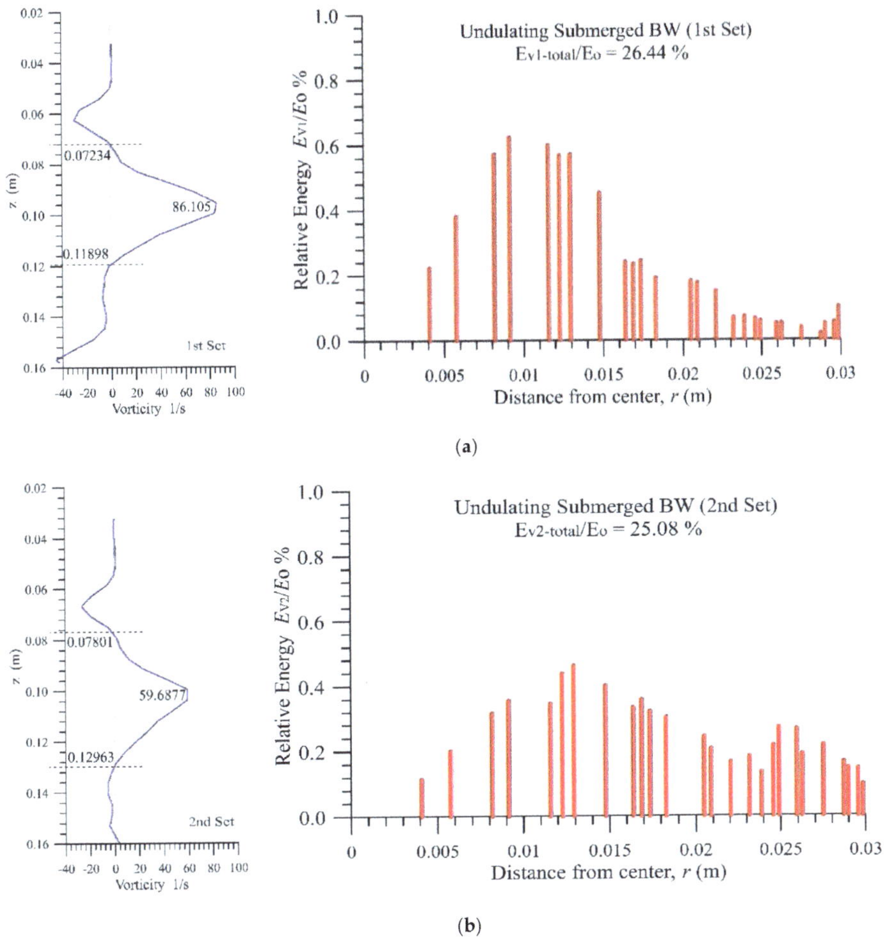

Figure 14. The vorticity distribution along the vertical line of the vortex center and relative vortex energy distribution of the maximum vortex induced by a solitary-like wave propagating over (**a**) the first set and (**b**) the second set of a composite set of undulating submerged breakwaters.

A comparison of the temporal variation and distribution of streamlines with the relative vorticity at three points in time, T1, T2, and T3, representing the generated solitary-like wave passing through the first set of the rectangular and undulating submerged BWs, respectively, is shown in Figures 15 and 16. The overall distributions of the streamlines of the generated solitary-like waves propagating over the two types of submerged BWs are similar. Still, similar to the observations in the single set case, more local small eddies occurred between the blocks of the undulating submerged BWs than between the rectangular submerged BWs. In addition, the variation in streamlines indicates that the movement of water particles was significantly hindered by the rectangular submerged BWs. When observing the vorticity distribution along the vertical line of the vortex center induced by the first BW set, as shown in Figures 15 and 16, it can be seen that the induced vortex of the rectangular BWs is relatively divergent compared with that of the undulating submerged BWs, for which the induced vortex is relatively concentrated. In addition, the magnitude of the maximum vorticity of the undulating submerged BWs appears to be larger than that of the rectangular BWs, which is the same as for the single set case.

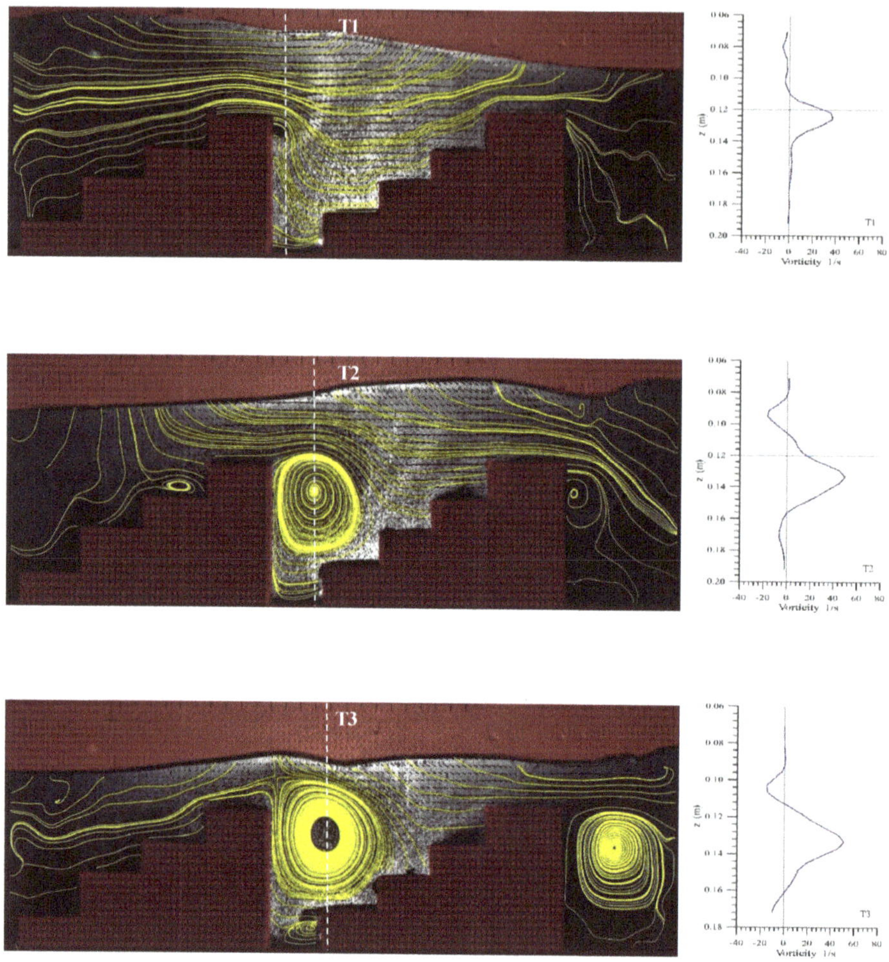

Figure 15. The temporal variation of streamlines and the vorticity distribution along the vertical line of the vortex center induced by a solitary-like wave propagating over a composite set of rectangular submerged breakwaters.

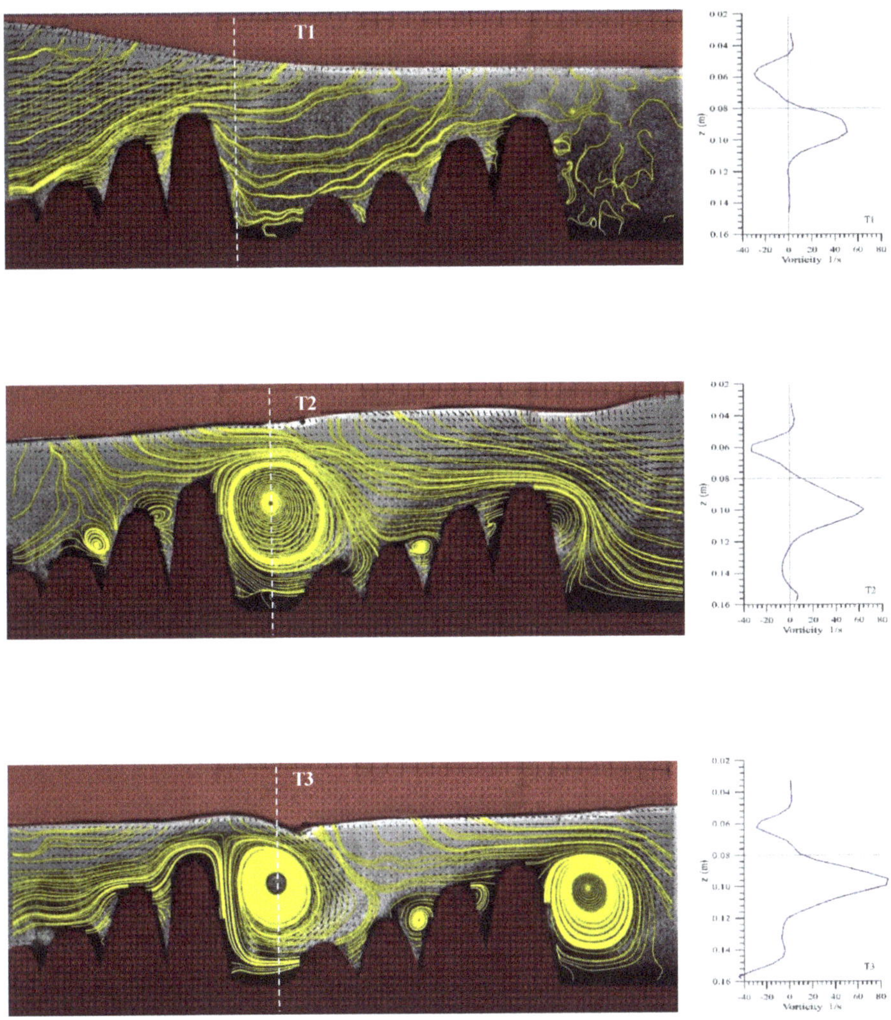

Figure 16. The temporal variation of streamlines and the vorticity distribution along the vertical line of the vortex center induced by a solitary-like wave propagating over a composite set of undulating submerged breakwaters.

5. Conclusions

In this study, a series of hydraulic model experiments was conducted to analyze the formation and variations of obstacle-induced vortices. These vortices were induced by landslide-generated solitary-like waves propagating over single and composite sets of rectangular and undulating submerged BWs. The induced vortices were further analyzed by PIV techniques using PIVlab, and the correlation between vortex energy and wave energy attenuation induced by different types of submerged BWs was investigated. The results showed that the sum of the relative vortex energy generated behind the composite submerged BW sets is greater than that of the single submerged BW sets. The trend of the estimated total relative vortex energy of the maximum vortices is in agreement with the sum of the energy attenuation, K_L, indicating that the attenuated wave energy is related to the induced vortices. Moreover, the total relative vortex energy of the rectangular submerged BW sets is approximately 1.29 (the single set) and 1.27 (the composite set) times larger than that of the undulating submerged BW sets.

This shows that the rectangular submerged BW sets have stronger effects on the energy attenuation of the solitary-like wave. The magnitude of the maximum vorticity of the undulating BW sets appears to be larger than that of the rectangular BW sets; however, the total vortex energy is shown to be slightly smaller, owing to the narrower vortex area range. Finally, the induced vortex of the rectangular submerged BW sets is relatively divergent when compared with that of the undulating submerged BW sets, which is relatively concentrated.

Author Contributions: Conceptualization, methodology, R.-S.S.; resources, W.-K.W.; formal analysis, C.-Y.L. and R.-S.S.; writing—original draft, C.-Y.L. and R.-S.S.; data curation, C.-Y.L.; writing—review and editing, C.-Y.L. and R.-S.S. All authors have read and agreed to the published version of the manuscript.

Funding: This research received no external funding.

Conflicts of Interest: The authors declare no conflict of interest. The funders had no role in the design of the study; in the collection, analyses, or interpretation of data; in the writing of the manuscript, or in the decision to publish the results.

References

1. Davies, A.G.; Heathershaw, A.D. Surface-wave propagation over sinusoidally varying topography. *J. Fluid Mech.* **1984**, *144*, 419–443. [CrossRef]
2. Dalrymple, R.A.; Kirby, J.T. Water Waves Over Ripples. *J. Waterw. Port Coast. Ocean Eng.* **1986**, *112*, 309–319. [CrossRef]
3. O'Hare, T.J.; Davies, A.G. A new model for surface wave propagation over undulating topography. *Coast. Eng.* **1992**, *18*, 251–266. [CrossRef]
4. Kar, P.; Sahoo, T.; Behera, H. Effect of Bragg scattering due to bottom undulation on a floating dock. *Wave Motion* **2019**, *90*, 121–138. [CrossRef]
5. Shih, R.-S.; Weng, W.-K. A Study of Long Wave Attenuation over Composite Undulating Breakwaters. *J. Coast. Res.* **2016**, *32*, 78–90. [CrossRef]
6. Shih, R.-S.; Weng, W.-K. Experimental determination of the performance characteristics of an undulating submerged obstacle. *Ships Offshore Struct.* **2016**, *11*, 129–141. [CrossRef]
7. Shih, R.-S.; Weng, W.-K.; Li, C.-Y. Experimental Study on the Generation and Attenuation of Landslide Tsunamis. *J. Coast. Res.* **2018**, *34*, 864–876. [CrossRef]
8. Hino, M.; Yamazaki, T. Reflection and Transmission Coefficient, and Energy Loss of Surface Wave by a Vertical Barrier. *Proc. Jpn. Soc. Civ. Eng.* **1971**, *1971*, 75–80. [CrossRef]
9. Ikeda, S.; Asaeda, T.; Nomoto, K.; Kimura, E. Vortex Structure and Energy Dissipation Near a Vertical Plate under Wave Action. *Doboku Gakkai Ronbunshu* **1985**, *1985*, 87–96. [CrossRef]
10. Hokamura, T.; Ejima, T.; Imamura, I.; Tsushima, S.; Tsujimoto, G.; Yuhi, M.; Oki, K.; Yamada, F. Wave energy dissipation mechanisms of the double barrier floating breakwater. *Proc. Coast. Eng. JSCE* **2008**, *55*, 896–900. [CrossRef]
11. Umeyama, M. Investigation of Single and Multiple Solitary Waves Using Superresolution PIV. *J. Waterw. Port Coast. Ocean Eng.* **2013**, *139*, 304–313. [CrossRef]
12. Zaghian, R.; Tavakoli, M.R.; Karbasipour, M.; Nili Ahmadabadi, M. Experimental study of flow structures of a solitary wave propagating over a submerged thin plate in different angles using PIV technique. *Int. J. Heat Fluid Flow* **2017**, *66*, 18–26. [CrossRef]
13. Li, J.; Zheng, Y.-X.; Gong, P.-H.; Guan, C.-T. Numerical simulation and PIV experimental study of the effect of flow fields around tube artificial reefs. *Ocean Eng.* **2017**, *134*, 96–104. [CrossRef]
14. Chang, C.-H.; Lin, C.; Wang, K.-H.; Jaf, J.M. Numerical simulations and experimental visualizations of the vortex characteristics for a solitary wave interacting with a bottom-mounted vertical plate. *J. Hydro-Environ. Res.* **2018**, *19*, 88–102. [CrossRef]
15. Lin, C.; Wong, W.-Y.; Raikar, R.V.; Hwung, H.-H.; Tsai, C.-P. Characteristics of Accelerations and Pressure Gradient during Run-Down of Solitary Wave over Very Steep Beach: A Case Study. *Water* **2019**, *11*, 523. [CrossRef]
16. Thielicke, W.; Stamhuis, E. PIVlab—Towards User-friendly, Affordable and Accurate Digital Particle Image Velocimetry in MATLAB. *J. Open Res. Softw.* **2014**, *2*, e30. [CrossRef]

17. Thielicke, W. The Flapping Flight of Birds: Analysis and Application. Ph.D. Thesis, Rijksuniversiteit Groningen, Groningen, The Netherlands, 2014.
18. Katell, G.; Eric, B. Accuracy of solitary wave generation by a piston wave maker. *J. Hydraul. Res.* **2002**, *40*, 321–331. [CrossRef]
19. Renouard, D.P.; Santos, F.J.S.; Temperville, A.M. Experimental study of the generation, damping, and reflexion of a solitary wave. *Dyn. Atmos. Oceans* **1985**, *9*, 341–358. [CrossRef]
20. Ippen, A.T.; Kulin, G. The shoaling and breaking of the solitary wave. *Coast. Eng. Proc.* **1954**, *1*, 4. [CrossRef]
21. Monaghan, J.J.; Kos, A. Scott Russell's wave generator. *Phys. Fluids* **2000**, *12*, 622. [CrossRef]
22. Enet, F.; Grilli, S.T. Experimental Study of Tsunami Generation by Three-Dimensional Rigid Underwater Landslides. *J. Waterw. Port Coast. Ocean Eng.* **2007**, *133*, 442–454. [CrossRef]
23. Watts, P. Tsunami Features of Solid Block Underwater Landslides. *J. Waterw. Port Coast. Ocean Eng.* **2000**, *126*, 144–152. [CrossRef]
24. Najafi-Jilani, A.; Ataie-Ashtiani, B. Estimation of near-field characteristics of tsunami generation by submarine landslide. *Ocean Eng.* **2008**, *35*, 545–557. [CrossRef]
25. Hino, M.; Yamazaki, T. Vortex formation and energy dissipation from a vertical plate. In Proceedings of the 18th Coastal Engineering, Cape Town, South Africa, 14–19 November 1982; pp. 61–65.
26. Coastal Engineering Research Center (U.S.). *Shore Protection Manual*; U.S. Army Coastal Engineering Research Center: Fort Belvoir, VA, USA, 1973.
27. Svendsen, I.A. *Introduction to Nearshore Hydrodynamics*; Advanced Series on Ocean Engineering; World Scientific: Singapore, 2006; Volume 24, ISBN 978-981-256-142-8.
28. Song, T.; Chiew, Y.M. Turbulence Measurement in Nonuniform Open-Channel Flow Using Acoustic Doppler Velocimeter (ADV). *J. Eng. Mech.* **2001**, *127*, 219–232. [CrossRef]
29. Ruonan, B.; Liekai, C.; Xingkui, W.; Danxun, L. Comparison of ADV and PIV Measurements in Open Channel Flows. *Procedia Eng.* **2016**, *154*, 995–1001. [CrossRef]
30. Goda, Y.; Suzuki, T. Estimation of incident and reflected waves in random waves. In Proceedings of the 15th Conference on Coastal Engineering, Honolulu, HI, USA, 11–17 July 1976; Volume 1, pp. 828–845. [CrossRef]
31. Mansard, E.P.D.; Funke, E.R. The Measurement of Incident and Reflected Spectra Using a Least squares Method. In Proceedings of the 17th Conference on Coastal Engineering ASCE, Sydney, Australia, 16–21 July 1980; pp. 154–172. [CrossRef]

Journal of
Marine Science and Engineering

MDPI

Article

Dynamic Stability of an Ocean Current Turbine System

Shueei-Muh Lin [1], Yang-Yih Chen [2], Hung-Chu Hsu [2,*] and Meng-Syue Li [3]

[1] Department of Mechanical Engineering, Kun Shan University and Green Energy Technology Research Centre (GETRC), Tainan 710303, Taiwan; smlin45@gmail.com
[2] Department of Marine Environment and Engineering, National Sun Yat-sen University, Kaohsiung 80424, Taiwan; yichen@faculty.nsysu.edu.tw
[3] Marine Science and Information Research Center, National Academy of Marine Research, Kaohsiung 80661, Taiwan; lms@naor.oac.gov.tw
* Correspondence: hchsu@mail.nsysu.edu.tw; Tel.: +886-7-525-5172

Received: 8 July 2020; Accepted: 1 September 2020; Published: 6 September 2020

Abstract: This paper presents a theoretical solution for the dynamic stability of the ocean current turbine system developed in Taiwan. This system is tethered to the sea floor and uses the Kuroshio Current to produce electricity. To maintain the performance of the turbine system in the presence of the Kuroshio Current, the stability of the surfaced turbine needs to be considered. The proposed system is composed of a turbine, a buoyance platform, a traction rope, and a mooring foundation. The two-dimensional theoretical solutions treat the turbine as a rigid body with a movable structure that is moored with two cables. In this model, the gravity, buoyancy, and drag force generated by the wave on the turbine structure are considered. In addition, an analytical solution is proposed for the general system. Finally, the effects of the wave on the pitch motion and dynamical stability of the ocean current turbine system are investigated.

Keywords: stability; ocean current power system; surface type; buoyance platform; mooring foundation

1. Introduction

The application of traditional energy resources (e.g., fossil fuels) causes serious environmental pollution. Renewable energy resources, such as wind, sunlight, ocean currents, waves, and tidal current, have been introduced as alternatives for achieving a low-carbon society. Ocean current energy is a potential power source that must be developed. Various forms of ocean energy are being investigated as potential sources of power generation [1–4]. For example, the Kuroshio Current—a strong current passing through the east of Taiwan—is expected to be an excellent energy resource. It has a mean velocity of 1.2–1.53 m/s near the surface and the potential electricity capacity of the Taiwan Current is approximately 4 GW [5]. This ocean energy source is stable and abundant, with the potential to be developed and utilized.

Tidal current—which can be extracted from the rise and fall of sea levels under the gravitational force exerted by the Moon and Sun as well as Earth's rotation—is one of the most valuable resources. Moreover, tidal current energy is more predictable than wind and wave energies [6]. Tidal current turbine (TCT) can be categorized into horizontal- and vertical-axis tidal turbines [7,8]. These inventions can harness the kinetic energy of tides and principally convert it to electricity. Horizontal-axis TCTs are the most common device, with their rotation axis parallel to the current stream direction [9,10]. By contrast, vertical-axis TCTs rotate about a vertical axis perpendicularly to the current stream [11]. Chen and Lam [6] reviewed the survivability of tidal power devices used to harness tidal power.

Zhou et al. [7] presented up-to-date information on large tidal turbine projects with a power exceeding 500 kW as well as their achievements and development histories. Most industrialized marine current turbine (MCT) devices are horizontal-axis turbines, with the rotation axis parallel to the current flow direction. The main disadvantages associated with vertical-axis turbines include their relatively low self-starting capability, high torque fluctuations, and generally lower efficiency than horizontal-axis turbines. The power of industrialized MCT devices rigidly fixed at a seabed below 80 m depth, such as Atlantis AR1000 turbine, Voith Hydro turbine, and GE-Alstom tidal turbine, is over 1 MW at a current speed of approximately 2.4–4 m/s. These are called seabed-mounted turbines. Some devices, such as the Scotrenewables SR250 turbine and Ocean Renewable Power Company (ORPC) turbine, are flexibly moored at deep seabeds.

A flexible moored device is an important tool for deployment in deep water. The majority of moored tidal current turbine developers agree that by using a flexibly moored system, the device will be automatically self-aligned to the direction of current flow [7,12–15]. A traditional design uses the gravity foundations or piles in deep water, which are complex and expensive. The flexible mooring lines and anchors can be deployed in deep water, where the other designs may by impractical [16]. Thus, it is important to develop a mathematical model for a flexible mooring system for ocean current energy systems. Muliawan et al. [17] determined the extreme responses in the mooring lines of a two-body floating wave energy converter with four catenary cables. Angelelli et al. [18] investigated the behavior of a wave energy convertor mooring system with four spread cables by using Ansys AQWA software. Chen et al. [4] investigated the wave-induced motions of a floating wave energy converter (WEC) with mooring lines by using the smoothed particle hydrodynamics method. Davidson and Ringwood [19] reviewed the mathematical models for wave energy convertor mooring systems. The marine energy developer Minesto [20] developed a floating subsea kite with flexible mooring.

Cribbs [7] proposed a conceptual design for the flexible mooring of a current turbine fixed to a 300-m-deep seabed. The system included a mooring chain, a mooring line, a flounder plate, two lines for the turbine and the platform, a marine turbine, and a rotating turbine using blade-estimated airfoils. However, this system has not been practically applied thus far. Chen et al. [4] successfully moored a 50 kW ocean current turbine supplied by Wanchi Company to an 850-m-deep seabed at the offshore area of Pingtung County, Taiwan. At a current speed of 1 m/s, the output power of the system was 26 kW. IHI and NEDO [21] conducted a demonstration experiment of the ocean current turbine located off the coast of Kuchinoshima Island, Kagoshima Prefecture, and obtained data for commercialization; the demonstration experiment was conducted for seven days. The turbine comprised a combination of three cylindrical floats, called pods, having a total length of approximately 20 m, width of approximately 20 m, and turbine rotor diameter of approximately 11 m. The turbine system was moored from the anchor installed on a 100-m-deep seabed.

Besides the current rotating-type turbine, the Wanchi Company is also developing an ocean current turbine with a translational blade. To produce more power from the ocean, it is configured with several matrix-array turbines. The system comprises a turbine, a buoyance platform, a traction rope, and a mooring foundation, as shown in Figure 1. The ocean current turbine is tethered to an ~900-m-deep seabed and the ocean current flows perpendicular to the turbine plane. This system might be more unstable than the horizontal rotational turbine. The system is composed of a turbine, a buoyance platform, a traction rope, and a mooring foundation. The system stability is important for the practical operation. So far, only a few studies have investigated the stability of the ocean current turbine system. No literature is devoted to the mathematical model of the system about the coupled heaven, surge and pitch motions. No analytical solution of the system is also presented. Because the system must be sufficiently stable under the effect of wave, the mathematical model is developed and the analytical solutions are presented in this study. Moreover, the effects of several parameters on the system stability are investigated.

In Section 2, a two-dimensional model for the motions of the ocean current turbine system and floater is developed. The turbine system and floater are treated as rigid bodies and the cables are divided into two sections. Theoretical solutions of the motions are presented in Section 3. As detailed in Section 4, a series of simulations were conducted for evaluating the dynamic stability of the system under various wave conditions. Section 5 summarizes the present study and provides some concluding remarks.

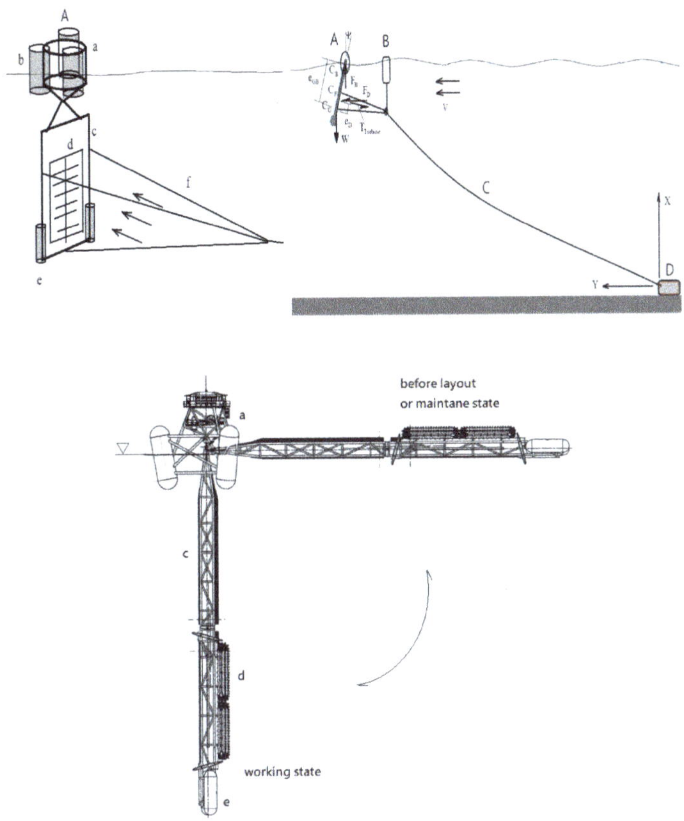

Figure 1. Current turbine system composed of turbine (**A**) (a: Generator; b: Float; c: Structure; d: Translational blade; e: Balanced weight; f: Cable; the current energy transferred by the blade to generator through the high-pressure oil); buoyance platform (**B**); traction rope (UHMWPE) (**C**); and anchor (**D**).

2. Governing Equations

According to Figure 1, the analysis of the motion and dynamic stability of the ocean current turbine system with a flexible mooring cable is complex. Assume that the structures of turbine and carrier subjected the force due to wave and current are kept during the motion. Because the displacement and pitching motion of the system are significantly concerned, these two components are considered rigid bodies. Moreover, the inertia effect of the elastic cable is neglected. Therefore, the system is simulated as a discrete one. The overall system comprises an anchored mooring, a carrier, and a vertical ocean turbine system driven by a translational blade with a gravity anchor. A mooring system that comprises a long main cable with a sub-cable connecting the carrier and ocean turbine is deployed. Due to the complexity of system stability, the flow field is considered to be in steady state and the ocean current velocity is assumed to be constant and uniform. The horizontal force applied to the turbine and its structure during electricity generation is also assumed to be constant. The coupled motion of the system includes the horizontal, vertical, and pitching oscillations. As shown in Figures 2 and 3, owing to the wave fluctuation, the buoyance forces applied on the floating platform and turbine can be expressed as follows:

$$F_{B1} = (H_1 - X_1)A_1\rho g \quad (1)$$

$$F_{B2} = (H_2 - X_2)A_2\rho g \quad (2)$$

where the subscripts "1" and "2" denote the carrier and turbine, respectively, and A_1 and A_2 denote the corresponding hydrodynamic areas. The wave heights at the carrier and turbine are $H_1 = H_0 \sin \Omega t$ and $H_2 = H_0 \sin(\Omega t + \Phi)$, where H_0 is the wave amplitude. ($H_i - X_i$, I = 1~2) indicates the vertical displacement, Ω is the wave frequency, and Φ is the phase lag due to the propagation delay from the carrier to the turbine. On the basis of linear wave theory, the phase is expressed as $\Phi = -kL_2$, where k indicates the wave number, L_2 is the horizontal distance between the carrier and turbine, $\rho(=1025\ \mathrm{kg \cdot m^{-3}})$ is the density of water, and g $(=9.81\ \mathrm{m \cdot s^{-2}})$ is the gravitational acceleration.

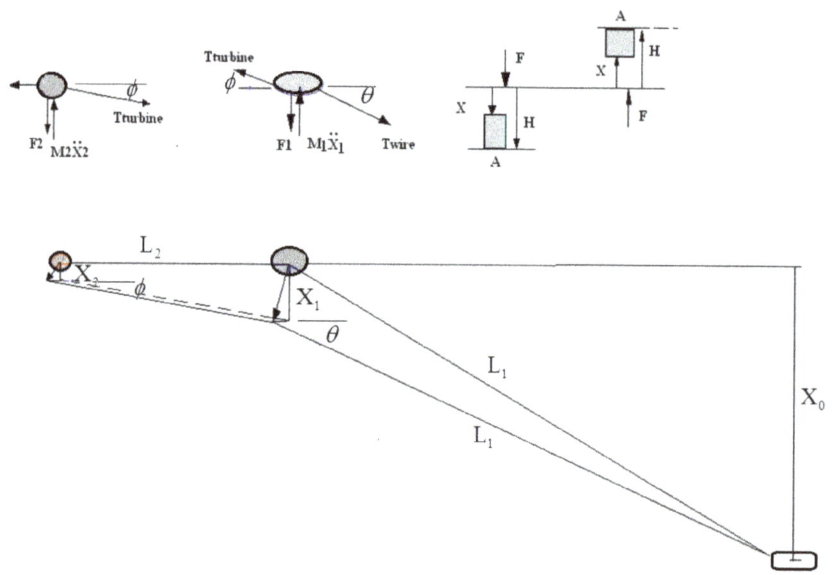

Figure 2. Concentrated mass model of the ocean current turbine system subjected to wave currents.

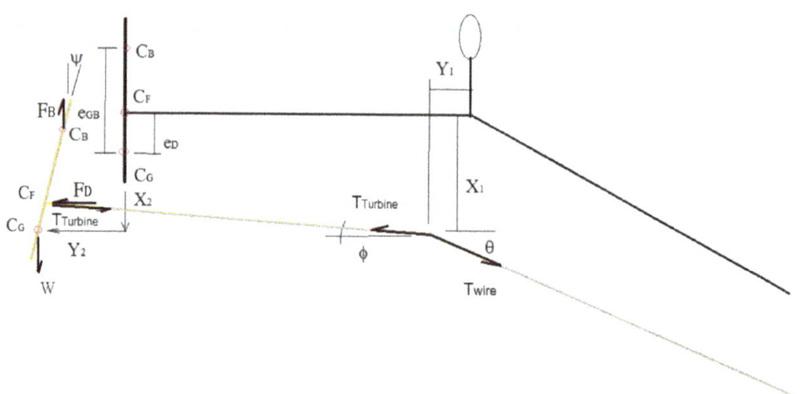

Figure 3. Coordinate and force distribution of the system.

According to the static equilibrium, the force in the horizontal direction at the joint o is expressed as follows:

$$T_{wire}\cos\theta = T_{turbine}\cos\phi \tag{3}$$

According to the dynamic equilibrium, the equations of motion of the floating carrier and turbine platform in the vertical direction are expressed as

$$M_1\ddot{X}_1 - F_{B1} + T_{tur}\sin\phi - T_{wire}\sin\theta = 0 \tag{4}$$

and

$$M_2\ddot{X}_2 - F_{B2} - T_{tur}\sin\phi = 0 \tag{5}$$

where $\ddot{X}_1$ and $\ddot{X}_2$ are the vertical accelerations of the floater and turbine, respectively; T_{tur} is the tension of the wire between the turbine and floater; and T_{wire} is that between the cable and anchor. If the vertical displacement is substantially smaller than the connecting cable between the floating and turbine platforms, the angle ϕ is very small, as shown in Figure 2, and can be approximated as follows:

$$\sin\phi \approx \frac{X_1 - X_2}{L_2} \tag{6}$$

The cable made by polyethylene dyneema connecting the floating platform and mooring foundation is considered. The material is the ultra-high molecular weight polyethylene (UHMWPE). Because the material properties are great strength, light weight and flexible, the mooring cable is likely straight during the turbine subjected to the ocean current force. Moreover, because the vertical displacement is significantly smaller than the water depth, the following approximate relation can be obtained:

$$\sin\theta \approx \frac{X_0}{L_1} \tag{7}$$

Through substituting Equations (3), (6) and (7) into Equations (4) and (5), we obtain the coupled equations of motion in terms of vertical displacements $\{X_1, X_2\}$ for the carrier and turbine:

$$\begin{aligned} &M_1\ddot{X}_1 + \left[A_1\rho g + T_{tur}\tfrac{1}{L_2}\right]X_1 - T_{tur}\tfrac{1}{L_2}X_2 \\ &= T_{tur}\frac{X_0}{\sqrt{L_1^2 - X_0^2}} + H_1A_1\rho g = T_{tur}\frac{X_0}{\sqrt{L_1^2 - X_0^2}} + A_1\rho g H_0\sin\Omega t \end{aligned} \tag{8}$$

and

$$M_2\ddot{X}_2 - \frac{T_{tur}}{L_2}X_1 + \left(A_2\rho g + \frac{T_{tur}}{L_2}\right)X_2 = A_2\rho g H = A_2\rho g H_0\sin(\Omega t + \Phi) \tag{9}$$

Equations (8) and (9) can be expressed as the equation of vertical motion in the matrix form

$$\begin{bmatrix} M_1 & 0 \\ 0 & M_2 \end{bmatrix}\begin{bmatrix} \ddot{X}_1 \\ \ddot{X}_2 \end{bmatrix} + \begin{bmatrix} K_{11} & K_{12} \\ K_{21} & K_{22} \end{bmatrix}\begin{bmatrix} X_1 \\ X_2 \end{bmatrix} = \begin{bmatrix} T_{Tur}\frac{X_0}{\sqrt{L_1^2 - X_0^2}} + A_1\rho g H_0\sin\Omega t \\ A_2\rho g H_0\sin(\Omega t + \Phi) \end{bmatrix} \tag{10}$$

where the first 2×2 matrix is the mass one, the second 2×2 matrix is the stiffness one, the last term is the forcing one due to the wave and the force of the turbine and

$$K_{11} = A_1\rho g + T_{Tur}\frac{1}{L_2},\ K_{12} = K_{21} = \frac{-T_{Tur}}{L_2},\ K_{22} = A_2\rho g - \frac{T_{Tyr}}{L_2},$$

where X0 is the depth of the seabed, {X1, X2} indicate the vertical displacements, Φ is the phase lag angle, and {M1, M2} are the masses.

Because the distance e_{GB} between the centers of gravity and buoyance of the ocean turbine set is almost constant during the pitching motion, the righting moment is '$We_{GB}\sin\psi$' and the heeling angle curve can be expressed as '$e_{GB}\sin\psi$'. Based on the principle of dynamic equilibrium, the equation of horizontal translational motion for the turbine can be derived

$$M_2\ddot{Y}_2 - F_D + T_{turbine}\cos\phi = 0, \tag{11a}$$

where $F_D = C_D\frac{1}{2}\rho A\left(V - \dot{Y}_2 - \dot{\psi}l\right)^2$. Considering the velocity resulting from the oscillation $\left(\dot{Y}_2 + \dot{\psi}l\right) << V$, the drag force becomes $F_D \approx C_D\frac{1}{2}\rho A\left(V^2 - 2V\left(\dot{Y}_2 + \dot{\psi}l\right)\right)$.

Because the angle ϕ approaches zero and $C_D\frac{1}{2}\rho AV^2 \approx T_{turbine}$, Equation (11a) becomes equivalent to the equation of horizontal motion:

$$M_2\ddot{Y}_2 + C_D\rho AV\dot{Y}_2 + C_D\rho AVl\dot{\psi} = 0 \tag{11b}$$

In the dynamic equilibrium equation, the pitching motion of the turbine is expressed as

$$I\ddot{\psi} + We_{GB}\sin\psi + F_De_D\cos\psi = (T_{Turbine}\cos\phi)(e_D\cos\psi) - (T_{Turbine}\sin\phi)(e_D\sin\psi) \tag{12a}$$

Considering the pitching angle ψ to be small and based on Equation (6), the equation of the pitching motion becomes

$$I\ddot{\psi} - C_D\rho AVle_D\dot{\psi} + \left[We_{GB} + T_{Turbine}e_D\left(\frac{X_1 - X_2}{L_2}\right)\right]\psi - C_D\rho AVe_D\dot{y} = 0 \tag{12b}$$

where y indicates the horizontal displacement, ψ indicates the pitch angle of the turbine, A is the area of drag, CD is the drag coefficient, e_{GB} is the distance between the centers of gravity and buoyance, V is the current velocity, and l is the radius of rotation.

3. Theoretical Solutions

3.1. Solution of the Vertical Motion

The solutions of vertical displacement of the turbine and carrier comprise static and dynamic components and can be expressed as

$$X_1(t) = X_{10} + X_{11}(t), X_2(t) = X_{20} + X_{21}(t) \tag{13}$$

where $\{X_{10}, X_{20}\}$ are the static displacements and $\{X_{11}, X_{21}\}$ are the dynamic displacements. By substituting Equation (13) into Equation (10) and dividing it into the static and dynamic subsystems, we obtain the following:

Static subsystem:

$$\begin{bmatrix} K_{11} & K_{12} \\ K_{21} & K_{22} \end{bmatrix}\begin{bmatrix} X_{10} \\ X_{20} \end{bmatrix} = \begin{bmatrix} T_{Tur}\frac{X_0}{\sqrt{L_1^2 - X_0^2}} \\ 0 \end{bmatrix} \tag{14}$$

It should be noted that the Equation (14) demonstrates the relation between the vertical displacements of the turbine and carrier and the current velocity without the wave effect.

Dynamic subsystem:

$$\begin{bmatrix} M_1 & 0 \\ 0 & M_2 \end{bmatrix}\begin{bmatrix} \ddot{X}_{11} \\ \ddot{X}_{21} \end{bmatrix} + \begin{bmatrix} K_{11} & K_{12} \\ K_{21} & K_{22} \end{bmatrix}\begin{bmatrix} X_{11} \\ X_{21} \end{bmatrix} = \begin{bmatrix} A_1\rho gH_0\sin\Omega t \\ A_2\rho gH_0\sin(\Omega t + \Phi) \end{bmatrix} \tag{15}$$

It should be noted that the Equation (15) demonstrates the dynamic behavior subjected to the wave effect under the static equilibrium. From Equation (14), the theoretical solutions of the static equilibrium can be easily derived:

$$X_{10} = \frac{T_{Tur}\frac{X_0}{\sqrt{L_1^2 - X_0^2}}}{\left(K_{11} - K_{12}\frac{K_{21}}{K_{22}}\right)},\ X_{20} = -\frac{K_{21}}{K_{22}}X_{10} \tag{16}$$

Further, the harmonic response of the system under the wave effect will be obtained. The solutions of the dynamic state of Equation (15) are assumed as

$$\begin{bmatrix} X_{11} \\ X_{21} \end{bmatrix} = \begin{bmatrix} X_{11c} \\ X_{21c} \end{bmatrix}\cos\Omega t + \begin{bmatrix} X_{11s} \\ X_{21s} \end{bmatrix}\sin\Omega t \tag{17}$$

Substituting Equation (17) into Equation (15), and after some manipulations, the theoretical solutions of the dynamic state can be derived as follows:

$$\begin{bmatrix} X_{11c} \\ X_{21c} \end{bmatrix} = \frac{\beta_c}{|A|}\begin{bmatrix} -K_{21} \\ K_{11} - \Omega^2 M_1 \end{bmatrix}, \tag{18}$$

and

$$\begin{bmatrix} X_{11s} \\ X_{21s} \end{bmatrix} = \frac{1}{|A|}\begin{bmatrix} \alpha\left(K_{22} - \Omega^2 M_2\right) - \beta_s K_{21} \\ -\alpha K_{12} + \beta_s\left(K_{11} - \Omega^2 M_1\right) \end{bmatrix}, \tag{19}$$

where

$$|A| = \begin{vmatrix} K_{11} - \Omega^2 M_1 & K_{12} \\ K_{21} & K_{22} - \Omega^2 M_2 \end{vmatrix},\ \alpha = A_1\rho g H_0, \beta_c = A_2\rho g H_0 \sin\Phi, \beta_s = A_2\rho g H_0\cos\Phi. \tag{20}$$

As the determinant function of the frequency is equal to zero, $|A| = 0$, the resonance condition applies. The two resonant frequencies can be derived as

$$\Omega_{1,2}^2 = \frac{1}{2}\left[\left(\frac{K_{22}}{M_2} + \frac{K_{11}}{M_1}\right) \pm \sqrt{\left(\frac{K_{22}}{M_2} + \frac{K_{11}}{M_1}\right)^2 - 4\left(\frac{K_{11}}{M_1}\frac{K_{22}}{M_2} - \frac{K_{12}K_{21}}{M_1 M_2}\right)}\right] \tag{21}$$

It is well known that if the wave frequency approaches to the natural frequency of the system, the resonance or instability will occur. However, based on Formula (21) one can tune the natural frequencies of system away the wave frequency to avoiding the instability of vertical vibration.

3.2. Solutions of the Horizontal and Pitching Motions

Consider the horizontal and pitching motions of the static and dynamic systems. The solutions of the horizontal displacement and pitching angle are expressed as

$$\begin{aligned} y &= y_0 + y_c\cos\Omega t + y_s\sin\Omega t \\ \psi &= \psi_0 + \psi_c\cos\Omega t + \psi_s\sin\Omega t \end{aligned} \tag{22}$$

where $\{y_0, \psi_0\}$ are the static horizontal translational and pitching displacements without the wave effect $\{y_c, y_s, \psi_c, \psi_s\}$ are the harmonic solutions. By substituting Equation (22) into Equation (11), we can obtain the following two equations:

$$o_1 y_c + o_2 y_s + o_3\psi_s = 0 \tag{23}$$

$$p_1 y_s + p_2 y_c + p_3\psi_c = 0 \tag{24}$$

where

$$\begin{aligned}o_1 = -M_2\Omega^2, o_2 = \Omega C_D\rho AV, o_3 = \Omega C_D\rho AVl.\\ p_1 = M_2\Omega^2, p_2 = \Omega C_D\rho AV, p_3 = \Omega C_D\rho AVl.\end{aligned} \tag{25}$$

By Substituting Equations (22) and (13) into Equation (12), after some manipulation, we obtain

$$\begin{aligned}&We_{GB}\psi_0\\ &+\left(\begin{array}{l}-\Omega^2 I\psi_c - \Omega C_D\rho AVle_D\psi_s + T_{Tur}e_D\left(\frac{X_{11c}-X_{21c}}{L_2}\right)\psi_0\\ +\left(We_{GB} + T_{Tur}e_D\frac{X_{10}-X_{20}}{L_2}\right)\psi_c - \Omega C_D\rho AVe_D y_s\end{array}\right)\cos\Omega t\\ &+\left(\begin{array}{l}-\Omega^2 I\psi_s + \Omega C_D\rho AVle_D\psi_c + T_{Tur}e_D\left(\frac{X_{11s}-X_{21s}}{L_2}\right)\psi_0\\ +\left(We_{GB} + T_{Tur}e_D\frac{X_{10}-X_{20}}{L_2}\right)\psi_s + \Omega C_D\rho AVe_D y_c\end{array}\right)\sin\Omega t\\ &+T_{Tur}e_D\left(\frac{X_{11c}-X_{21c}}{L_2}\right)\psi_c\frac{1+\cos 2\Omega t}{2}\\ &+T_{Tur}e_D\left(\frac{X_{11s}-X_{21s}}{L_2}\right)\psi_s\frac{1-\cos 2\Omega t}{2}\\ &+\left[T_{Tur}e_D\left(\frac{X_{11s}-X_{21s}}{L_2}\right)\psi_c + T_{Tur}e_D\left(\frac{X_{11c}-X_{21c}}{L_2}\right)\psi_s\right]\sin 2\Omega t = 0\end{aligned} \tag{26}$$

Equation (26) indicates that the horizontal and pitching displacements depend on the resonant frequencies Ω and vertical displacements $\{X_1, X_2\}$, which are time-dependent functions. Therefore, the characteristic equations can be derived using the orthogonality relation of $\{\sin n\Omega t, \cos m\Omega t\}$ as follows:

First, integrating Equation (26) from 0 to period T (=$2\pi/\Omega$), Equation (26) can be expressed as

$$q_0\psi_0 + q_1\psi_c + q_2\psi_s = 0 \tag{27a}$$

where

$$q_0 = We_{GB}, q_1 = \frac{1}{2}T_{Tur}e_D\left(\frac{X_{11c}-X_{21c}}{L_2}\right), q_2 = \frac{1}{2}T_{Tur}e_D\left(\frac{X_{11s}-X_{21s}}{L_2}\right). \tag{27b}$$

Second, multiplying Equation (26) with $\cos\Omega t$ and integrating it from 0 to T, Equation (26) can be expressed as

$$r_0\psi_0 + r_1\psi_c + r_2\psi_s + r_3 y_s = 0 \tag{28a}$$

where

$$\begin{aligned}r_0 = -T_{Tur}e_D\left(\frac{X_{11c}-X_{21c}}{L_2}\right),\ r_1 = -\left(-\Omega^2 I + We_{GB} + T_{Tur}e_D\frac{X_{10}-X_{20}}{L_2}\right)\\ r_2 = \Omega C_D\rho AVle_D,\ r_3 = \Omega C_D\rho AVe_D.\end{aligned} \tag{28b}$$

Finally, multiplying Equation (26) with $\sin\Omega t$ and integrating it from 0 to T, Equation (26) can be expressed as

$$s_0\psi_0 + s_1\psi_c + s_2\psi_s + s_3 y_c = 0 \tag{29a}$$

where

$$\begin{aligned}s_0 = T_{Tur}e_D\left(\frac{X_{11s}-X_{21s}}{L_2}\right),\ s_1 = \Omega C_D\rho AVle_D,\\ s_2 = \left(-\Omega^2 I + We_{GB} + T_{Tur}e_D\frac{X_{10}-X_{20}}{L_2}\right),\ s_3 = \Omega C_D\rho AVe_D\end{aligned} \tag{29b}$$

Equations (23), (24) and (27)–(29) can be further expressed as

$$\begin{bmatrix}0 & 0 & o_3 & o_4 & o_5\\ 0 & 0 & p_3 & p_4 & p_5\\ q_1 & q_2 & q_3 & 0 & 0\\ r_1 & r_2 & r_3 & 0 & r_5\\ s_1 & s_2 & s_3 & s_4 & 0\end{bmatrix}\begin{bmatrix}\psi_0\\ \psi_c\\ \psi_s\\ y_c\\ y_s\end{bmatrix} = 0 \tag{30a}$$

The eigenvalues of Equation (30a) provide a measure of system stability. These values are characterized using a matrix, which is the Jacobian of the state of the system. The eigenvalues of the matrix are the characteristic roots of the state equation and can be determined using the roots of characteristic equations of the system.

$$\begin{vmatrix} 0 & 0 & o_3 & o_4 & o_5 \\ 0 & 0 & p_3 & p_4 & p_5 \\ q_1 & q_2 & q_3 & 0 & 0 \\ r_1 & r_2 & r_3 & 0 & r_5 \\ s_1 & s_2 & s_3 & s_4 & 0 \end{vmatrix} = 0 \quad (30b)$$

or

$$o_3(q_1 r_2 s_4 p_5) - o_4(q_1 r_2 s_3 p_5 - s_1 r_2 q_3 p_5) + o_5(q_1 r_2 s_3 p_4 - s_1 r_2 q_3 p_4 - r_1 q_2 p_3 s_4) = 0 \quad (30c)$$

From the characteristic Equation (30c), we can determine the system's resonance under the effects of various environmental parameters.

4. Numerical Results

It is important to design the current turbine system to be stable in order to obtain good performance and control requirement. The theoretical solutions, including static and dynamic stability systems, are introduced in Section 3. Here, the behaviors of the turbines and floaters under the effects of wave and ocean currents are numerically analyzed. Figures 4–7 show the effects of wave frequency f, drag force T_{Tur}, and turbine area A_2 on the amplitudes $\{X_{11}, X_{21}\}$ of vertical vibration.

These amplitudes under the condition of $\left\{T_{Tur} = 40 \text{ tons}, A_1 = 29.9 \text{ m}^2, A_2 = 80 \text{ m}^2\right\}$ at various wave frequencies are shown in Figure 4, where the dynamic motions of the turbine and floater change with the wave frequency. A lower wave frequency causes a lager vertical displacement of the turbine and floater, and the excitation of the floater is larger than that of the turbine. The vertical vibrations have two peaks at the natural frequencies {0.751 Hz, 1.22 Hz}, which are almost the same for the turbine and floater. This is attributed to the resonance vertical motion. Figure 5 shows the effect of wave frequency on $\{X_{11}, X_{21}\}$ under the condition of $\left\{T_{Tur} = 100 \text{ tons}, A_1 = 29.9 \text{ m}^2, A_2 = 80 \text{ m}^2\right\}$. The vertical vibrations have two peaks at the natural frequencies {0.761 Hz, 1.21 Hz}. The results in Figures 4 and 5 show that the pretension T_{Tur} has only a slight influence on the resonance condition.

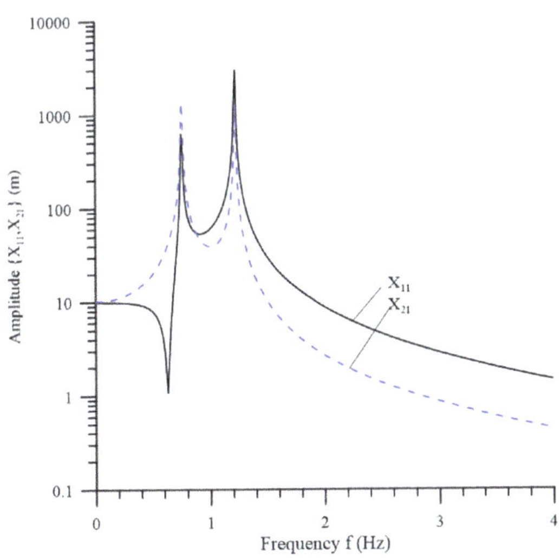

Figure 4. Effect of the wave frequency f on the amplitudes $\{X_{11}, X_{21}\}$ of vertical vibration. ($A_1 = 29.9$ m^2, $A_2 = 80$ m^2, $M_1 = 13.26$ tons, $M_2 = 100$ tons, $L_1 = 2780$ m, $L_2 = 50$ m, $X_0 = 850$ m, $H_0 = 10$ m, $T_{Tur} = 40$ tons).

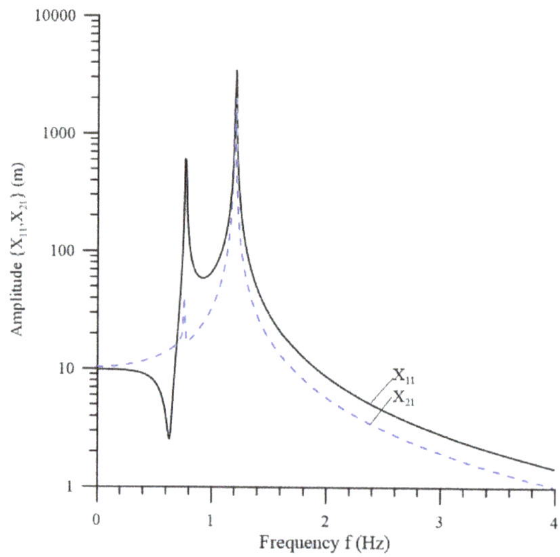

Figure 5. Effect of wave frequency f on $\{X_{11}, X_{21}\}$. ($T_{Tur} = 100$ tons; other parameters are the same as those in Figure 4).

The effects of the hydrodynamic area of the turbine and floater on vertical excitation are shown in Figures 6 and 7, respectively. Figure 6 demonstrates the effect of wave frequency on $\{X_{11}, X_{21}\}$ at $\{T_{Tur} = 40$ tons, $A_1 = 29.9\ \mathrm{m}^2, A_2 = 120\ \mathrm{m}^2\}$. The two natural frequencies at the resonance are {0.751 Hz, 1.49 Hz}. Compared to Figure 4, this figure shows that the turbine area has a substantial effect on the second natural frequency. However, the effect on the first natural frequency is negligible. The effect of the wave frequency on $\{X_{11}, X_{21}\}$ at $\{T_{Tur} = 40$ tons, $A_1 = 40\ \mathrm{m}^2$, $A_2 = 80\ \mathrm{m}^2\}$ is shown in Figure 7. The two natural frequencies are {0.871 Hz, 1.49 Hz}, which differ from the results shown in Figure 4. The resonance effect occurs at higher wave frequencies, which shows that the floater hydrodynamic area has a more substantial influence on the two natural frequencies. We can thus conclude that the larger the floater and turbine areas are, the higher the natural frequencies are. Thus, the effect of the floater area is more significantly greater than that of the turbine area.

It is well known that when dynamic stability is being considered, a larger- e_{GB} between the gravity and buoyance centers is preferred. The critical distance e_{GB} about the instability of pitching motion is investigated here. Figure 8 shows the variation in the critical distance e_{GB} with the wave frequencies under three drag forces T_{Tur} at the moment of inertial $I = 8.33 \times 10^8\left(kg - m^2\right)$. For the wave frequency over 0.03 Hz, the larger the drag force T_{Tur}, the longer the critical distance $e_{GB,critical}$. In other word, for the larger the drag force T_{Tur} the longer distance is required for the dynamic stability. Moreover, for a wave frequency over 0.03 Hz, the critical distance, e_{GB}, is less than one, except at a frequency of 1.9 Hz. Thus, if e_{GB} is larger than 1 m, the instability will not occur at a wave frequency above 0.03 Hz. However, for a wave frequency under 0.03 Hz, e_{GB} is larger than 2 m, and thus the ocean current system will be unstable. According to Figure 9, at $I = 1.67 \times 10^6\left(kg - m^2\right)$, for the wave frequency under 1 Hz, the critical distance $e_{GB,critical}$ is larger than 2 m and the entire system cannot maintain dynamic stability. It is concluded from Figures 8 and 9 that the larger the moment of inertial I, the better the stability of system.

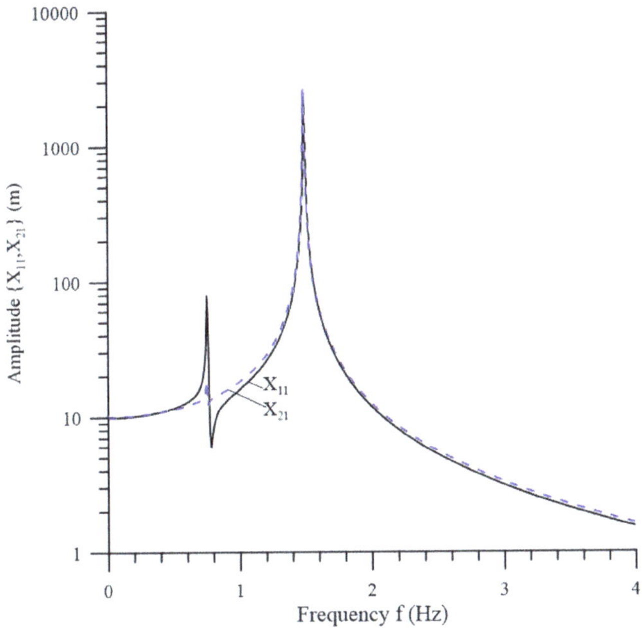

Figure 6. Effect of wave frequency f on $\{X_{11}, X_{21}\}$. ($A_2 = 120$ m^2; other parameters are the same as those in Figure 4).

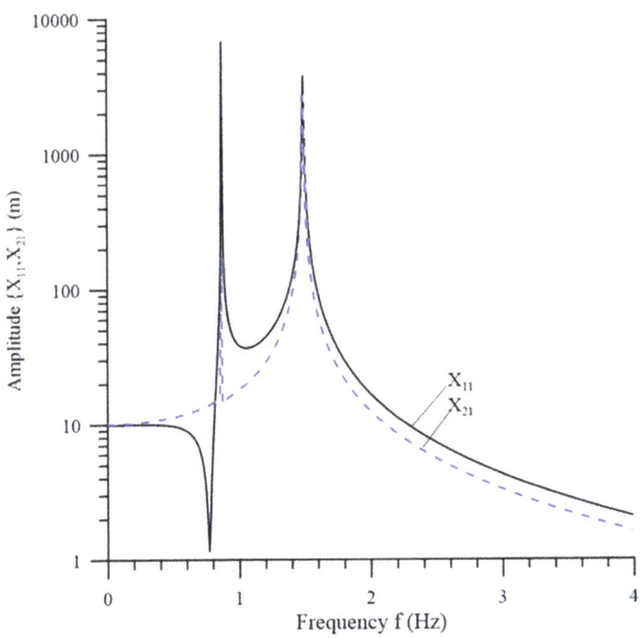

Figure 7. Effect of the wave frequency f on $\{X_{11}, X_{21}\}$. ($A_1 = 40$ m^2; other parameters are the same as those in Figure 4).

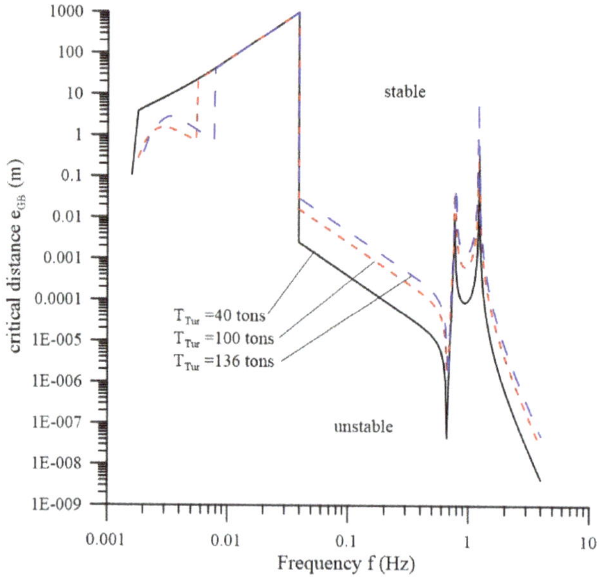

Figure 8. Effect of the wave frequency f and drag force T_{Tur} on the critical distance between the centers of gravity and buoyance for resonance. ($C_D = 1.0$, $V = 1$ m/s, $e_D = 1$ m, $\ell = 1$ m, $I = 8.33 \times 10^8$ (kg − m^2); other parameters are the same as those in Figure 4).

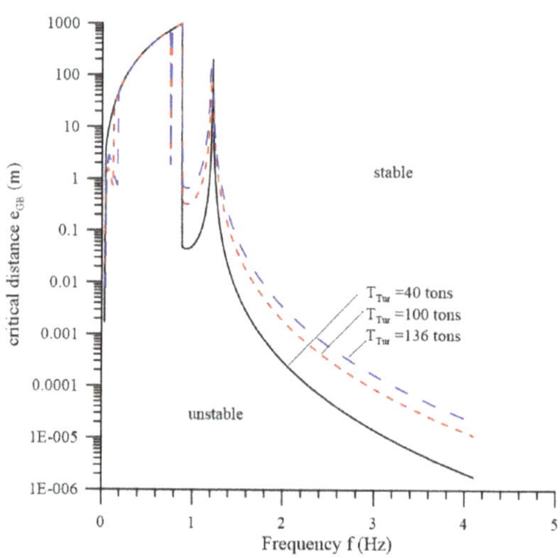

Figure 9. Effect of wave frequency f and drag force T_{Tur} on the critical distance between the centers of gravity and buoyance for resonance. ($I = 1.67 \times 10^6$(kg − m^2); other parameters are the same as those in Figure 8).

5. Conclusions

In this study, a mathematical model for a system comprising a turbine, buoyance platform, traction rope, and mooring foundation was developed. A theoretical solution of the dynamic stability analyses of the system is proposed. Based on the presented Formula (21) one can easily tune the natural frequencies of the system away from the wave frequency to avoid the instability of vertical motion. In addition, the critical distance $e_{GB,critical}$ of pitching stability is investigated. Several trends about the stability due to wave excitation are obtained as follows:

(1) The effect of the pretension T_{Tur} of rope on the natural frequencies of vertical motion is negligible.
(2) The larger the areas of the floater and turbine, the higher the natural frequencies of vertical motion.
(3) The larger the moment of inertial I, the better the stability of system.

In subsequent work the influences of nonlinear wave and current interaction will be investigated.

Author Contributions: Conceptualization, S.-M.L., Y.-Y.C., H.-C.H. and M.-S.L.; methodology, H.-C.H., S.-M.L. and Y.-Y.C.; software, H.-C.H. and M.-S.L.; validation, S.-M.L.; formal analysis, H.-C.H., S.-M.L.; investigation, H.-C.H.; resources, H.-C.H.; data curation, S.-M.L. and M.-S.L.; writing—original draft preparation, H.-C.H., S.-M.L. and M.-S.L.; writing—review and editing, H.-C.H.; visualization, S.-M.L.; supervision, Y.-Y.C.; project administration, H.-C.H.; funding acquisition, H.-C.H. All authors have read and agreed to the published version of the manuscript.

Funding: This research was funded bythe Ministry of Science and Technology of Taiwan, R. O. C., grant number MOST 107-2221-E-110-077-MY3.

Acknowledgments: The authors would like to thank the referees for their helpful comments and suggestions. The support of the Ministry of Science and Technology of Taiwan, R. O. C., is gratefully acknowledged (MOST106-3113-E-110-001-CC2, MOST107- 2221-E-110-077-MY3).

Conflicts of Interest: The authors declare no conflict of interest.

Nomenclature

A	cross-sectional area
C_D	drag coefficient
e_D	distance between the centers of gravity and drag force
e_{GB}	distance between the centers of gravity and buoyance
H_0	wave height
g	gravity acceleration
k	wave number
L_1	distance between the carrier and foundation
L_2	distance between the carrier and turbine
M_1, M_2	masses of carrier and turbine, respectively
$T_{Turbine}$	tension force of cable between the carrier and turbine
T_{wire}	tension force of cable between the carrier and anchor
T	time variable
V	ocean current velocity
X_0	depth of bed
X_1, X_2	vertical displacements of the carrier and turbine, respectively
Y_1, Y_2	horizontal displacements of the carrier and turbine, respectively
l	radius of rotation of turbine about the z-axis
ρ	density of water
Ω	wave angular frequency
Φ	phase due to the delay of propagation from the carrier to the turbine, $-kL_2$
ϕ, θ	angles of wires at the carrier
ψ	pitch angle of the turbine
Subscript:	
1, 2	the carrier and turbine, respectively

References

1. Lago, L.I.; Ponta, F.L.; Chen, L. Advances and trends in hydrokinetic turbine systems. *Energy Sustain. Dev.* **2010**, *14*, 287–296. [CrossRef]
2. Barbarelli, S.; Amelio, M.; Castiglione, T.; Florio, G.; Scornaienchi, N.M.; Cutrupi, A.; Zupone, G.L. Analysis of the equilibrium conditions of a double rotor turbine prototype designed for the exploitation of the tidal currents. *Energy Convers. Manag.* **2014**, *87*, 1124–1133. [CrossRef]

3. Barbarelli, S.; Castiglione, T.; Florio, G.; Scornaienchi, N.M.; Zupone, G.L. Design and numerical analysis of a double rotor turbine prototype operating in tidal currents. *Energy Procedia* **2016**, *101*, 1199–1206. [CrossRef]
4. Chen, D.; Nagata, S.; Imai, Y. Modelling Wave-Induced Motions of a Floating WEC with Mooring Lines Using the SPH Method. In Proceedings of the 3rd Asian Wave and Tidal Energy Conference, Singapore, 24–28 October 2016.
5. Yang, Y.J.; Jan, S.; Chang, M.-H.; Wang, J.; Mensah, V.; Kuo, T.-H.; Tsai, C.-J.; Lee, C.-Y.; Andres, M.; Centurioni, L.R.; et al. Mean structure and fluctuations of the Kuroshio East of Taiwan from in situ and remote observations. *Oceanography* **2015**, *28*, 74–83. [CrossRef]
6. Chen, L.; Lam, W.H. A review of survivability and remedial actions of tidal current turbines. *Renew. Sustain. Energy Rev.* **2015**, *43*, 891–900. [CrossRef]
7. Zhou, Z.; Benbouzid, M.; Charpentier, J.F.; Scuiller, F.; Tang, T. Developments in large marine current turbine technologies—A review. *Renew. Sustain. Energy Rev.* **2017**, *71*, 852–858. [CrossRef]
8. Chen, Y.Y.; Hsu, H.C.; Bai, C.Y.; Yang, Y.; Lee, C.W.; Cheng, H.K.; Shyue, S.W.; Li, M.S.; Hsu, C.J. Evaluation of Test Platform in the Open Sea and Mounting Test of KW Kuroshio Power-Generating Pilot Facilities. In Proceedings of the 2016 Taiwan Wind Energy Conference, Keelung, Taiwan, 2016. (In Chinese)
9. Coiro, D.P.; Marco, A.D.; Nicolosi, F.; Melone, S.; Montella, F. Dynamic behaviour of the patented kobold tidal current turbine: Numerical and experimental aspects. *Acta Polytech.* **2005**, *45*, 77–84.
10. Coiro, D.P.; Maisto, U.; Scherillo, F.; Melone, S.; Grasso, F. Horizontal Axis Tidal Current Turbine: Numerical and Experimental Investigations. In Proceedings of the Owemes, Civitavecchia, Italy, 20–22 April 2006.
11. Shiono, M.; Suzuki, K.; Kiho, S. Output Characteristics of Darrieus Water Turbine with Helical Blades for Tidal Current Generations. In Proceedings of the 12th International Offshore and Polar Engineering Conference, Kitakyushu, Japan, 26–31 May 2002.
12. Cribbs, A.R. Model Analysis of a Mooring System for an Ocean Current Turbine Testing Platform. Master's Thesis, Florida Atlantic University, Boca Raton, FL, USA, December 2010.
13. Benelghali, S.; Benbouzid, M.; Charpentier, J.F. Generator systems for marine current turbine applications: A comparative study. *IEEE J. Ocean. Eng.* **2012**, *37*, 554–563. [CrossRef]
14. Bowie, A.E.D. Flexible Moorings for Tidal Current Turbines. Master's Thesis, University of Strathclyde, Glasgow, Scotland, 2012.
15. Li, Y. On the definition of the power coefficient of tidal current turbines and efficiency of tidal current turbine farms. *Renew. Energ.* **2014**, *68*, 868–875. [CrossRef]
16. Nautricity. Green Energy: Powered by the Sea, Harnessed by Nautricity. 2012. Available online: http://www.nautricity.com/ (accessed on 3 July 2012).
17. Muliawan, M.J.; Gao, Z.; Moan, T. Application of the contour line method for estimating extreme responses in the mooring lines of a two-body floating wave energy converter. *J. Offshore Mech. Arct. Eng.* **2013**, *135*, 031301. [CrossRef]
18. Angelelli, E.; Zanuttigh, B.; Martinelli, L.; Ferri, F. Physical and Numerical Modelling of Mooring Forces and Displacements of a Wave Activated Body Energy Converter. In Proceedings of the ASME 2014 33rd International Conference on Ocean, Offshore and Arctic Engineering, San Francisco, CA, USA, 8–13 June 2014; p. V09AT09A044.
19. Davidson, J.; Ringwood, J.V. Mathematical modelling of mooring systems for wave energy converters—A review. *Energies* **2017**, *10*, 666. [CrossRef]
20. Swedish Marine Energy Developer Minesto. Available online: https://www.offshore-energy.biz/the-launch-of-the-worlds-first-subsea-tidal-kite-video/ (accessed on 27 December 2018).
21. IHI and NEDO. The Demonstration Experiment of the IHI Ocean Current Turbine Located off the Coast of Kuchinoshima Island, Kagoshima Prefecture. *IHI Eng. Rev.* **2019**, *52*.

Journal of
Marine Science and Engineering

MDPI

Article

Wave Induced Effects on the Hydrodynamic Coefficients of an Oscillating Heave Plate in Offshore Wind Turbines

Krish Thiagarajan [1,*,†] and Javier Moreno [2,‡]

1 Department of Mechanical and Industrial Engineering, University of Massachusetts Amherst, Amherst, MA 01007, USA
2 Marflet Marine, 28014 Madrid, Spain; javiermorenodesalas@gmail.com
* Correspondence: kthiagarajan@umass.edu
† Legal name: Krish Thiagarajan Sharman; Also Adjunct Faculty, Advanced Structures and Composites Center, University of Maine.
‡ Iberdrola Scholar, University of Maine.

Received: 19 July 2020; Accepted: 14 August 2020; Published: 18 August 2020

Abstract: The research problem discussed in this paper is of relevance to floating offshore wind turbine design, where heave plates are attached to the columns of a semi-submersible in order to improve vertical plane stability and the power output. Because of the shallow draft of these structures, the heave plates are proximal to the water surface. When subject to vertical plane motions the flow around a plate is altered by the presence of the free surface, resulting in changes in added mass and damping forces. In this paper, we present the experimental results for the added mass and damping coefficients for circular heave plates attached to a column, when oscillating in heave in the presence of oncoming waves. The results tend to indicate that applying the hydrodynamic coefficients obtained from still water experiments for a structure moving in waves may only be an approximation. For different relative phases of the wave and the motion, large variations could occur. We define a modified Keulegan—Carpenter (KC) number that depends on the relative amplitude of motion with respect to the wave. With this definition, the added mass and damping values are seen to be closer to the still water trends. However, at lower KC values, the added mass coefficients could differ by 30%, which can affect natural frequency estimates. Thus, caution needs to be exerted in the selection of hydrodynamic coefficients for heave plates oscillating in proximity to the free surface.

Keywords: heave plate; free surface effect; floating offshore wind turbine; hydrodynamic coefficients; added mass; damping coefficient; forced oscillation in waves; Keulegan Carpenter number

---

1. Introduction

Heave plates play an important role in the hydrodynamic behavior of floating offshore wind turbine (FOWT) structures. The concept of heave plates arose from their application in offshore spar production platforms, where their characteristics of increasing heave added mass and damping are exploited in order to maintain heave motion within acceptable limits. In the case of a FOWT, heave plates provide increased added mass in the vertical plane that shifts the platform resonance period away from the wave and wind-induced excitation periods and increase the total damping of the platform by enhancing the vortex shedding process [1]. Some prototype designs, e.g., Windfloat [2] (Figure 1) or a spar [3] use heave plates to stabilize the platform in pitch, thus improving the power output of the wind turbine.

Experimental research is the main means to study the hydrodynamic characteristics of heave plates, although some computational fluid dynamics solutions have also been reported (see e.g., [4,5]).

Scaled model tests can provide a good understanding on the behavior of floating platforms at various stages of design and development. This reduces risks and helps to optimize the design of the prototype platform. Hydrodynamic data on heave plates are usually reported using two non-dimensional characteristic parameters (Keulegan–Carpenter number KC and frequency parameter β) that represent the amplitude and frequency of oscillation of any heave plate configuration [6–10]. These dimensionless parameters of relevance are defined as [11]:

$$KC = \frac{2 \cdot \pi \cdot Z_0}{D_d} \tag{1}$$

$$\beta = \frac{D_d^2 \cdot f}{\nu} \tag{2}$$

Here, D_d is the disk diameter, Z_0 and f are, respectively. the heave amplitude and frequency of oscillation, and ν is the kinematic viscosity of the fluid.

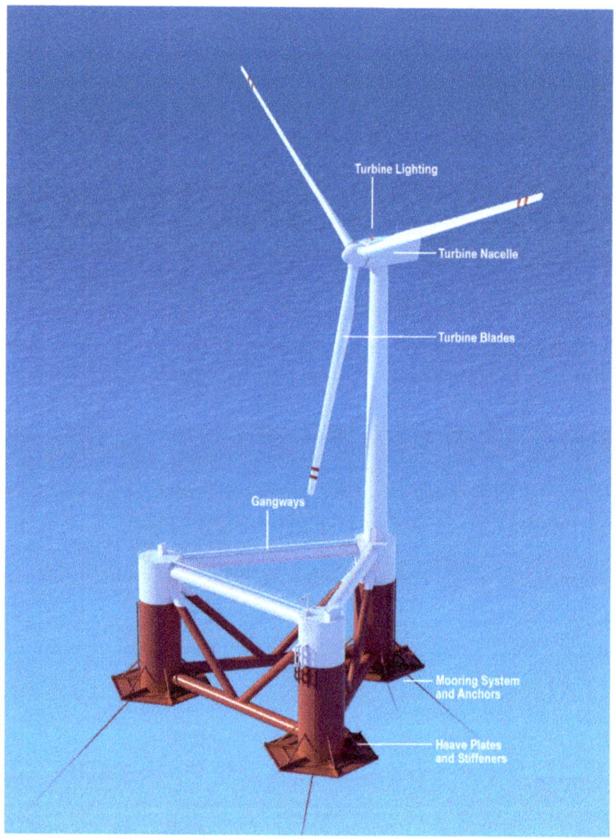

Figure 1. A floating offshore wind turbine showing heave plates attached to the structural columns [2]. Attributed to Untrakdrover/CC BY-SA (https://creativecommons.org/licenses/by-sa/3.0).

Studies on the hydrodynamic coefficients of structures with heave plates have focused on obtaining values of these coefficients at parametric ranges of importance, as well as on behavioral trends with respect to KC and β. Thiagarajan and Troesch in [7] conducted model tests on circular columns of a TLP platform. It was found that, while the added mass coefficient was invariant at low

KC numbers at a fixed oscillation frequency, the heave damping coefficient was found to be linear when plotted versus KC, and was made up of two components:

- Friction drag: damping obtained by integrating the shear stress over the wet surface of the body. This damping is dependent on viscosity, hence on Reynolds number or β. It is negligible except at very small KC numbers.
- Form drag: term due to flow separation and vortex shedding at the edges. It is obtained by integrating the normal stresses over the wet surface of the body. It is highly influenced by the geometry of the body. They found this term to be linear with the amplitude of oscillation.

The same linear tendency of damping when plotted against KC was also found in [12]. In this study, experimental tests at low KC were performed on a circular heave plate of 0.609 m diameter attached to a column of 0.457 m diameter. The experiments were performed at 1:75 scale by [8] on a spar platform with two circular heave plates of 0.68 m and 0.60 m diameter. They also found that the heave damping was primarily form drag for low β numbers. The effect of β was found to be small for small KC.

On the matter of vortex shedding flows, Tao and Thiagarajan [5] studied the viscous flow around an oscillating cylinder with a heave plate by direct numerical simulations. Flows at KC numbers ranging from 7.5×10^{-4} to 0.75 and at $\beta = 1.585 \times 10^5$ were studied. Three different shedding modes were found, i.e. independent, interactive, and uni-directional vortex shedding. The occurrence of these modes was shown to be dependent on KC and the aspect ratio represented by the ratio between the disk thickness t_d and the disk diameter D_d . The vortex shedding was found to be uni-directional for thinner disks at low KC. A quantitative method of identifying the vortex shedding flow regimes based on KC and aspect ratio was presented in [6]). A distinct increase in damping, depending on the vortex shedding regime, was also observed.

The added mass effect was examined in detail in a number of publications. The added mass coefficient was found to have a weakly linear trend as the range of KC was increased. The added mass of a circular plate attached to and separated from a column was studied by [13]. The added mass coefficient was found to double when the plate was separated from the disk, but still largely invariant with KC. A similar observation was also found by [14], whose experiments covered a range of separation distances between the column and the plate. The authors also provided theoretical formulations for the added mass coefficient as a function of the separation distance.

The above observations largely apply to solid heave plates submerged in water and oscillating in isolation. On the other hand, proximity to a boundary, thickness to width ratio, the shape of the edge, and porosity of the heave plates can alter the behavior of the hydrodynamic coefficients. In [9], experiments with 0.4 m rectangular heave plates were performed with different submergence from the free surface ranging from 0.4 to 1.2 m. The results showed that the added mass increased linearly with KC but showed indifferent trend over the range of submergence tested (whose minimum value was one diameter). The drag coefficient—which is related to damping coefficient over KC—was found to vary inversely with KC, but showed similar invariant behavior with submergence. In all these cases, the effect of changing β was negligible over the ranges tested. Furthermore, the effect of ambient currents on the hydrodynamic coefficients of a plate attached to a column has also been studied (see e.g., [12]).

When a heave plate is brought closer to a surface either by increasing the KC value or by reducing the draft of the attached column, then some changes become apparent. Numerical studies by [10] have shown that vortices shed by heave plates when executing large amplitude oscillations can disturb an otherwise quiescent free surface. They also showed that both coefficients exhibited dramatic variations with increasing KC, which depended on the distance from the free surface. A similar behavior was also observed when a plate was moved close to a solid surface like a seabed. Energy dissipation arguments were used to explain damping variations that were observed when amplitude of oscillation was changed or when the free surface was proximal.

This paper arose out of an interest to know how waves on the free surface will alter the hydrodynamic coefficients of heave plates. To study this problem, we consider the forced heave motion of a column with a heave plate in the presence of waves. The problem is of relevance to floating offshore wind turbine design, where heave plates are attached to the columns of a semi-submersible. Because of the shallow draft of these structures, the heave plates are proximal to the water surface and, hence, wave induced water motions could affect the hydrodynamic behavior. In the next section, we present some theoretical background, followed by details of an experimental study that was conducted for this research.

2. Theoretical Model

We initially consider a structure that is forced to harmonically oscillate in the vertical direction, Z (heave) in still water, Figure 2. Using Newton's second law, the force F_{33} that is required to move the structure is shown as Equation (3):

$$F_{33}(t) = (M + A_{33})\ddot{Z} + B_{33}\dot{Z} + K_{33}Z \tag{3}$$

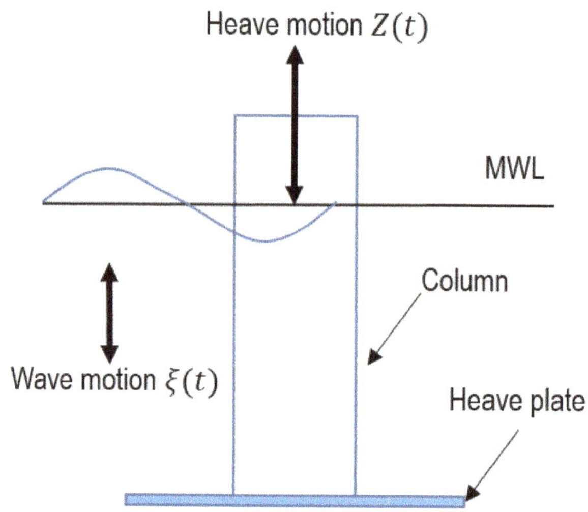

Figure 2. Schematic of a column with heave plate showing motions of relevance.

Subscript 3 denotes the heave direction, M and A_{33} are the mass and heave added mass of the body, respectively, and B_{33} is the linearized heave damping coefficient. $K_{33} = \rho g A_w$ is the heave hydrostatic restoring coefficient that depends on the water plane area A_w. The non-dimensional hydrodynamic coefficients are defined as:

$$A'_{33} = \frac{A_{33}}{A_{33_{th}}}; B'_{33} = \frac{B_{33}}{2\pi f \cdot A_{33_{th}}} \tag{4}$$

where $A_{33_{th}} = \frac{1}{12}\rho(2D_d^3 + 3\pi D_d^2 z - \pi^3 z^3 - 3\pi D_c^2 z)$ is the theoretical added mass for a column with a disk attached at the bottom [15]. D_c and D_d are the column and disk diameters, respectively, and $z = \frac{1}{\pi}\sqrt{D_d^2 - D_c^2}$.

The damping forces typically have a linear and a quadratic component [12]. By using a linearized damping coefficient, the nonlinear effects are translated into a varying dependence on the coefficients KC and β. A typical forced oscillation experiment in still water can be conducted in order to evaluate this dependence. A least squares method [11] can be used to find the optimum hydrodynamic coefficients, A_{33} and B_{33}, which minimizes the error between the measured force during experiments (F_{exp}) and the heave force (F_{33}) over n samples (Equation (3)):

$$E^2 = \sum_{i=1}^{n}(F_{exp_i} - F_{33_i})^2 \tag{5}$$

$$\frac{dE^2}{dA_{33}} = 0 \tag{6}$$

$$\frac{dE^2}{dB_{33}} = 0 \tag{7}$$

The only unknowns in Equation (5) are A_{33} and B_{33}. The displacement and the force are measured during the experiments. The velocity and acceleration can be obtained by numerically differentiating the displacement. By solving Equations (6) and (7), we obtain a system of two equations and two unknowns (A_{33} and B_{33}), which can be solved numerically.

A windowing method was used during the least squares process. In this method, the added mass and damping are evaluated in different cycles. First, the acceleration signal is divided into different windows with each one containing an acceleration peak. For each window, the added mass and damping are evaluated using the least squares method. The obtained damping coefficient is rejected at this stage, since it is poorly evaluated at acceleration peaks. This process is repeated over 32 cycles yielding the mean and the standard deviation of each coefficient through statistical analysis. The same procedure is done with the damping coefficient while using the velocity signal. This time the added mass is rejected, keeping the damping coefficient. This method was also used in [16].

The relative kinematics between the platform and the water particles need not be considered when the platform is oscillating in still water. When the platform is oscillating in waves, the presence of the wave field alters the water particle kinematics near the edge of a heave plate. The resulting effect is complicated by the fact that the wave field itself is altered by the presence of the oscillating object. Furthermore, it is apparent that the motion of the object is caused by the forcing due to waves. In a linear sense, one can distinguish between radiation and incident/diffraction problems. In order to obtain suitable hydrodynamic coefficients for solving the radiation problem, one can use a linearized version of the relative velocity model described in [17],

$$F_{33}(t) = M\ddot{Z} + A_{33}(\ddot{Z} - \ddot{\xi}) + B_{33}(\dot{Z} - \dot{\xi}) + K_{33}Z + \frac{\pi D_d^2}{4}P_b - \frac{\pi}{4}(D_d^2 - D_c^2)P_t \tag{8}$$

where P_t and P_b, respectively, represent the wave dynamic pressure acting on the top and bottom of the heave plate, and $\dot{\xi}$ and $\ddot{\xi}$ are the water particle vertical velocity and acceleration respectively at the mean position of the plate. It can be verified that accounting for dynamic pressures on both surfaces of a disk is equivalent to using a relative motion in the restoring force term, i.e., $K_{33}(Z - \xi)$.

If the wave kinematics at the mean position of the plate can be measured or estimated satisfactorily using a wave theory, then one can replace the Equation (3) with Equation (8) in the least squares analysis described above. One can then imagine two different approaches to evaluate the added mass and damping coefficients in the presence of waves:

- Absolute model: this model is the same as the one used to evaluate the coefficients in still water. In this model the waves are not taken into account explicitly. Therefore, a variation is expected when comparing the obtained coefficients in waves with the ones obtained in still water. This model is akin to the default approach, and consistent with the linear superposition of radiation and incident wave problems.
- Relative model: this model takes into account the presence of waves interacting with the heave plate. It includes the Froude–Krylov forces as well as the relative kinematics between the heave plate and the water particles. Here, the coefficients are explicitly made dependent on

wave kinematics. This approach may be considered to be more appropriate for evaluating suitable coefficients.

In the vicinity of a heave plate, the phase difference between the plate motion and the wave-induced water particle motion becomes a key parameter in the relative model. Let us, for example, consider the relative motion as:

$$Z_{rel}(t) = Z(t) - \xi(t) \tag{9}$$

where

$$\xi(t) = \xi_0 \cos(wt + \alpha_w) \tag{10}$$

$$Z(t) = Z_0 \cos(wt + \alpha_p) \tag{11}$$

By simple rearrangement, we can obtain

$$Z_{rel}(t) = A_{rel} \cdot cos(wt + \alpha) \tag{12}$$

where,

$$A_{rel} = \sqrt{Z_0^2 + \xi_0^2 - 2Z_0\xi_0 \cos(\alpha)} \tag{13}$$

and $\alpha = \alpha_w - \alpha_{hp}$. Thus, one could define a "relative" KC number

$$KC_w = \frac{2 \cdot \pi \cdot A_{rel}}{D_d} \tag{14}$$

This is similar to the relative velocity based KC number mentioned in [17].

3. Experimental Setup

A circular heave plate of diameter 0.25 m and thickness 4.3 mm attached to a column of diameter 0.088 m and draft 0.19 m is considered. This model is a 1:80 scaled version of a demonstration prototype off the coast of Spain reported in [18,19]. The experiments were performed at the Marine Ocean and Offshore Research (MOOR) wave tank facility at the University of Maine, which is 8 m long and 1 m wide. The water depth for the experiments was kept at 0.7 m. A wedge-shaped plunger type wave maker was installed at one end, and a passive energy absorbing beach at the other end. The wave maker is capable of producing regular waves from 0.5–2 s periods and amplitudes ranging from 0.002–0.132 m. The beach design was optimized to produce reflection of 5–10% over most of the range of testing.

Forced harmonic oscillation of the models in the vertical direction was achieved while using a Parker ETH032 linear actuator driven by a 750 W Parker servo motor. Two Omega force sensors were attached by two slender rods to measure the vertical forces (Figure 3). The heave displacement was measured by a string potentiometer. Output signals were amplified, sampled, and acquired at 1kHz. Using the least squares approach, the optimum hydrodynamic coefficients, A_{33} and B_{33}, which minimize the error between the measured force during experiments (F_{exp}) and heave force (F_{33}) are found. A 32-cycle windowing method described in the previous section was used in the added mass and damping evaluation.

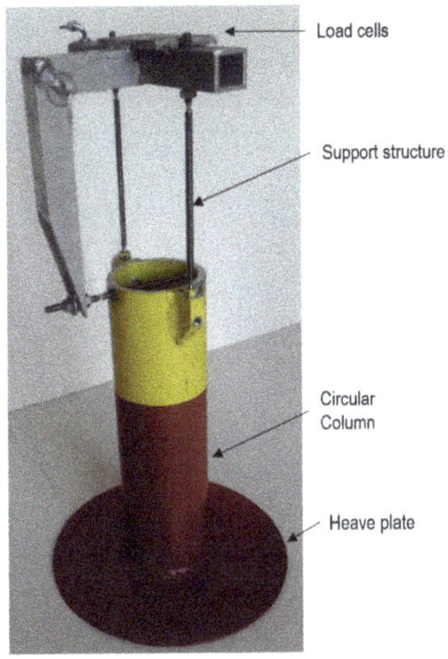

Figure 3. Column with a circular heave plate attached to a frame.

The first set of experiments was conducted in still water. The model was forced to oscillate over a range of KC values from 0.05–1.2 at a frequency of 1 Hz ($\beta = 62251$). At larger oscillation amplitudes, disturbance caused by the motion on the free surface radiated outwards. Two triangular wave deflectors were located on each of the tank walls at the heave plate location to reduce reflection from the side walls arising from the disturbance (Figure 4). This simple device performed satisfactorily, as evidenced by Fourier analysis of the force time histories. This showed that several spurious peaks were present in the time histories recorded without the deflectors.

Figure 4. Photo showing the Marine Ocean and Offshore Research (MOOR) wave tank with the model and actuator in the foreground. Wave deflectors on either side for still water runs.

The second set of experiments were conducted in waves. The model was forced to oscillate at a frequency of 1 Hz and two KC values of 0.5 and 0.84. The wave frequency was set at 1 Hz to match the heave plate oscillation frequency. The wave steepness varied from $H/L = 0.018 - 0.02$. Table 1 presents the experimental test matrix. The phase difference between the wave and the platform motion was introduced manually by visual observation of the first three waves measured by a probe located adjacent to the model. The heave plate motion was triggered when the third wave crest reached a desired distance from a zero phase mark. This approach resulted in several runs at different phases ranging from $\alpha = 0°$ to $\alpha = 360°$. Each experiment was repeated for different phases in between this range in order to study the effects of the phase between the platform and the wave. In order to calculate the phase between the wave and the heave plate signal, a frequency analysis was performed using the platform displacement and the wave elevation signals, such as shown in Figure 5.

Table 1. Wave-heave plate interaction test matrix. H/L is the wave steepness.

f (Hz)	H/L	*KC*
1	0.020	0.84
1	0.018	0.84
1	0.018	0.5

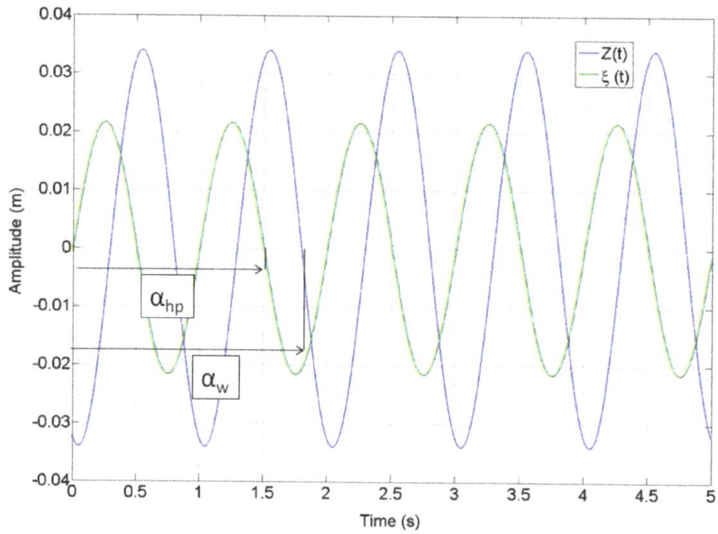

Figure 5. Wave and heave plate phase.

The quality of the generated wave was tested with a repeatability test. The platform was replaced with a third wave probe to measure the wave field at the exact location of the platform. Each wave case was repeated three times. For each case, a statistical analysis was performed that yielded the mean and standard deviation of the wave amplitude and frequency.

4. Results

The added mass and damping coefficients for the wave experiments were obtained by two different approaches. In the "Absolute model" approach, Equation (3) is used in the least squares evaluation. This model is identical to the still water case, and all wave-induced variations were visible in the trends of the coefficients with the phase angle. In the "Relative model" approach, Equation (8) which incorporates relative kinematics is used. Sample added mass and damping results using the two equation models are shown in Figures 6 and 7 for different phase angles at $H/L = 0.02$ and $KC = 0.84$. At this KC value, still water added mass and damping values are 1.42 and 0.85, respectively.

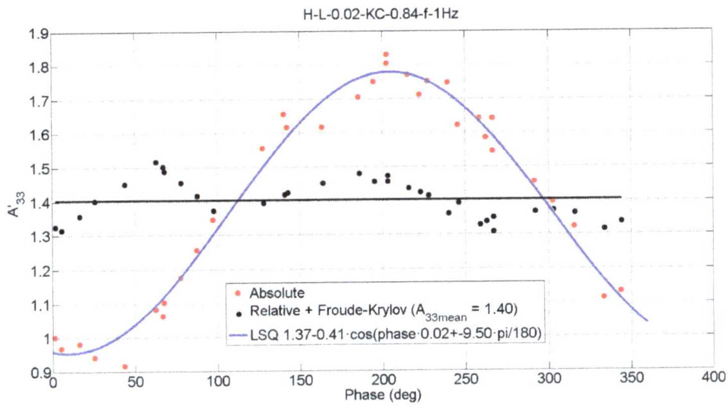

Figure 6. Added mass coefficient vs. phase angle for $H/L = 0.02, KC = 0.84$.

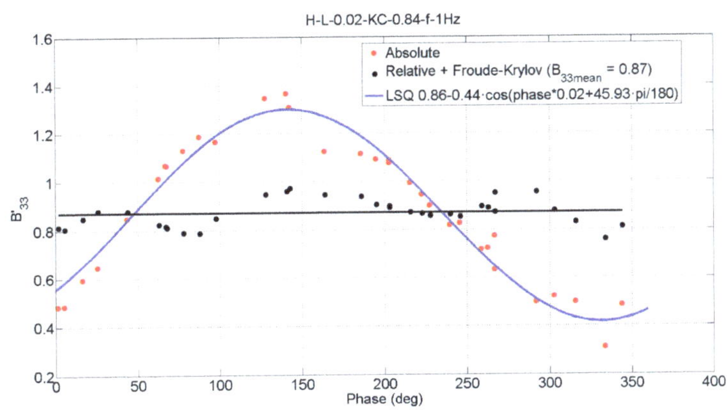

Figure 7. Damping coefficient vs. phase angle for $H/L = 0.02, KC = 0.84$.

When using the absolute model approach, a clear sinusoidal trend is observed with respect to the phase angle. Interestingly, the mean value of this sinusoidal variation matches with the corresponding still water added mass and damping values to within 3%. When the relative flow approach is used, the trend of both coefficients with the phase is much flatter, tending towards a constant value that matches the still water value to within 4%.

The relative phase between the plate and the wave gives rise to a relative change in the KC, although the amplitude of oscillation is kept constant. Figures 8 and 9 present the added mass and damping coefficients obtained using the relative model against KC_w. Additionally shown are the results obtained in still water for the added mass and damping coefficients vs. KC. The results are presented for the cases $KC = 0.84$ and $KC = 0.5$ for a frequency of oscillation of 1Hz and for $H/L = 0.018$ and $H/L = 0.02$. The observed linear trend in the coefficients is remarkable. It can be seen that the added mass and damping coefficients increase as the relative displacement between the plate and the wave particles increases. The added mass coefficients in waves show a steeper linear trend when compared with the still water coefficients. For small KC, the added mass coefficients in still water are higher. As KC increases, the coefficients in waves become slightly higher than the ones in still water. The damping coefficients in still water and in waves are very similar in slope, with the zero offset showing a difference.

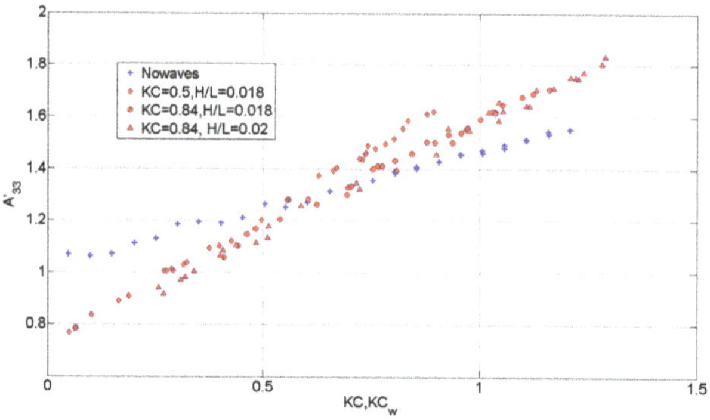

Figure 8. Comparison of the added mass coefficients in waves vs. KC_w with those in still water vs. KC.

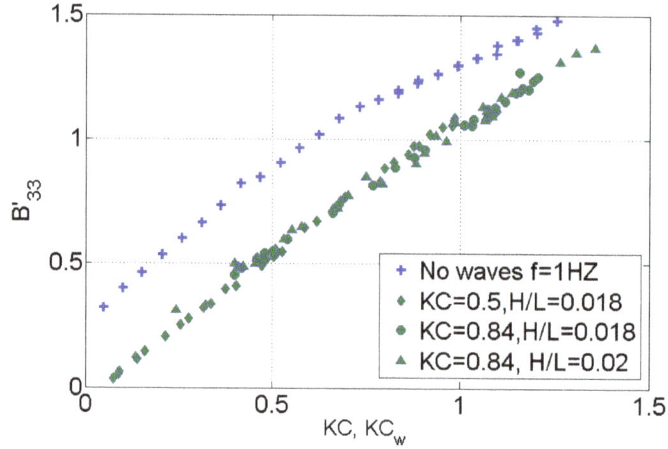

Figure 9. Comparison of the damping coefficients in waves vs. KC_w with those in still water vs. KC.

5. Discussion and Conclusions

This paper has focused on the effect of ambient wave motion on the hydrodynamic forces acting on an oscillating heave plate. When compared to the plate oscillating in still water, large differences in the values of the added mass and damping coefficients are observed. These differences are quite pronounced when the relative motion between the water and the plate are not taken into account. The results from Figures 6 and 7 tend to indicate that applying the added mass and damping coefficients obtained from still water experiments for simulating the motion of a structure in waves may lead to inconsistent results. However, due to scarcity of data on oscillating plates in waves, one method of getting reliable added mass and damping values would be by using the newly defined KC_w, which depends on the relative amplitude of motion with respect to the wave. As seen in Figures 8 and 9, the trends between the results in waves are somewhat closer to those that were obtained in still water.

Because damping values are more critical in estimating the maximum motions around resonance, a relative phase angle of $\pi/2$ may be used for KC_w. This could be used iteratively along with motion magnitude to find the optimum damping coefficient. On the other hand, it is seen in Figure 6 that, at around a phase angle of 90°, the added mass coefficients in waves and in still water are similar in magnitude. However, added mass coefficients are of relevance in all motion ranges. From Figure 8, it is seen that, at lower KC values, the added mass coefficients could differ by 30%, which can affect inertial load calculations. Thus, caution needs to be exerted in selection of hydrodynamic coefficients for heave plates oscillating in proximity to the free surface. More data would support better estimates of hydrodynamic coefficients for use in simulation of offshore wind turbine platform motions. Future work by the researchers would include a broader range of wave parameters and oscillation ranges. It is also envisaged that currents could be added to the environment in order to understand the combined effect of waves and currents.

Author Contributions: Conceptualization, K.T.; methodology, K.T. and J.M.; experimentation and validation, J.M.; analysis, J.M. and K.T.; resources, K.T.; writing–original draft preparation, K.T.; writing–review and editing, K.T. and J.M.; project administration, K.T.; All authors have read and agreed to the published version of the manuscript.

Funding: This research received no external funding.

Acknowledgments: The paper is a development of an earlier version presented at the 36th International Workshop on Water Waves and Floating Bodies (IWWWFB36) held in Plymouth, MI. The authors acknowledge the support of Stephen Abbadessa, Matthew Cameron and Raul Urbina, Department of Mechanical Engineering, University of Maine in support of the MOOR laboratory. The second author acknowledges the financial support of Iberdrola Foundation for his fellowship during the study.

Conflicts of Interest: The authors declare no conflict of interest.

Abbreviations

The following abbreviations are used in this manuscript:

α	Phase difference between the wave and the plate motion
β	Frequency parameter
ν	Kinematic viscosity of water
ξ	Water particle vertical displacement
ξ_0	Water particle vertical displacement amplitude
ρ	Mass density of water
A_{33}	Added mass in heave
$A_{33_{th}}$	Theoretical added mass in heave
A'_{33}	Added mass coefficient in heave
A_rel	Relative motion between wave and plate
A_w	Water plane area of the column
B'_{33}	Non-dimensional damping coefficient in heave
B_{33}	Damping coefficient in heave
D_c	Column diameter
D_d	Heave plate diameter
f	frequency of oscillation
F_{33}	Heave exciting force
F_{exp}	Heave force measured in an experiment
H	Wave height
K_{33}	Restoring force coefficient in heave
KC	Keulegan-Carpenter Number
KC_w	Relative Keulegan-Carpenter Number
L	Wave length
M	Mass of the oscillating structure
P	wave-induced dynamic pressure
Z	Heave displacement
Z_0	Heave motion amplitude
Z_{rel}	Relative displacement between wave and plate motion in the vertical direction

References

1. Subbulakshmi, A.; Sundaravadivelu, R. Heave damping of spar platform for offshore wind turbine with heave plate. *Ocean Eng.* **2016**, *121*, 24–36. [CrossRef]
2. Roddier, D.; Cermelli, C.; Aubalt, A.; Weinstein, A. WindFloat: A floating foundation for offshore wind turbines—Part I: Design basis and qualification. *J. Renew. Sustain. Energy* **2010**, *2*, 33104. [CrossRef]
3. Liu, L.; Guo, Y.; Jin, W.; Yuan, R. Motion performances of a 5 MW VAWT supported by spar floating foundation with heave plates. In Proceedings of the ASME 2017 36th International Conference on Offshore Mechanics and Arctic Engineering, Trondheim, Norway, 25–30 June 2017; Paper 62625.
4. Zhang, S.; Ishihara, T. Numerical study of hydrodynamic coefficients of multiple heave plates by large eddy simulations with volume of fluid method. *Ocean Eng.* **2018**, *163*, 583–598. [CrossRef]
5. Tao, L.; Thiagarajan, K. Low *KC* flow regimes of oscillating sharp edges I. Vortex shedding observation. *Appl. Ocean Res.* **2003**, *25*, 21–35. [CrossRef]
6. Tao, L.; Thiagarajan, K. Low *KC* flow regimes of oscillating sharp edges. II. Hydrodynamic forces. *Appl. Ocean Res.* **2003**, *25*, 53–62. [CrossRef]
7. Thiagarajan, K.P.; Troesch, A.W. Hydrodynamic heave damping estimation and scaling for tension leg platforms. *J. Offshore Mech. Arctic Eng.* **1994**, *116*, 70–76. [CrossRef]
8. Thiagarajan, K.P.; Datta, I.; Ran, A.Z.; Tao, L.; Halkyard, J.E. Influence of heave plate geometry on the heave response of classic spars. In Proceedings of the ASME 21st International Conference on Offshore Mechanics and Arctic Engineering, Oslo, Norway, 23–28 June 2002; pp. 621–627.
9. Li, J.; Liu, S.; Zhao, M.; Teng, B. Experimental investigation of the hydrodynamic characteristics of heave plates using forced oscillation. *Ocean Eng.* **2013**, *66*, 82–91. [CrossRef]
10. Garrido-Mendoza, C.A.; Thiagarajan, K.P.; Souto-Iglesias, A.; Bouscasse, B.; Colagrossi, A. Numerical investigation of the flow features around heave plates oscillating close to a free surface or seabed. In Proceedings of the ASME 2014 33rd International Conference on Offshore Mechanics and Arctic Engineering, San Francisco, CA, USA, 8–13 June 2014.
11. Sarpkaya, T.; Isaacson, M.d.S.Q. *Mechanics of Wave Forces on Offshore Structures*; Van Nostrand Reinhold Company: New York, NY, USA, 1981.
12. Thiagarajan, K.P.; Troesch, A.W. Effects of appendages and small currents on the hydrodynamic heave damping of TLP columns. *J. Offshore Mech. Arctic Eng.* **1998**, *120*, 37–42. [CrossRef]
13. Lake, M.; Troesch, A.W.; Perlin, M.; Thiagarajan, K.P. Scaling effects in hydrodynamic coefficient estimation of TLP and spar structures. *J. Offshore Mech Arctic Eng.* **2000**, *122*, 118–124. [CrossRef]
14. Zhu, L.; Lim, H.-C. Hydrodynamic characteristics of a separated heave plate mounted at a vertical circular cylinder. *Ocean Eng.* **2017**, *131*, 213–223. [CrossRef]
15. Tao, L.; Molin, B.; Scolan, Y.-M.; Thiagarajan, K. Spacing effects on hydrodynamics of heave plates on offshore structures. *J. Fluids Struct.* **2007**, *23*, 1119–1136. [CrossRef]
16. Beatty, S.; Buckham, B. Experimental comparison of self-reacting point absorber WEC designs. In Proceedings of the 10th European Wave and Tidal Conference (EWTEC), Aalborg, Denmark, 2–5 September 2013.
17. Chakrabarti, S.K. *Hydrodynamics of Offshore Structures*; WIT Press/Comp Mechanics: Wessex, UK, 1987.
18. Lopez Pavon, C.; Watai, R.A.; Ruggeri, F.; Simos, A.; Souto-Iglesias, A. Influence of wave induced second-order forces in semi-submersible FOWT mooring design. *J. Offshore Mech. Arctic Engng.* **2015**, *137*, 31602–31612. [CrossRef]
19. Lopez Pavon, C.; Souto-Iglesias, A. Hydrodynamic coefficients and pressure loads on heave plates for semi-submersible floating offshore wind turbines: A comparative analysis using large scale models. *Renew. Energy* **2015**, *81*, 864–881. [CrossRef]

Journal of
Marine Science and Engineering

MDPI

Article

Power Prediction of Wind Farms via a Simplified Actuator Disk Model

Yen-Cheng Chiang, Yu-Cheng Hsu and Shiu-Wu Chau *

Department of Engineering Science and Ocean Engineering, National Taiwan University, Taipei 10617, Taiwan; r06525016@ntu.edu.tw (Y.-C.C.); r07525006@ntu.edu.tw (Y.-C.H.)

* Correspondence: chausw@ntu.edu.tw

Received: 15 June 2020; Accepted: 13 August 2020; Published: 16 August 2020

Abstract: This paper aims to demonstrate a simplified nonlinear wake model that fills the technical gap between the low-cost and less-accurate linear formulation and the high-cost and high-accuracy large eddy simulation, to offer a suitable balance between the prediction accuracy and the computational cost, and also to establish a robust approach for long-term wind farm power prediction. A simplified actuator disk model based on the momentum theory is proposed to predict the wake interaction among wind turbines along with their power output. The three-dimensional flow field of a wind farm is described by the steady continuity and momentum equation coupled with a k-ε turbulence model, where the body force representing the aerodynamic impact of the rotor blade on the airflow is uniformly distributed in the Cartesian cells within the actuator disk. The characteristic wind conditions identified from the data of the supervisory control and data acquisition (SCADA) system were employed to build the power matrix of these typical wind conditions for reducing the computation demands to estimate the yearly power production. The proposed model was favorably validated with the offshore measurement of Horns Rev wind farm, and three Taiwanese onshore wind farms were forecasted for their yearly capacity factors with an average error less than 5%, where the required computational cost is estimated about two orders of magnitude smaller than that of the large eddy simulation. However, the proposed model fails to pronouncedly reproduce the individual power difference among wind turbines in the investigated wind farm due to its time-averaging nature.

Keywords: power prediction; capacity factor; actuator disk; wind farm

---

1. Introduction

The growing demand for wind energy has greatly enhanced the fast development of large-scale wind farms, especially in the offshore sectors. The weather research and forecasting (WRF) model [1,2] germinated in the late 1990s is a widely adopted tool to evaluate the atmospheric condition for a given area, and to assess its corresponding wind energy potential [3,4]. After the site for a planned wind farm has been determined via a proper wind resource assessment and evaluation, the decision around wind turbine location in the wind farm becomes the next key task before the related construction and installation work can begin. However, the wind turbine array installed generally suffers from various degrees of power loss due to the wake interaction among neighboring wind turbines. In order to efficiently harvest the wind energy captured by wind turbines installed in a wind farm, the wake interaction between the upstream and downstream wind turbines has to be accurately forecasted for maximizing the power production of a wind farm through an appropriate wind turbine siting. Three major numerical models have been proposed to identify the impact of the upstream wake on the downstream wind turbines, alongside the resulting power deficiency. The analytical model based on the conservation of momentum that arose in the 1980s is the most simple and efficient model to predict the downstream wind speed evolution of a single wind turbine [5,6], and this linear approach is further applied to the estimation of the complex shadow effects behind multiple upstream wind turbines [7]. One major drawback of this approach is to assume a linear wake decay along the flow direction that is rarely observed in the field cases characterized by non-trivial turbulence. Additionally, this model is also criticized for not being self-consistent because it uses an empirical constant to describe the downstream wake expansion behavior as well as the prediction accuracy being heavily dependent on a reasonable estimation of the wake recovery behavior. When the wind turbine is not aligned with the wind direction, the substantial flow angle leads to a downstream wake with high degree of nonlinearity, where the linear model obviously loses its validity. Despite these drawbacks, this approach has been broadly employed in the early planning stage of wind farm development due to its efficiency. A computational fluid dynamics (CFD) model with the detailed wind turbine geometry resolved in the flow domain, i.e., a full CFD model, also serves as a popular approach to investigate the nonlinear flow around wind turbines. Owing to being capable of delivering the temporal flow details in space, this approach is widely adopted to study the aerodynamic and aeroacoustic characteristics of single wind turbines [8–10]. The full CFD model is able to model the wake flow in a self-consistent way because the flow nonlinearity, together with the turbulence transport, is well considered in the related governing equations. However, it apparently becomes impractical in the flow prediction of massive wind farms with large number of wind turbines due to the complexity in grid generation and the accompanied computational cost [11]. Nowadays, the most prevalent nonlinear approach to simulate wind farm flow is the CFD approach incorporated with a body force model where the rotor geometry is fully neglected in a Cartesian grid system and replaced by a force distribution in space to account for the aerodynamic force of the rotor blades exerted on the incoming airflow [12,13]. In this approach, the blade element momentum (BEM) theory [14] is adopted to estimate the body force distribution in space by using the flow condition at the blade section location to obtain the corresponding lift and drag coefficient from a pre-defined aerodynamics table. Large eddy simulation (LES) [15,16] is a state-of-art model in this category, where a precursor calculation is first conducted for the flow domain to create the required unsteady turbulent inflow condition within a fixed period followed by a transient wind farm modeling with the designated body force distribution from the BEM model to represent the aerodynamic impact of the wind turbines on the airflow. In addition to a self-consistent nature in turbulence, the ability to offer a fine temporal resolution of turbulence in space, especially the contribution of large eddies, serves as the main cause of its great popularity. By contrast, the length modeling time of its unsteady process together with the high computational cost to resolve the eddies embedded in the flow field is regarded as the main technical barrier to prevent it from serving as an affordable design tool. Due to this numerical feature, LES is prone more employed in the diagnostic or validation cases rather than in the early or design stage of wind farm development. The motivation of this paper is to bridge the low-cost and less-accurate linear model and the high-cost and high-accuracy nonlinear model via a proposed simplified nonlinear model that provides a reasonable balance between the prediction accuracy and the computational cost. When an average power output of wind farms or wind turbines is of main interest, such as in the design stage or in a long-term prediction for estimating the capacity factor, a nonlinear steady actuator-disk model offers a very promising solution to acquire sufficient accuracy with economical computation expense. In this study, a simplified actuator disk model based on the momentum theory is proposed to model the flow around and across wind turbines while an accompanied approach to efficiently forecast the yearly capacity factor of a wind farm is further demonstrated. The proposed approach is first validated by the power measurements of Horns Rev offshore wind farm under different wind conditions. Three Taiwanese onshore wind farms are then employed to verify the accuracy of the proposed approach in the forecast of yearly capacity factor.

2. Mathematical Model

2.1. Governing Equations

It is assumed that the flow field in a wind farm is incompressible and steady, where gravity and temperature effects are fully neglected. The continuity and Reynolds-Averaged Navier-Stokes (RANS) equations are given as follows:

$$\frac{\partial u_i}{\partial x_i} = 0, \tag{1}$$

$$u_j \frac{\partial u_i}{\partial x_j} = -\frac{1}{\rho}\frac{\partial p}{\partial x_i} + \upsilon \frac{\partial^2 u_i}{\partial x_i \partial x_j} - \frac{\partial \overline{u_i' u_j'}}{\partial x_j} + f_i, \tag{2}$$

where ρ denotes the density, p the pressure, $\upsilon(= \upsilon_m + \upsilon_t)$ the effective kinematic viscosity, υ_m the kinematic viscosity, υ_t the turbulent kinematic viscosity, $\overline{u_i' u_j'}$ the Reynolds stress, f_i the body force representing the aerodynamic characteristic of rotor. A k-ε turbulence model was adopted in this study. The transport equation of turbulent kinetic energy (k) and its dissipation rate (ε) are shown as follows [17]:

$$\frac{\partial}{\partial x_j}\left(k u_j\right) = \frac{\partial}{\partial x_j}\left(\left(\upsilon + \frac{\upsilon_t}{\sigma_k}\right)\frac{\partial k}{\partial x_j}\right) + P_k - \varepsilon, \tag{3}$$

$$\frac{\partial}{\partial x_j}\left(\varepsilon u_j\right) = \frac{\partial}{\partial x_j}\left(\left(\upsilon + \frac{\upsilon_t}{\sigma_\varepsilon}\right)\frac{\partial \varepsilon}{\partial x_j}\right) + c_{\varepsilon 1} P_k \frac{\varepsilon}{k} - c_{\varepsilon 2}\frac{\varepsilon^2}{k}, \tag{4}$$

$$\upsilon_t = c_\mu \frac{k^2}{\varepsilon}, \tag{5}$$

$$P_k = -\overline{u_i' u_j'}\frac{\partial u_j}{\partial x_j} + P_b, \tag{6}$$

where P_k represents the total production of turbulent kinetic energy and P_b the production of turbulent kinetic energy due to rotor motion. The equation constants of the turbulence model $\left(c_\mu, \sigma_k, \sigma_\varepsilon, c_{\varepsilon 1}, c_{\varepsilon 2}\right)$ are given in Table 1. The wall function approach [17] was additionally adopted in this study to properly describe the region bridging the viscos layer and the turbulent region.

Table 1. The equation constants of turbulence model.

c_μ	σ_k	σ_ε	$c_{\varepsilon 1}$	$c_{\varepsilon 2}$
0.09	1.00	1.30	1.44	1.92

2.2. Actuator Disk Model

In the proposed simplified actuator disk model, the wind turbine geometry is fully neglected in the computational domain [18]. Instead, the blade force acting on the incoming airflow is modeled via an extra body force term, i.e., f_i, in the momentum equations in the swept region of rotor where the body force distribution following the aerodynamic characteristics of rotor under specific wind conditions is prescribed [19–21]. Similar to the body force term, a turbulence source term, i.e., P_b, representing the turbulence generated by the rotor motion is additionally applied to the Cartesian cells residing within the swept region in the solution of turbulent kinetic energy k. In the proposed actuator disk model, the geometric parameters describing the wind turbine are the rotor diameter (D) and the hub height (H). The operation parameters describing the wind turbine are the rotor angular velocity (ω), the output power (P), and the average axial velocity at rotor (U_{AVE}).

2.3. Body Force Distribution

The body force assigned to the cell in the swept region can mainly be decomposed into an axial component and a tangential component. The radial component is too small to be included in the body force term when compared to other two components. Due to the blades steadily sweeping over the space of the actuator disk at a specific speed, the body force distribution is therefore equally distributed in the azimuthal direction to represent a time-averaging effect of the rotor force acting on the incoming airflow. The body force distribution along the radial direction theoretically has to comply with the aerodynamic performance of airfoils varying along the radial direction. Prior knowledge of the blade geometry is therefore necessary to determine the correct radial variation of body force. However, this information is frequently inaccessible because of a planned power output instead of a wind turbine solution being generally available in the early design stage or the geometric details of the installed wind turbine being unattainable in many cases. Because the wake variation in the radial direction benefits from the turbulent mixing quickly attenuating along the downstream distance, the wake with a growing uniformity is expected to develop at the rotor disk of downstream locations. This wake behavior can be easily modelled by a uniform body force distribution in the radial distribution, provided that the far wakes, such as those taking place in a wind farm having multiple wind turbines arranged in rows or columns, are of main interest. Thus, a constant body force density results over the rotor disk.

Based on the momentum theory, the total axial force that the rotor disk exerts on the incoming airflow (F_a) is estimated from the power of wind turbine and the average axial velocity component of the airflow across the rotor:

$$U_{AVE} = \frac{1}{A}\int U dA, \tag{7}$$

$$F_a = \frac{P}{U_{AVE}}, \tag{8}$$

where U is the axial flow velocity and A the swept area of the rotor disk. The accompanied axial force density (f_a) is defined by the following equation:

$$f_a = \frac{4P}{\pi RD^2 U_{AVE}}, \tag{9}$$

where R denotes the rotor radius. The rotor torque (Q) contributed by the incoming airflow to drive the rotor operating at a rotor speed of ω is given as follows:

$$Q = \frac{P}{2\pi\omega}. \tag{10}$$

The accompanied tangential force density (f_t) is then obtained from the identity of the rotor torque with the surface integral contributed by the tangential body force:

$$f_t = \frac{6P}{\pi^2 D^3 \omega}. \tag{11}$$

The power curve of a wind turbine, Figure 1a, only depicts the correlation between the undisturbed incoming wind speed (V) and its power output where no related information of the axial velocity component across the rotor is disclosed. It is necessary to define the average axial velocity of airflow across the rotor as a function of power and rotor speed for the operation range before the evaluation of body force density via Equations (9) and (11) is possible in any wake prediction of the wind farm. Therefore, the correlation among U_{AVE}, ω and P has to be established in order to enable the body force calculation. A single wind turbine is initially adopted to discover the required correlation among U_{AVE}, ω and P. With a given inflow velocity, the corresponding rotor speed and power are obtained from the power curve. These two values are directly enforced in the flow simulation and the average axial velocity component is then evaluated by Equation (7) after a converged solution of the flow field is attained. A correlation among U_{AVE}, ω and P for the full operation range is subsequently established with the help of a comprehensive flow simulation under various inflow conditions defining a power curve. Figure 1b illustrates the correlation among U_{AVE}, ω and P for the wind turbine V80 with the aforementioned approach. In this study, the power curve governing the wind turbine power characteristics is explicitly replaced by the dependence of P and ω on U_{AVE} that is adopted to determine the power behavior of wind turbines for feasible wind conditions.

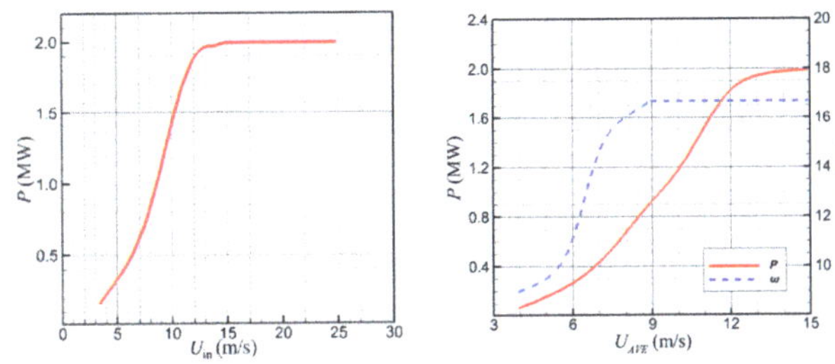

(a) (b)

Figure 1. Wind turbine V80: (**a**) Power curve; (**b**) the correlation among U_{AVE}, ω and P.

The body force is first distributed over the rotor disk based on a prescribe arrangement of disk elements with the abovementioned force density and the force on disk elements is further distributed to the neighboring Cartesian cells. In the definition of disk elements, the rotor disk is divided into I radial segments and J circumferential segments. The disk element is represented by its geometric center, appearing as the red point in Figure 2a. For simplicity, the disk element is homogenously distributed along the radial and circumferential direction over the rotor disk. The body force acting on a disk element is hence expressed as

$$f = (f_a S)\mathbf{e}_a + (f_t S)\mathbf{e}_t, \tag{12}$$

where S is the area of a disk element, $\mathbf{e}_a$ the unit normal vector in the axial direction and $\mathbf{e}_t$ the unit normal vector in the tangential direction. Due to numerical reasons, the body force of a disk element is only distributed within a finite region. Only the Cartesian cells with a distance (d) to the target disk element smaller than a given value (ϵ) is considered in the body force distribution. Figure 2b depicts the distribution of body force from a disk element to its neighboring Cartesian cells defined by the black crosses. The weighting of body force distribution among neighboring Cartesian cells is governed by a Gaussian distribution, $\eta_\epsilon(d)$:

$$\eta_\epsilon(d) = \frac{1}{\epsilon^2 \pi^{2/3}} \exp\left[-\left(\frac{d}{\epsilon}\right)^2\right]. \tag{13}$$

Therefore, the body force applied to a Cartesian cell f_ϵ is expressed as follows:

$$f_\epsilon = \sum\nolimits_k f_k \eta_\epsilon, \tag{14}$$

where f_k denotes the body force of a disk element that is distant less than ϵ from the Cartesian cell.

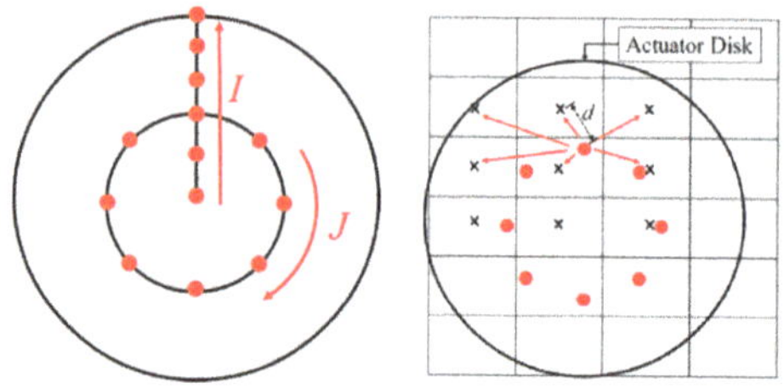

(a) (b)

Figure 2. (**a**) The arrangement of disk elements; (**b**) to distribute body force from a disk element to neighboring Cartesian cells.

3. Numerical Method

3.1. Solution Process

In the method proposed in this study, an in-house code is implemented for the wake flow prediction of wind farms based on the proposed mathematical model where a linear domain decomposition scheme is employed to parallelize the flow computation on a Cartesian gird via the MPI library [22]. A second-order finite volume method based on a deferred correction approach is adopted to discretize the governing equations and a semi-implicit method for pressure-linked equations (SIMPLE) algorithm is employed to decouple velocity and pressure. During the outer iteration of a flow computation, the velocity fields are predicted by solving the momentum equations. After the correction of mass flux at the outlet boundary, the velocity and pressure fields are updated through the pressure correction obtained from solving the continuity equation. The turbulence model is subsequently solved for the turbulence property of the flow field. In the inner iteration of a flow computation, the linearized equations are solved via a strongly implicit procedure (SIP) solver [23] to determine the field variables. Figure 3 shows the solution procedure of the proposed numerical method, as well as the corresponding in-house code, WIFA3D. The flow calculation is considered to be converged when the normalized residual of governing equations is smaller than 10^{-3} during an iterative solution process.

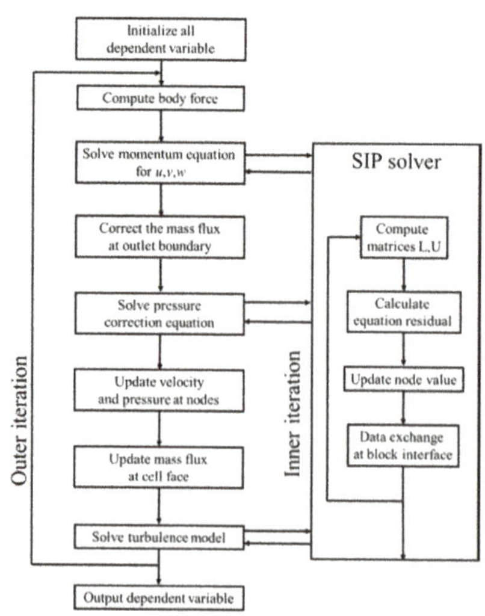

Figure 3. The solution procedure of the proposed numerical method.

3.2. Computational Domain

The typical computational domain adopted in this study has the shape of a rectangular box that principally targets offshore wind farms or inland wind farms on flat terrain, such as the onshore wind farms near coastlines demonstrated in the later sections. The computational domain is demonstrated in Figure 4, where the blue-line region explicitly defines the boundary of the interested wind farm. The origin is defined at the center of the wind farm. The symbols a to h label the corners of the computational domain that is featured with a width of L_1 alongside a height of L_2. The distance between the inflow boundary (*abcd*) and the studied wind farm is L_3, while the distance between the outflow boundary (*efgh*) and the investigated wind farm is L_4. The values of L_1, L_2, L_3, and L_4 is properly determined according to the size of a wind farm.

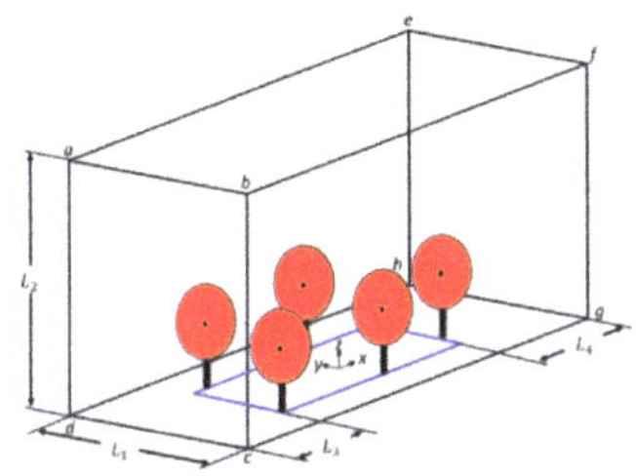

Figure 4. Computational domain adopted in the wind farm prediction.

3.3. Boundary Conditions

For the convenience of specifying boundary conditions, the inflow direction is always parallel to the x-direction of the coordinate system. The inflow condition is adopted on the boundaries *abcd* and *aefb*. The inflow wind speed varying with height is given in the following equation:

$$U_a = U_{\text{in}}\left(\frac{z}{H}\right)^{\delta} \tag{15}$$

where U_a denotes the inflow velocity, U_{in} the wind velocity at the hub height, z the height measured from the ground, H the hub height, and δ the index of ground roughness that is set 0.1 for smooth terrains appearing in the offshore or near shore applications [24]. A zero gradient condition is applied to the field variable at the outflow boundary *efgh*. A symmetry condition is employed on the boundaries *aefd* and *bfgc*. A no-slip condition is assigned to the boundary *cdhg*. The boundary conditions used in this study are summarized in Table 2 where U, V, and W are the velocity components in the x, y, and z directions, respectively, and n the normal direction of the boundary.

Table 2. The equation constants of turbulence model.

Boundary	Type	U	V	W
abcd, aefb	Inlet	(Equation (15))	0	0
aehd, bfgc	Slip wall	$\frac{\partial U}{\partial n} = 0$	0	$\frac{\partial W}{\partial n} = 0$
cdhg	No-slip wall	0	0	0
efgh	Outlet	$\frac{\partial U}{\partial n} = 0$	$\frac{\partial V}{\partial n} = 0$	$\frac{\partial W}{\partial n} = 0$

3.4. Grid Dependence

A Cartesian grid to discretize the computational domain with a grid number of about 8.2 million is schematically shown in Figure 5, where N_x, N_y, and N_z denotes the grid segments in the x, y, and z direction, respectively. The increase in the grid number apparently helps to deliver more accurate results, but with a high computation cost. Therefore, a grid dependency study was adopted to find the balance between grid number and prediction accuracy. In this study, five grid levels systematically refined from 0.2 million to 32 million cells were chosen to investigate the dependence of U_{AVE} on the grid number. Richardson's extrapolation was employed to estimate the grid-independent value (ϕ_∞) of a physical quantity (ϕ) in a second-order scheme:

$$\phi_\infty = \phi_n + \frac{\phi_n - \phi_{n-1}}{\left(\sqrt[3]{\frac{N_n}{N_{n-1}}}\right)^2 - 1}, \tag{16}$$

where ϕ_n and ϕ_{n-1} are the physical quantity of the nth and $(n-1)$th grid levels, respectively, and N_n and N_{n-1} are the grid number of the nth and $(n-1)$th grid levels, respectively. The average wind speed at the rotor disk of the wind turbine V80 under a wind speed of 5 m/s was surveyed for its grid dependency. The rated wind speed Figure 6a shows that the average wind speed at the rotor disk gradually converges to a grid-independent solution of 4.922 m/s as the number of grid ($N_x \times N_y \times N_z$) steadily grows. The dimensionless grid size (dx) along with the discretization error (E_ϕ) is given as follows:

$$x = \sqrt[3]{\frac{1}{N}}, \tag{17}$$

$$E_\phi = \left|\frac{\phi_\infty - \phi}{\phi_\infty}\right|, \tag{18}$$

where N is the grid number. Figure 6b shows the discretization error of U_{AVE}. on various grids. The third grid level with a grid number of 8.2 million gives a discretization error of U_{AVE} in 0.46%. This clearly suggests that the proposed approach has good numerical consistency and sufficient numerical stability. The role of grid spacing in three directions is governed by their individual purpose. The grid distribution of the vertical coordinate is focused on a good representation of the near-wall viscous layer where substantial velocity gradient along the vertical direction is expected, as well as a smoother velocity profile at higher altitude. The longitudinal and horizonal grid spacing is used to capture the wake variation across the rotor, while the former additionally keeps track of the downstream wake evolution. Based on this consideration, the grid is vertically clustered near the solid wall to adapt to the wall function employed in this study accompanied by an appropriate expansion rate to reach a nearly uniform size in the actuator disk for accurately resolving the airflow across the rotor. A similar grid spacing is generally used within the actuator disk for both horizontal and vertical direction and this grid size is also adopted in the longitudinal direction. This grid practice aimed to minimize the additional numerical diffusion caused by the excessive aspect ratio of grid size.

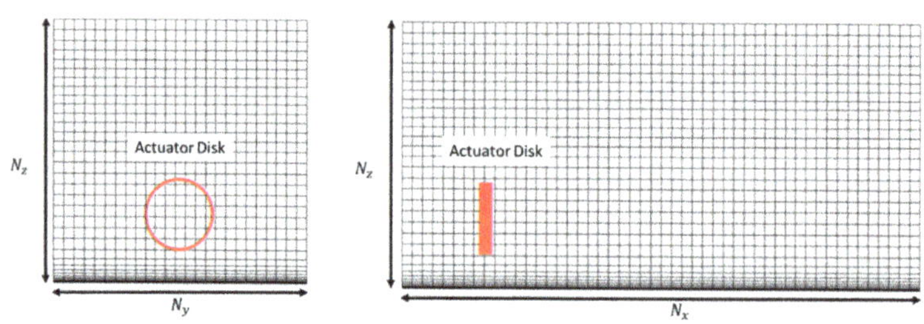

(**a**) (**b**)

Figure 5. A Cartesian grid schematically employed in the grid-dependent study: (**a**) front view; (**b**) side view.

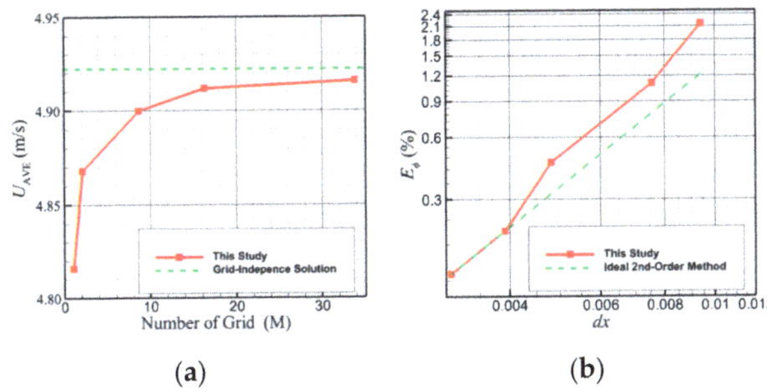

Figure 6. Grid dependency study: (**a**) U_{AVE}; (**b**) discretization error.

4. Numerical Results

4.1. Validation of Horns Rev Wind Farm

This study first predicts the power output of Horns Rev wind farm for a wind speed of 8 m/s with three distinct wind directions. The wind turbine layout of Horns Rev wind farm is shown in Figure 7a, where the red solid dots represent the locations of the installed Vestas V80 wind turbines. Three specific wind directions ($\alpha = 220°, 270°$, and $312°$) corresponding to a tower spacing (d) of $9.3D$, $7D$, and $10.4D$ are selected as the validation cases. Figure 7b illustrates the typical grid arrangement for the Horns Rev wind farm, where L_x is the longitudinal size of the wind farm, L_y the lateral size of the wind farm, $L_1 = L_y + 12D$, $L_2 = 12.5D$, $L_3 = 25D$ and $L_4 = 50D$. The numerical mesh was particularly clustered in the region where the rotor of a wind turbine is located. Two different domain sizes, i.e., a full model and a reduced one, are additionally compared to identify their computational advantages and drawbacks in the power prediction of wind farms. In the full model, a complete wind farm is simulated, Figure 8, whereas the reduced model only considers a cascading wind turbine arrangement due to the tower location symmetry of the investigated wind turbine array, Figure 9. Tables 3 and 4 summarize the grid parameters to create a numerical mesh for the full and reduced domains, respectively. The first group of parameters, i.e., dx, dy, dz, represent the grid spacing of the rotor region in the x, y, and z directions, respectively, where dz_0 is the first off-wall grid spacing for accommodating the adopted turbulence model. The second group of parameters, i.e., $N_{r,y}$ and $N_{r,z}$, denotes the grid numbers covering the rotor area in the spanwise and vertical directions, respectively. Figure 10 displays the predicted axial velocity distribution at the hub height for various wind directions. The airflow obviously decelerates across wind turbines and the wake gradually recovers its velocity along the downstream direction. A slow wake recovery was found as airflow passes through multiple wind turbines. It is mainly due to a successive energy transfer from airflow to the wind turbine that results in less chance for airflow to receive energy from neighboring high velocity streams. The normalized power of a turbine row (P_n) is defined as follows:

$$P_n = \frac{P_R}{P_1}, \tag{19}$$

where P_1 represents the power of the first wind turbine row and P_R the power of a wind turbine row defined in Figure 8. The normalized power P_n for three wind directions is forecasted in Figure 10, where the field measurement is labeled an average value alongside its upper and lower limits. For understanding the relative prediction accuracy of the proposed approach compared to other numerical approaches, a benchmark LES result [25] was chosen in this study to serve as a comparison basis of the numerical predictions. Figure 11 depicts the power forecast of the full model along with the LES prediction. In the case of $\alpha = 220°$, the prediction of the full model favorably falls within the measurement limits. Despite a small overprediction of the average power output, the full model successfully delivered a sharp power drop in the second row, accompanied by a very mild power decrease along the downstream direction that truthfully describes the field performance of wind turbine rows. By contrast, the LES approach delivers a similar conclusion to the full model but it suggests an almost unchanged power tendency of the downstream rows where the average power is underpredicted especially for the second and third rows. For the wind farm simulation of $\alpha = 270°$, the power forecast of the full model very well agreed with the measurement, except for a within-limits overprediction of the average power at the second row, whereas the LES prediction gives an obvious underprediction for all downstream wind turbine rows in addition to a deep power deficiency of the second row that is apparently inconsistent with the wind farm measurement. A large discrepancy between the measurement and the numerical result predicted by the full model, as well as the LES approach, was found in the power prediction of $\alpha = 312°$. The field data indicates a clear power decline growing with the increase of the row number that is unable to be reproduced by both numerical approaches. One cause of this large disagreement on the power prediction might come to the unsatisfactory modeling of the turbulence development in the atmosphere boundary by both methods. As the tower spacing increases, the wake recovery is expected to become more pronounced. The nonlinear behavior of strong wake recovery at large tower spacing should be further investigated in future studies to obtain an accurate power prediction of wind turbine array, especially for the wind farms employing high-power wind turbines. Another reason for this substantial underprediction is the experimental data representing a period of measurement that records the wind turbine power under time-varying wind direction and speed rather than a fixed wind direction and speed in the ideal case considered in the numerical calculation. From this viewpoint, the case of $\alpha = 222°$ clearly exhibits a more stable wind condition than the case of $\alpha = 312°$. Figure 11 also compares the numerical results between the full and reduced models. Interestingly, the reduced model gives a very close result to the full model. This implies a significant reduction up to one order of magnitude in the computation cost can be easily achieved by utilizing the geographical symmetry without bring substantial modeling errors.

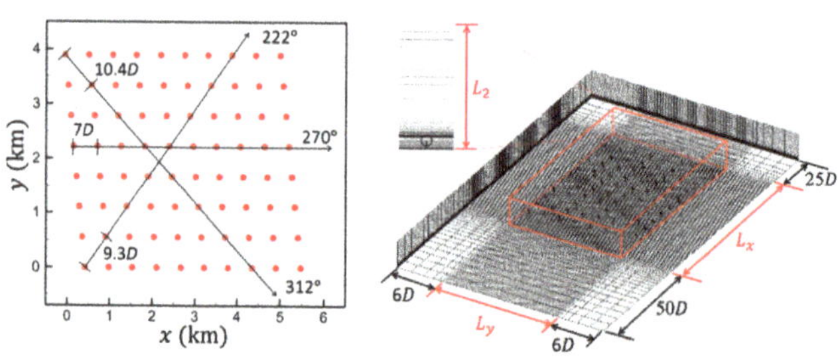

(a) (b)

Figure 7. Horns Rev wind farm: (**a**) definition of wind direction; (**b**) grid arrangement for $\alpha = 270°$

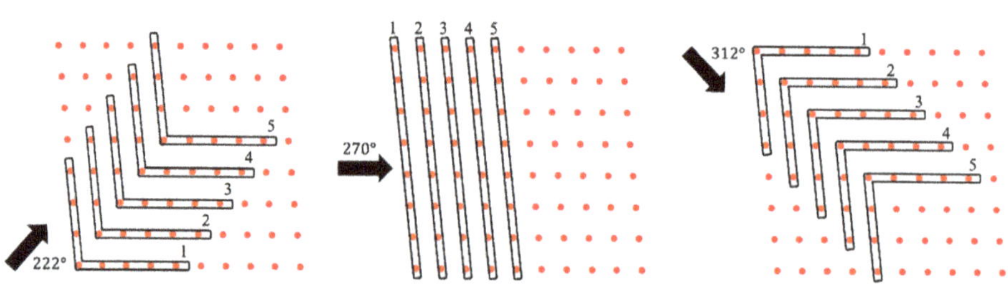

(a) (b) (c)

Figure 8. The definition of a turbine row in the full model: (**a**) $\alpha = 222°$; (**b**) $\alpha = 270°$; (**c**) $\alpha = 312°$.

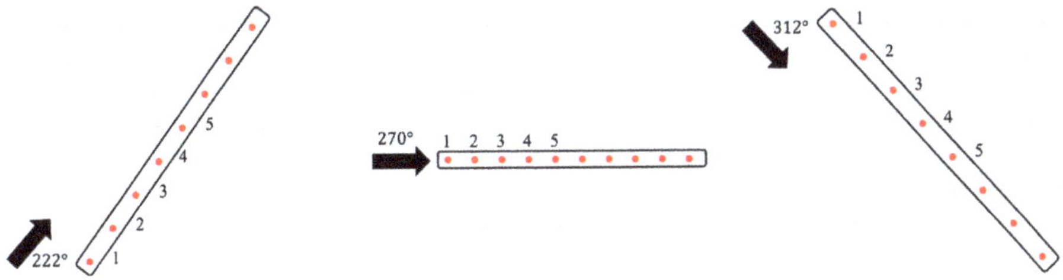

(a) (b) (c)

Figure 9. The definition of a turbine row in the reduced model: (**a**) $\alpha = 222°$; (**b**) $\alpha = 270°$; (**c**) $\alpha = 312°$.

Table 3. Grid parameters adopted for the full model of the Horns Rev wind farm.

α	L_x/D	L_y/D	dx (m)	dy (m)	dz (m)	dz_0 (cm)	$N_{r,y}$	$N_{r,z}$	$N_x \times N_y \times N_z$
222°	85	90							850×900×55
270°	81	55	8	8	8	10	10	10	810×550×55
312°	95	80							950×800×55

Table 4. Grid parameters adopted for the reduced model of Horns Rev wind farm.

α	L_x/D	L_y/D	dx (m)	dy (m)	dz (m)	dz_0 (cm)	$N_{r,y}$	$N_{r,z}$	$N_x \times N_y \times N_z$
222°	79								1185×60×72
270°	77	4	5.3	5.3	5.3	10	15	15	1155×60×72
312°	87								1305×60×72

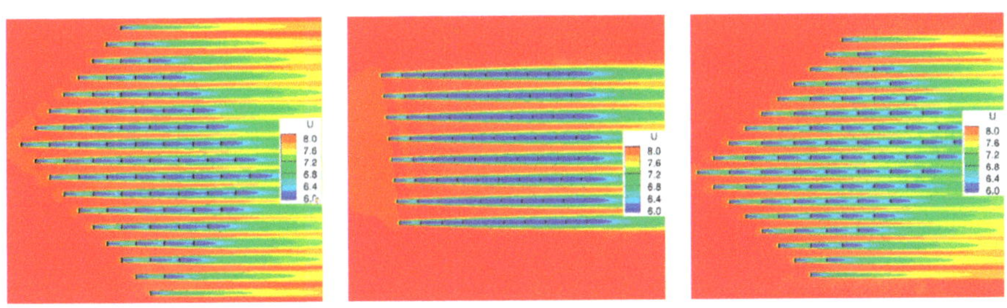

(a) (b) (c)

Figure 10. Axial velocity distribution at the hub height: (**a**) $\alpha = 222°$; (**b**) $\alpha = 270°$; (**c**) $\alpha = 312°$.

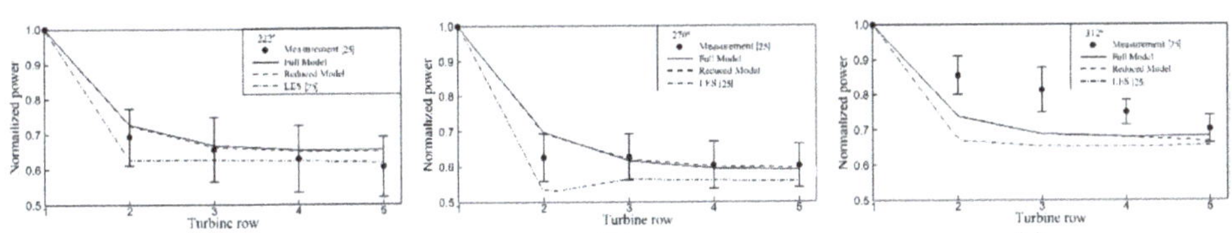

(a) (b) (c)

Figure 11. The comparison of power prediction: (**a**) $\alpha = 222°$; (**b**) $\alpha = 270°$; (**c**) $\alpha = 312°$.

4.2. Long-Term Power Prediction

The power captured by a wind turbine mainly depends on its local wind conditions [26]. Hence, the long-years SCADA measurement was adopted to describe the local wind conditions. The SCADA data, such as wind velocity, wind direction, and power output, used in this study is recorded on a ten-minutes basis. However, these data possibly contain incomplete or erroneous information when the wind turbine is under maintenance, limits its power output, or experiences a system failure. Therefore, pre-processing is required to remove the invalid information inconsistent with the design power curve. Power prediction of a wind farm is an essential target of this study. However, wind velocity and wind direction randomly change in real wind conditions. In the long-term power prediction, to model wind conditions with high data frequency, do provide good time accuracy in power, but it is very time-consuming as well as computationally expensive. Hence, it is necessary to use a more robust and efficient approach to long-term power prediction. With a statistical analysis of the wind condition and its corresponding wind turbine power, a number of characteristic wind conditions, i.e., inflow wind direction and inflow wind velocity, were first defined for the wind farm under investigation. The power output of each characteristic wind condition was then predicted through a numerical simulation of the proposed actuator disk model. A power matrix to define the wind turbine power under the selected combination of inflow direction and inflow velocity was then obtained for the investigated wind turbine, Figure 12a, where α is the inflow direction, U_{in} the inflow velocity and P the power output of a wind turbine. The long-term power output of a wind turbine is therefore easily calculated via a numerical interpolation within the power matrix for any given wind condition. In regard to the accuracy of power prediction via the power matrix, a sufficient number of wind directions and wind velocities is generally required to satisfactorily resolve the meteorological features of the wind farm under investigation. In this study, 36 wind directions alongside 18 wind speeds were employed to defined the characteristic wind conditions. For the yearly SCADA data set with a data frequency of 10-min mean, the simulation cases were reduced from 52,560 to 648, namely a computation reduction in two orders of magnitude. For the onshore wind farms surveyed in this study and other similar wind farms where no anemometer is available to provide relevant atmospheric information for the power prediction, the wind condition had to be determined from the available SCADA data of the installed wind turbines. Because each wind turbine records the wind velocity and wind direction that it experiences, it is necessary to select a virtual anemometer independent of wind turbine locations for representing a unified and reasonable inflow condition for the wind farm. The inflow wind direction of a wind farm for a given time span is defined as the average wind direction of all wind turbines in the wind farm at that time span, Figure 12b:

$$\alpha = \frac{\sum_{i=1}^{i=m} \alpha_i}{m} \tag{20}$$

where m denotes the total number of wind turbines installed in the wind farm. The inflow wind velocity is defined as the wind velocity recorded at the most upstream wind turbine with respect to the given inflow wind direction. Figure 12c schematically illustrates a wind farm with an inflow wind direction of 45°. In this case, the wind turbine WT2 is most upstream, so its wind velocity measurement was regarded as the inflow wind velocity of the wind farm for an inflow wind direction of 45°. Figure 13 shows the procedure for long-term wind farm power prediction. In this study, the capacity factor (CF) that represents the efficiency of a wind turbine over a time span is defined in Equation (21), where t is the time span of interest and P_r the rated power of the wind turbine studied.

$$\text{CF} = \frac{1}{T}\frac{\int P dt}{P_r}. \tag{21}$$

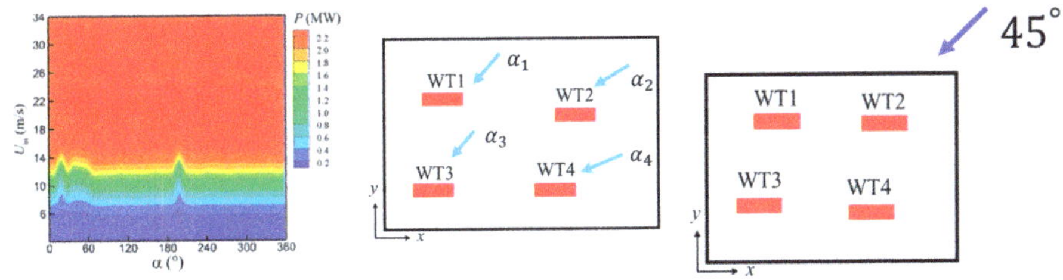

(**a**) (**b**) (**c**)

Figure 12. (**a**) The power matrix of a wind turbine to define correlation among U_{in}, α and P; (**b**) the definition of inflow wind direction for a wind farm; (**c**) the wind farm with an inflow wind direction of 45°.

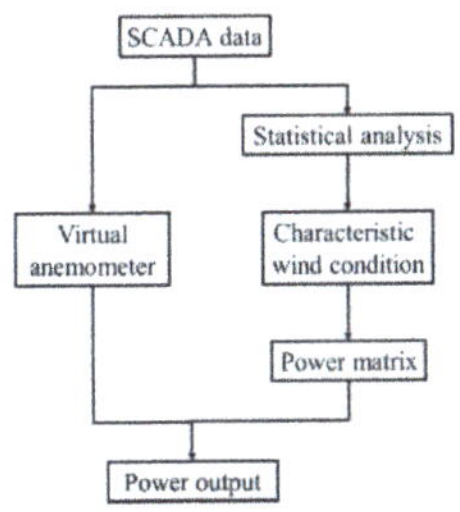

Figure 13. The procedure for long-term wind farm power prediction.

4.3. Description of Onshore Wind Farms

Three Taiwanese wind farms, labelled A, B, and C, were investigated for their long-term power prediction in this study. Wind farm A and wind farm B are onshore wind farms where the wind turbines reside along the coastline, while wind farm C is situated about 2 km away from a river mouth to the sea. The time span of SCADA data employed in this study for fixing the wind conditions was 54 months for wind farm A and 12 months for wind farms B and C. Figure 14 discloses the layouts of their wind turbine array, with the numbering of installed turbines where the direction of north is given in the upper right corner. Wind farm A has a wind turbine array of 8 Enercon E40-600 wind turbines. Figure 15a gives the power curve of the E40-600 wind turbine. The 600-kW wind turbine operates between a cut-in wind speed of 2.5 m/s and a cut-off wind speed of 25 m/s and has a rated wind speed of 13 m/s. The rotor speed of this wind turbine varies in the range of 18 rpm and 34.5 rpm. The wind conditions disclosed by SCADA are given in Figure 5b,c where Φ_p denotes the probability function. The main wind direction of the wind farm A lies between $\alpha = 20°$ and 35°, while a probability peak around 17% occurs around a wind speed of 5 m/s. Ten Enercon E70 wind turbines builds the wind turbine array of wind farm B. The power characteristics of the Enercon E70 wind turbine are shown in Figure 16a, where the rated power of 2.3 MW is achieved at the rated wind speed of 15 m/s. This wind turbine starts to rotate at a wind speed of 2.5 m/s and stops to operate at a wind speed of 34 m/s. The allowable range of rotor speeds is designated between 6 rpm and 21.5 rpm. Figure 16b displays the wind direction experienced by the wind turbine WT1, which is the only available SCADA data for the wind direction in wind farm B. As indicated by this figure, the major wind direction is between 15° and 20°. The maximum probability of the wind speed for wind farm B is about 11%, and the corresponding wind speed is approximately 3 m/s (Figure 16c). Wind farm C is the largest wind farm in wind turbine number, as well as in total power, compared to its two counterparts. There are 23 wind Vestas V80 turbines in total installed in wind farm C, which is designed in a two-row arrangement. The installed wind turbine features a rated power of 2 MW, a rated wind speed of 14.5 m, and a rated rotor speed of 16.7 rpm. The lower and upper limits of wind speed for a normal wind turbine operation are 3.5 m/s and 25 m/s, respectively. Figure 1a illustrates the power dependence on the wind speed of V80. Unlike the other two wind farms, wind farm C shows some disagreement in the wind direction among its wind turbines (Figure 17). This inconsistency in the wind direction was then technically removed in the statistical analysis in determining the inflow condition of a virtual anemometer, as described in the previous sections. The dominant wind direction nearly comes from the north with a variation of about 10°. In contrast to the wind direction, the wind velocity exhibits a high consistence among wind turbines. A probability of 13% was found for a wind speed of 3 m/s (Figure 18). Clearly, wind farm A seems to have better wind conditions than wind farm B and wind farm C. The superior wind condition of wind farm A comes from its geographical advantage, i.e., a wind turbine array on an island in the Taiwan strait. Table 5 summarizes the main geometry and operation parameters of the wind turbines installed at the wind farms investigated.

(a) (b) (c)

Figure 14. The wind turbine layout of the wind farm: (**a**) wind farm A; (**b**) wind farm B; (**c**) wind farm C.

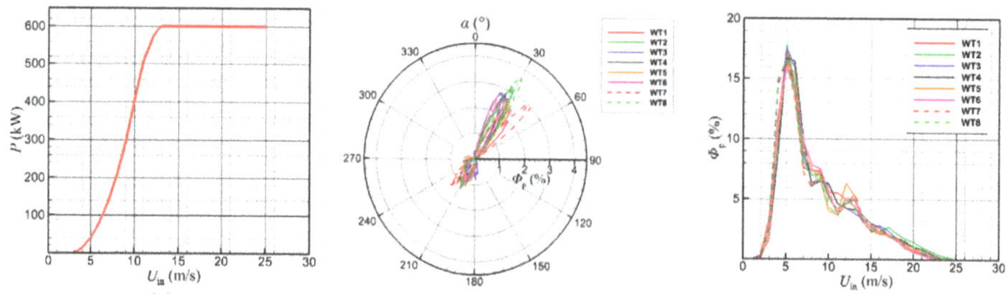

(a) (b) (c)

Figure 15. Wind farm A: (**a**) power curve; (**b**) wind direction probability; (**c**) wind velocity probability.

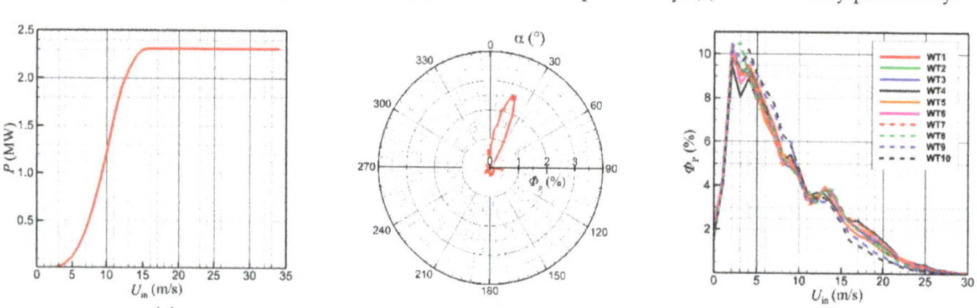

(a) (b) (c)

Figure 16. Wind farm B: (**a**) power curve; (**b**) wind direction probability; (**c**) wind velocity probability.

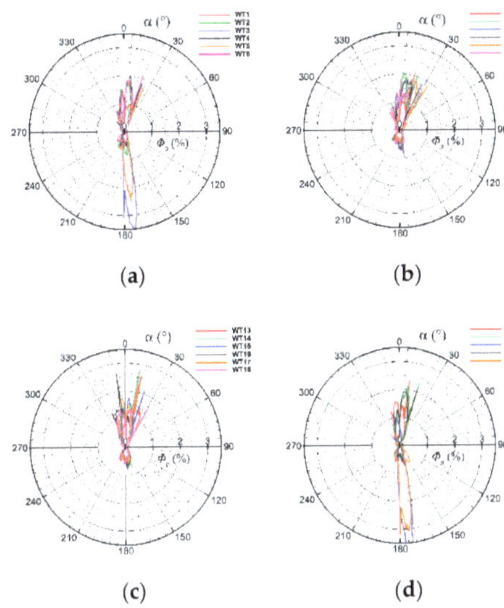

Figure 17. The wind direction probability of wind farm C: (a) Wind turbine (WT)1 to WT6; (b) WT7 to WT12; (c) WT13 to WT18; (d) WT19 to WT23.

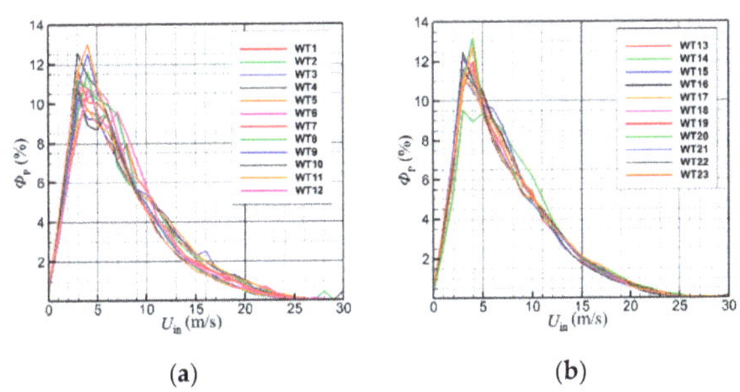

Figure 18. The wind velocity probability of wind farm C: (a) WT1 to WT12; (b) WT13 to WT23.

Table 5. Main parameters of the wind turbines installed at the investigated wind farms.

Wind Farm	Type	N_{WT}	P_r (MW)	D	H	ω_r (rpm)	U_r (m)
A	Enercon E40-600	8	0.6	40	46	34.5	13
B	Enercon E70	10	2.3	71	64	21.5	15
C	Vestas V80	23	2	80	70	16.7	14.5

4.4. Yearly Capacity Factor

Figure 19a,b displays the wind conditions delivered by the virtual anemometer of wind farm A. These wind conditions indicate a dominant wind direction from the northeast in winter and from the southeast in the summer, while wind speed in the range between 5 m/s and 13 m/s has a probability function decreasing from 15% to 5%. Figure 19c shows the predicted yearly capacity factor for all wind turbines installed in wind farm A. The measurement indicates that the capacity factor of individual wind turbines varies between 0.4 and 0.46. The numerical prediction of the proposed approach delivers a mean capacity factor of 0.423 that well agrees with the measured b mean value of 0.432. The variation in CF among wind turbines seems to be underestimated by the numerical model, despite giving a similar tendency where the wind turbine WT6 delivers the least power among its counterparts. The power deficiency of WT6 can be explained by the high probability of the wind direction between $\alpha = 15°$ and 25°. As the wind condition falls in this range, WT6 directly suffers from the wake interaction with WT1 that accounts for its low capacity factor. The underestimation of the power variation in the wind turbine array should stem from the time resolution of SCADA data, i.e., a ten-minute mean. The averaging of wind speed and wind direction undoubtedly eliminate the power spike within a ten-minute time span, and hence lead to a smoother power history than instantaneous measurement. Figure 20a presents the wind velocity estimated by the virtual anemometer of wind farm B while the wind direction was already given in Figure 16b. The wind mainly comes from the northeast direction ($15° \le \alpha \le 20°$) in the cold season and from the southeast direction ($200° \le \alpha \le 210°$) in the hot season, together with the wind speed growing from 2.5 m/s to 15 m/s accompanied by a probability function falling from 8% to 3%. The estimated yearly capacity factor for all wind turbines installed in wind farm B is depicted in Figure 20b. The measured capacity factor of individual wind turbines oscillates between 0.3 and 0.39, whereas the numerical prediction shows a smaller range of power variation, $0.35 \le CF \le 0.39$. The predicted mean capacity factor, 0.38, favorably agrees with the measured mean capacity factor of 0.35. The overprediction, alongside a small power variation of the wind farm, as explained earlier, originates from the large time span of SCADA data. The improvement in predication accuracy requires us to use the wind conditions of smaller time intervals. Additionally, the proposed approach was unable to identify WT7 as a leading power contributor in the wind farm, and instead suggested an almost uniform power performance from WT1 to WT8, except for correctly recognizing WT9 and WT10 to be the least power contributors. The unsatisfactory prediction of power tendency among wind turbines is also heavily impacted by using a single wind turbine SCADA date to determine the incoming wind direction. Figure 20c unveils the major reason resulting in their power deficiency. Under the principal wind direction, i.e., $15° \le \alpha \le 20°$, WT9 and WT10 strongly suffers from a low incoming wind speed that clarifies their low mean value of the capacity factor. Figure 21a presents the dependence of the wind direction on its probability function, where the main wind direction resides in a region $0° \le \alpha \le 20°$. The maximum probability of 16% occurs at a wind speed of 5 m/s, while the probability function drops to 3% as the wind velocity reduces to the rated wind speed of 14.5 m/s, Figure 21b. The capacity factor for the wind turbines in wind farm C is compared between the numerical prediction and SCADA measurement in Figure 21c. The comparison shows that the measured capacity factor lies between 0.22 and 0.32, whereas the predicted capacity factor varies from 0.34 to 0.39. In respect to the wind farm power performance, the average capacity factor of the wind farm is 0.271 based on the SCADA data, whereas 0.368 is estimated by the numerical approach. A large discrepancy is apparently observed in the power prediction of wind farm C. Figure 21d shows a typical example of the raw SCADA data obtained from a wind turbine in wind farm C together with the design power curve. Noticeably, the real-time power is almost on the right-hand side of the design power curve. It implies a power degradation of deterioration of the wind turbine due to some unknown or unavailable reasons, because the design power curve generally serves as a mean line of the real-time power, as depicted in Figure 21e. This uncovers the main cause of large CF errors for wind farm C. A remedy is to obtain a corrected power curve based on the SCADA data to reflect the realistic power behavior of the wind turbine installed in the wind farm. The corrected power curve that truthfully delivers the correlation between the wind speed and the wind turbine power is given as the solid blue curve in Figure 21e. With this revision, the updated CF prediction is shown in Figure 21f, where the capacity factor of wind farm C is predicted to be 0.268. Nevertheless, the distinct CF variation among the installed wind turbines (~0.1) suggested by the SCADA data was unsurprisingly flattened into a much smaller value of 0.02, although the higher power output of the first-row turbines than the second-row turbines was still distinguishable from the numerical simulation. Table 6 shows a comparison of the yearly capacity factor of the investigated wind farms where the subscript S denotes the simulation result and the subscript E represents the SCADA measurement. The prediction error of the yearly capacity factor for a wind farm is less than 9% for a wind farm with a limited SCADA data set, whereas the prediction error can be reduced to 2% and less provided a more comprehensive SCADA data are available.

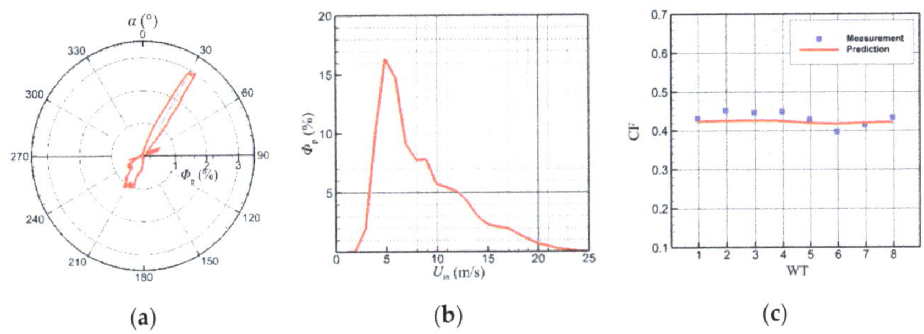

Figure 19. Wind farm A: (**a**) wind direction probability; (**b**) wind velocity probability; (**c**) yearly capacity factor.

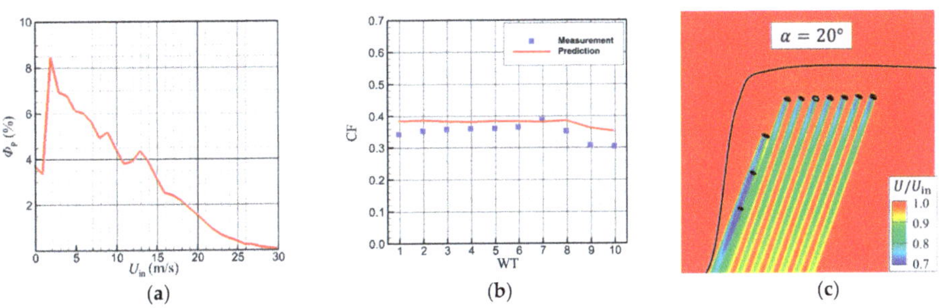

Figure 20. Wind farm B: (**a**) wind velocity probability; (**b**) yearly capacity factor; (**c**) axial velocity distribution at the hub height for $\alpha = 20°$ and $U_{\text{in}} = 10$ m/s.

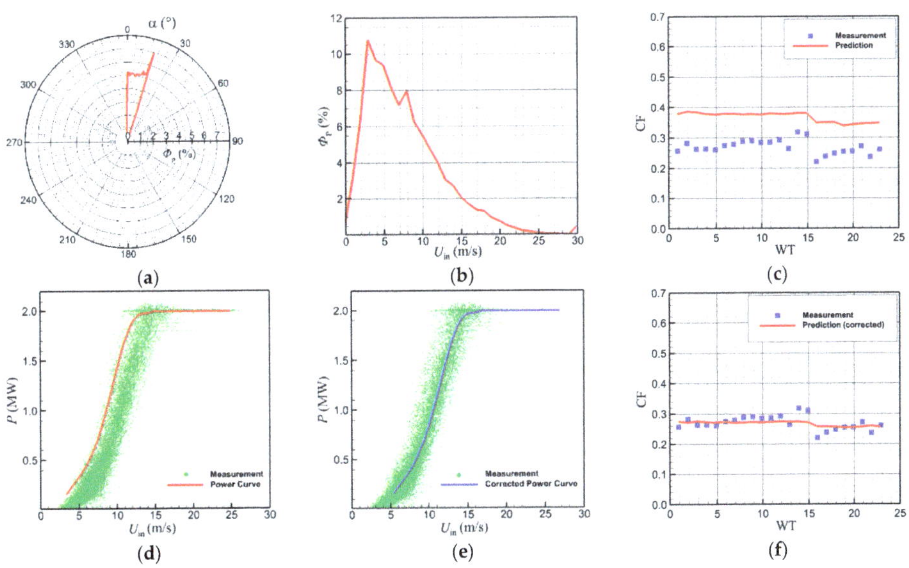

Figure 21. Wind farm C: (**a**) wind direction probability; (**b**) wind velocity probability; (**c**) yearly capacity factor; (**d**) power curve; (**e**) corrected power curve; (**f**) yearly capacity factor with the corrected power curve.

Table 6. Comparison of yearly capacity factor among the investigated wind farms.

Wind Farm	*A*	*B*	*C*
CF_E	0.432	0.350	0.271
CF_S	0.423	0.380	0.268
Error (%)	2.1	8.6	1.1

4.5. Performance Assessment of the Proposed Model

One important aim of this study is to establish a numerical model that can be easily applied in the early stages of wind farm development to verify or select an appropriate site for wind turbines in the planned wind farm, along with the capability to accurately estimate the long-term capacity factor required in financial investment in a wind farm project. To meet this target, a simplified nonlinear wake model based on the momentum theory is correspondingly proposed in this paper and an in-house code, WIFA3D, is accordingly implemented. Therefore, the momentum change of airflow across the wind turbine is directly governed by the power curve where no geometry details of blade section are required. In this way, the simplified actuator disk model can be directly used to verify the layout of a wind turbine array in the initial phase of a wind farm project when only the principal operation parameters, such as power curve and rotor speed, are available. Another advantage of this model is to offer sufficient prediction accuracy without demanding excessive computation expense. This model practically links the technical gap between the low-cost and less-accurate linear formulation and the high-cost and high-accuracy large eddy simulation. The prediction accuracy of the yearly capacity factor was clearly displayed in the cases of three onshore wind, where the individual prediction error ranged from 1.1% to 8.6% and the average prediction error was less than 4%. A recent study using the LES approach to validate the case of Horns Rev wind farm [27] was referenced to illustrate the computational benefit of the proposed approach. In the case of simulating a wind farm with eight cascading wind turbines, i.e., the reduced wind farm model with $\alpha = 270°$, 6000 outer iterations are required by WIFA3D to achieve the numerical convergence. In the referenced LES study, 12,000 time steps for the wind farm simulation, alongside 200,000 time steps for the precursor computation, were reported with a comparable grid number. Because LES is a time-accurate computation, a converged solution of each time step commonly requires tens of outer iterations when the marching time step is moderate. Assuming that the computational cost of an outer iteration for both approaches to solve single scalar governing equation is approximately equal, the typical LES approach demands a computation cost about two orders of magnitude higher than the proposed approach. This highlights the computational efficiency of the proposed model. Despite delivering good prediction accuracy with low computation expense, the simplified actuator disk model has the following drawbacks. First, the proposed model shows a good prediction performance in the mean power, but it is prone to smooth the power variation among wind turbines in the same wind farm due to an averaging process in time and space. Second, the dynamic effect of power fluctuation contributed by the moving rotor, as well as by the unsteady incoming wind, is also unable to be reflected in its steady nature. Third, the uniform body force distribution based on the momentum conservation applied to the actuator disk could bring unfavorable impacts to the near wake because of the negligence of the blade aerodynamics. These issues also form the limitations of the proposed approach.

5. Conclusions

To meet the demand bridging the gap between low-cost linear formulation with unsatisfactory accuracy and the high-accuracy nonlinear models, such as large eddy simulation, featuring excessive computational costs, a simplified nonlinear wake model based on the momentum theory is proposed in this study to offer a reasonable balance between accuracy requirements and numerical expense. The numerical robustness of the proposed model further enables the establishment of an economic approach for long-term wind farm power prediction that provides the required information for the financial evaluation of wind farm development. The proposed model was first validated with the offshore measurement of Horns Rev wind farm for a given wind speed of 8 m/s alongside three wind directions. Except for the wind direction of 312°, the proposed approach delivered good agreement with the measurements, as well as a comparable accuracy level to the LES results. The in-house code WIFA3D only requires about one percent of the computational cost of the LES approach to predict the mean power delivered by a wind farm. This study also verifies that a reduced computational domain based on the symmetry of wind turbine arrays gives a significant computational advantage without sacrificing the prediction accuracy. This paper also proposes an efficient approach to predict the yearly capacity factor based on the characteristic wind conditions, where the required computational cost is possibly reduced by two orders of magnitude when compared with the traditional serial approach. The proposed approach can favorably forecast the yearly capacity factor of a wind farm with an average error of less than 5%, but the power variation across the wind turbine array is inevitably smoothed due to an averaging nature in time and space. The advantage of the simplified actuator disk model is demonstrated in its numerical robustness, as well as in the prediction accuracy of the mean power, whereas the time-accurate power behavior and the significant power variation among the wind turbines is unable to be captured by the proposed model due to its steady nature.

Author Contributions: Methodology, S.-W.C.; validation, Y.-C.H.; formal analysis, Y.-C.C.; writing—original draft preparation, Y.-C.C.; writing—review and editing, S.-W.C.; visualization, Y.-C.C., Y.-C.H. All authors have read and agreed to the published version of the manuscript.

Funding: This research was funded by the Ministry of Science and Technology, Taiwan, grant number MOST 107-3113-E-002-010- and MOST 108-3116-F-006-004-CC1.

Acknowledgments: The authors thank the Taiwan Power Company for providing relevant operation information of the investigated wind farms.

Conflicts of Interest: The authors declare no conflict of interest.

References

1. Skamarock, W.C.; Klemp, J.B.; Dudhia, J.; Gill, D.O.; Barker, D.; Duda, M.G.; Huang, X.; Wang, W.; Powers, J.G. *A Description of the Advanced Research WRF Version 3 (No. NCAR/TN-475+STR)*; University Corporation for Atmospheric Research: Boulder, CO, USA, 2008.
2. Powers, J.G.; Klemp, J.B.; Skamarock, W.C.; Davis, C.A.; Dudhia, J.; Gill, D.O.; Coen, J.L.; Gochis, D.J.; Ahmadov, R.; Peckham, S.E.; et al. The Weather Research and Forecasting Model: Overview, System Efforts, and Future Directions. *Bulletin Am. Meteorol. Soc.* **2017**, *98*, 1717–1737. [CrossRef]
3. Mahoney, W.P.; Parks, K.; Wiener, G.; Liu, Y.B.; Myers, W.L.; Sun, J.; Monache, L.D.; Hopson, T.; Johnson, D.; Haupt, S.E. A Wind Power Forecasting System to Optimize Grid Integration. *IEEE Trans. Sustain. Energy* **2012**, *3*, 670–682. [CrossRef]
4. Haupt, S.E.; Mahoney, W.P. Taming Wind Power with Better Forecasts. *IEEE Trans. Spectr.* **2015**, *52*, 47–52. [CrossRef]
5. Ainslie, J.F. Calculating the Flow Field in the Wake of Wind Turbines. *J. Wind Eng. Ind. Aerodyn.* **1988**, *27*, 213–224. [CrossRef]
6. Sørensen, J.N.; Shen, W.Z. Numerical Modeling of Wind Turbine Wakes. *J. Fluids Eng.* **2002**, *124*, 393–399. [CrossRef]
7. González-Longatt, F.; Wall, P.; Terzija, V. Wake Effect in Wind Farm Performance: Steady-State and Dynamic Behavior. *Renew. Energy* **2012**, *39*, 329–338. [CrossRef]
8. Lin, Y.; Lee, N.Z.; Chau, S.W. Flow-Structure Interaction Modelling of Rotating Wind Turbine with Deformable Rotor Blades. In Proceedings of the 9th International Workshop on Ship and Marine Hydrodynamics, Glasgow, UK, 24–26 August 2015.
9. Chuang, Y.H.; Chau, S.W. Aerodynamic and Aeroacoustic Prediction of Wind Turbine Rotor for a 2MW Horizontal-Axis Design under the Rated Condition. In Proceedings of the 6th International Symposium on Energy Challenges and Mechanics, Inverness, Scotland, 14–18 August 2016.

10. Yang, C.Y.; Kouh, J.S.; Chau, S.W. Study of Aerodynamic Loads Acting on Wind Turbines under the Typhoon Conditions in Taiwan. In Proceedings of the Advanced Maritime Engineering Conference, Hong Kong, China, 13–14 October 2016.
11. Choi, N.J.; Nam, S.H.; Jeong, J.H.; Kim, K.C. Numerical Study on the Horizontal Axis Turbines Arrangement in a Wind Farm: Effect of Separation Distance on the Turbine Aerodynamic Power Output. *J. Wind Eng. Ind. Aerodyn.* **2013**, *117*, 11–17. [CrossRef]
12. Troldborg, N. Actuator Line Modeling of Wind Turbine Wakes. Ph.D. Thesis, Technical University of Denmark, Kongens Lyngby, Denmark, 2008.
13. Sørensen, J.N.; Mikkelsen, R.F.; Henningson, D.S.; Ivanell, S.; Sarmast, S.; Andersen, S.J. Simulation of Wind Turbine Wakes Using the Actuator Line. *Tech. Phil. Trans. R. Soc. A* **2015**, *373*, 20140071. [CrossRef] [PubMed]
14. Lanzafame, R.; Messina, M. Fluid Dynamics Wind Turbine Design: Critical Analysis, Optimization and Application of BEM Theory. *Renew. Energy* **2007**, *32*, 2291–2305. [CrossRef]
15. Porte-Agel, F.; Wu, Y.T.; Lu, H.; Conzemius, R.J. Large-eddy Simulation of Atmospheric Boundary Layer Flow through Wind Turbines and Wind Farms. *J. Wind Eng. Ind. Aerodyn.* **2011**, *99*, 154–168. [CrossRef]
16. Jha, P.K.; Churchfield, M.J.; Moriarty, P.J.; Schmitz, S. Guidelines for Volume Force Distributions within Actuator Line Modeling of Wind Turbines on Large-Eddy Simulation-Type Grids. *J. Sol. Energy Eng.* **2014**, *136*, 031003. [CrossRef]
17. Launder, B.E.; Spalding, D.B. Numerical Computation of Turbulent Flows. *Comput. Methods Appl. Mech. Eng.* **1974**, *3*, 269–289. [CrossRef]
18. Mikkelsen, R.F.; Sørensen, J.N. Actuator Disc Methods Applied to Wind Turbines. Doctoral Dissertation, Technical University of Denmark, Kongens Lyngby, Denmark, 2004.
19. Lee, K.C.; Chau, S.W. Wind Farm Wake Modeling Using an Actuator Disk Model. In Proceedings of the Advanced Maritime Engineering Conference, 9–12 October 2018.
20. Crasto, G.; Gravdahl, A.R.; Castellani, F.; Piccioni, E. Wake Modeling with the Actuator Disc Concept. *Energy Procedia* **2012**, *24*, 385–392. [CrossRef]
21. Castellani, F.; Vignaroli, A. An Application of the Actuator Disc Model for Wind Turbine Wakes Calculations. *Appl. Energy* **2013**, *101*, 432–440. [CrossRef]
22. Chau, S.W.; Hsu, K.L. Modeling Steady Axis-Symmetric Thermal Plasma Flow of Air by a Parallelized Magneto-Hydrodynamic Flow Solver. *Comput. Fluids* **2011**, *45*, 109–115. [CrossRef]
23. Stone, H.L. Iterative Solution of Implicit Approximations of Multidimensional Partial Differential Equations. *SIAM J. Numer. Anal.* **1968**, *5*, 530–558. [CrossRef]
24. Hsu, S.A.; Meindl, E.A.; Gilhousen, D.B. Determining the Power-Law Wind-Profile Exponent under Near-Neutral Stability Conditions at Sea. *J. Appl. Meteorol.* **1994**, *33*, 757–765. [CrossRef]
25. Wu, Y.T.; Porté-Agel, F. Modeling Turbine Wakes and Power Losses within a Wind Farm Using LES: An Application to the Horns Rev Offshore Wind Farm. *Renew. Energy* **2015**, *75*, 945–955. [CrossRef]
26. Jamil, M.; Parsa, S.; Majidi, M. Wind Power Statistics and an Evaluation of Wind Energy Density. *Renew. Energy* **1995**, *6*, 623–628. [CrossRef]
27. Dekos, G.; Piggot, M.D.; Laizet, S. Development and Validation of the High-Order Finite Difference Wind Farm Simulator, WInc3D. In Proceedings of the Third Conference on Renewable Energies Offshore, Lisbon, Portugal, 8–10 October 2018.

Journal of
Marine Science and Engineering

MDPI

Article

Bragg Reflections of Oblique Water Waves by Periodic Surface-Piercing and Submerged Breakwaters

I-Fan Tseng [1], Chi-Shian You [1] and Chia-Cheng Tsai [1,2,3,*]

1 Department of Marine Environment and Engineering, National Sun Yat-sen University, Kaohsiung 80424, Taiwan; ifan@mail.nsysu.edu.tw (I.-F.T.); a0970926908@gmail.com (C.-S.Y.)
2 Department of Marine Environmental Engineering, National Kaohsiung University of Science and Technology, Kaohsiung 80778, Taiwan
3 Center of Excellence for Ocean Engineering, National Taiwan Ocean University, Keelung 20224, Taiwan
* Correspondence: tsaichiacheng@nkust.edu.tw

Received: 2 June 2020; Accepted: 13 July 2020; Published: 16 July 2020

Abstract: The Bragg reflections of oblique water waves by periodic surface-piercing structures over periodic bottoms are investigated using the eigenfunction matching method (EMM). Based on the assumption of small wave amplitude, the linear wave theory is employed in the solution procedure. In the step approximation, the surface-piercing structures and the bottom profiles are sliced into shelves separated by abrupt steps. For each shelf, the solution is composed of eigenfunctions with unknown coefficients representing the wave amplitudes. Upon applying the conservations of mass and momentum, a system of linear equations is obtained and is then solved by a sparse-matrix solver. The proposed EMM is validated by several examples in the literature. Then, the method is applied to solve Bragg reflections of oblique water waves by various surface-piercing structures over periodic bottoms. From the numerical experiments, Bragg's law of oblique waves was used to predict the occurrences of Bragg resonance.

Keywords: eigenfunction matching method; oblique wave; Bragg reflection; step approximation; surface-piercing structure; periodic bottom

1. Introduction

When wind waves generated in the deep-water approach coastal regions, they experience various physical phenomena caused by inferences with structures, bathymetric variations, nonlinear wave interactions, etc. To prevent coasts from huge wave attacks, either floating or submerged breakwaters are usually installed in the coastal area. Although the nonlinear effects become significant as the waves approach the shoreline, consistent linear solutions are still valuable and provide extensive information concerning the wave impact on the nearshore and coastal environments. Furthermore, the linear solution usually serves as the starting point for a weakly nonlinear model [1].

Both submerged and floating breakwaters are typically designed in coastal regions. Submerged breakwaters are conventional structures that often rest on the sea floor. In addition, they are built as submerged types to satisfy the requirements of coastal landscapes and ecologies. Submerged breakwaters are heavy and large, and are designed by engineers for different purposes [2]. On the other hand, floating breakwaters have the advantage of lower construction costs compared to submerged breakwaters. Floating structures may be used in fishing farms for ecology conservation, tourism, and leisure. Moreover, they may be implemented at the ocean engineering working stations, such as oil exploration stations [3]. Engineers also use assembled floating structures to construct an airport on the sea [4], or to provide hospitable environments on the surface of the water.

The Bragg reflection caused by the periodic breakwaters can help to effectively attenuate waves. For example, Mei et al. [5] considered using a series of submerged sinusoidal sandbars to protect the ocean platforms in Ekofisk of the North Sea, and Bailard et al. [6] found that the Bragg reflection of submerged bars can protect U.S. Gulf Coasts from storm-induced waves. In addition, Tsai and Wen [7] indicated that the Bragg reflections of submerged breakwaters were effective for protecting the Mi-Tuo Coast, Taiwan. The vortex generation and dissipation accompanying the Bragg scattering of water waves propagating over a series of submerged rectangular breakwaters were investigated by Hsu, et al. [8]. Recently, the Bragg reflections of floating breakwaters were studied by Ouyang et al. [9] and Ding et al. [10]. In this study, the combined Bragg reflections of periodic submerged and floating breakwaters are considered.

Numerical solutions are inevitable for solving water wave-scattering problems as analytic solutions are rare [11]. Berkhoff [12] derived the mild-slope equation (MSE) by integrating the governing equation over the vertical interval of water depth. Subsequently, the MSE was modified and improved in various studies [13,14]. In addition to the prescribed one-equation models, further improvements were made by including the evanescent modes, resulting in a system of hierarchical MSEs [15,16]. Athanassoulis and Belibassakis [17] additionally included a sloping-bottom mode to formulate the consistent coupled-mode system (CCMS), which has been applied to many water wave problems [18–20]. The MSE has been applied to solve problems of nonlinear waves [18,21], three dimensions [22], wave–current interactions [21,23], time evolutions [24], Bragg reflections [13], floating structures [25,26], etc. A comprehensive review can be found in a recent article [27].

Alternately, Takano [28] developed the eigenfunction marching method (EMM) for solving normal incident wave scattering over an elevated sill and a fixed surface obstacle. Subsequently, Kirby and his coauthors [29,30] applied the EMM to solve problems of wave scattering over a trench of oblique incidences. For waves propagating over an arbitrary bottom profile, Devillard et al. [31], O'Hare and Davies [32,33], and Tsai et al. [34,35] decomposed the bottom profiles into a sequence of flat shelves separated by steps. The EMM has been applied to problems of viscous wave scattering [36–38], water wave scattering by tension-leg structures [39] and thin floating plates [40]. The accuracy of the EMM solutions was shown to be comparable with that of the MSE solutions [41]. In addition, the mathematical formulation is EMM is simpler as there are no requirements for the spatial derivatives of the eigenfunctions; however, they are needed in the MSE. However, the applications of EMM to three-dimensional, nonlinear, and/or time-dependent problems require further investigation.

In 1966, Katō, et al. [42] conducted laboratory experiments on the reflections of wave scattering using four simple floating structures, including a rectangular structure. Through numerical method, the diffraction of oblique waves scattering by a surface-piercing rectangular structure was studied by Bai [43]. Sequentially, Kanoria et al. [44] derived analytical solutions for normally incident wave scattering by a surface-piercing rectangular structure in water of uniform finite depth. The analytical solutions were then extended to oblique waves by Söylemez and Gören [45]. For an arbitrary cross-section, Garrison [46] developed a Green's function procedure to compute oblique wave scattering. Using numerical methods, Ouyang et al. [9] recently studied the Bragg reflections of normal waves by fixed rectangular structures. Ding et al. [10] studied the Bragg reflections of normal waves by structures of different shapes using the boundary element method. All the prescribed studies consider scattering problems with different configurations of various structures over a flat bottom.

Manisha et al. [47] recently developed a model considering the effects of bottom undulations for oblique wave interaction with a surface-piercing rectangular structure behind a submerged breakwater or a trench. They connected the solutions of the MSE and EMM for the regions of the undulated bottom and rectangular structure over the flat bottom, respectively. In this study, the EMM model is developed for analyzing the combined phenomena of oblique incidence, surface-piercing structures of different shapes, Bragg reflections, and undulated bottoms. In the solution procedure, the surface-piercing structures and bottom topography are sliced into successive flat shelves separated by abrupt steps. The matching conditions of the normal flow flux and the continuity of pressure are imposed on the interface boundaries. The EMM model is validated by comparisons with analytical solutions in the literature [10,45,47].

Bragg's law is usually used to predict the wavelengths at which the X-rays are intensively reflected by crystalline solids [48]. For water-wave problems, Bragg's law is applied for scatterings by floating [9,10] and submerged [5–7,23] structures at normal incidence. For oblique incidence, Mei [49] and Dalrymple et al. [50] applied Bragg's law for water wave scattering by bottom ripples. In this study, numerical experiments were conducted to study the Bragg reflections by the combined effects of floating structures, bottom variations, and oblique incidence. In addition, the numerical results are compared with those predicted by Bragg's law.

This paper is organized as follows: the wave problem is mathematically modeled and the EMM solution is developed in Section 2, and the EMM model is validated in Section 3. Discussions on oblique Bragg reflections by surface-piercing and submerged breakwaters are provided in Section 4. Finally, conclusions are presented in Section 5.

2. Materials and Methods

2.1. The Mathematical Model

We consider the problem of oblique monochromatic water wave scattering by surface-piercing structures over uneven bottoms. The wave amplitude is assumed to be small enough that the linear wave theory is applicable. The wave motion is assumed to be time-harmonic, $e^{-\mathrm{i}\sigma t}$, where $\sigma = 2\pi/T$ is the angular frequency, T is the wave period, t is the time, and $\mathbf{i}$ is the unit of complex numbers. Figure 1 shows a schematic representation of the wave scattering problem induced by a surface-piercing structure over uneven bottoms. In the figure, the surface-piercing structures and sea bottom are discretized into a series of M shelves in the intervals of $x_{m-1} \le x \le x_m$ for $m = 1, 2, 3\ldots, M$, with a water depth $h_m > d_m$, where $d_m > 0$ is the submergence depth of the structure. Alternatively, $d_m = 0$ is designated if there is no structure in the interval. Furthermore, $x_0 = -\infty$ and $x_M = \infty$ are assumed.

Considering the solution on the m-th shelf in the interval $x_{m-1} \le x \le x_m$, the velocity of the fluid is defined by

$$\mathbf{u}_m = \nabla\phi_m, \tag{1}$$

where $\nabla = (\partial/\partial x, \partial/\partial y, \partial/\partial z)$ is the three-dimensional del operator with respect to the three-dimensional Cartesian coordinates (x, y, z) and ϕ_m is the velocity potential. According to the linear wave theory, the velocity potential is governed by the Laplace equation as

$$\nabla^2\phi_m = 0. \tag{2}$$

It should be noted that ϕ_m is only related to the spatial part of the velocity potential for the remainder of this work. If there is no structure in the interval ($d_m = 0$), the problem is subjected to the kinematic and dynamic free-surface boundary conditions, respectively, as

$$-\mathbf{i}\sigma\eta_m - \frac{\partial\phi_m}{\partial z} = 0 \tag{3}$$

and

$$-\mathbf{i}\sigma\phi_m + g\eta_m = 0 \text{ on } z = 0. \tag{4}$$

Equations (3) and (4) can be combined to obtain

$$\frac{\partial\phi_m}{\partial z} - \frac{\sigma^2}{g}\phi_m = 0 \text{ on } z = 0. \tag{5}$$

On the other hand, if there is a structure in the interval ($d_m > 0$), the boundary condition on the bottom of the surface-piercing structure is given by

$$\frac{\partial\phi_m}{\partial z} = 0 \text{ on } z = -d_m. \tag{6}$$

In addition, the boundary condition on the sea bottom can be expressed in the form of

$$\frac{\partial \phi_m}{\partial z} = 0 \text{ on } z = -h_m. \tag{7}$$

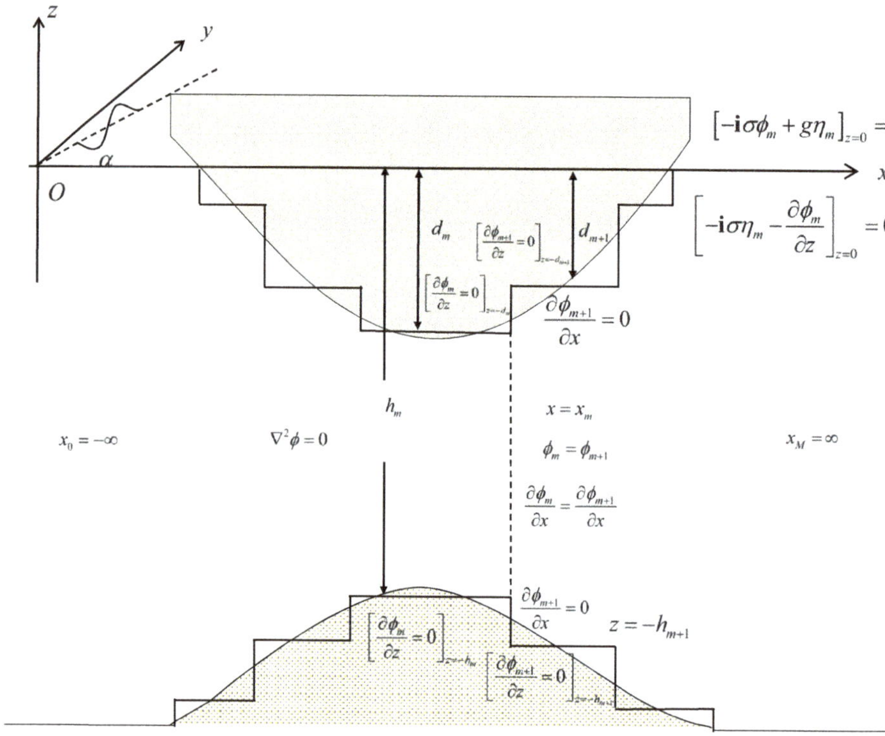

Figure 1. Schematic representation of the boundary-value problem of water-wave-scattering by the surface-piercing structure over even bottom. The study domain divided into different regions with M shelves separated by $M-1$ steps.

Equations (2)–(7) are sufficient to construct the complete solution by eigenfunctions and will be given in the next section.

Then, connection conditions are required to match the solutions ϕ_m and ϕ_{m+1} as

$$\phi_m = \phi_{m+1} \tag{8}$$

and

$$\frac{\partial \phi_m}{\partial x} = \frac{\partial \phi_{m+1}}{\partial x} \tag{9}$$

on $-d_m^{\max} < z < -h_m^{\min}$ and $x = x_m$ with $d_m^{\max} = \max(d_m, d_{m+1})$ and $h_m^{\min} = \min(h_m, h_{m+1})$. In addition, no-penetration conditions are needed on the bottom and structure side-walls, respectively, as

$$\frac{\partial \phi}{\partial x} = 0 \text{ on } -h_m^{\max} < z < -h_m^{\min} \text{ and } x = x_m \tag{10}$$

and

$$\frac{\partial \phi}{\partial x} = 0 \text{ on } -d_m^{\max} < z < -d_m^{\min} \text{ and } x = x_m. \tag{11}$$

The definitions of $h_m^{\max}$ and $d_m^{\min}$ are similar, and thus neglected here. In Equations (10) and (11), ϕ stands for either ϕ_m or ϕ_{m+1} depending on the water side of the wall.

Then, considering a monochromatic wave train with incidence angle α, amplitude a, frequency σ, and wavelength λ, which propagates towards the surface-piercing structures over an uneven bottom. Therefore to make the solution unique, the following far-field conditions are required

$$\eta = a\left(e^{\mathrm{i}\hat{k}_{1,0}x} + K_R e^{i\theta_R} e^{-\mathrm{i}\hat{k}_{1,0}x}\right)e^{\mathrm{i}k_y y} \text{ as } x \to -\infty \tag{12}$$

and

$$\eta = aK_T e^{\mathrm{i}\theta_T} e^{\mathrm{i}\hat{k}_{M,0}x} e^{\mathrm{i}k_y y} \text{as } x \to \infty. \tag{13}$$

where K_R, θ_R, K_T, and θ_T are real numbers, such that $K_R e^{\mathrm{i}\theta_R}$ and $K_T e^{\mathrm{i}\theta_T}$ are the reflection and transmission coefficients, respectively. In Equations (12) and (13), $\hat{k}_{1,0}$, $\hat{k}_{M,0}$, and k_y are positive real wavenumbers defined by

$$\hat{k}_{m,n} = \sqrt{k_{m,n}{}^2 - k_y{}^2}, \tag{14}$$

and

$$k_y = k_{1,0} \sin \alpha. \tag{15}$$

where $k_{1,0} = 2\pi/\lambda > 0$ and $k_{M,0} > 0$ are the progressive wavenumbers obtained from the dispersion relation

$$\frac{\sigma^2}{g} = k_{m,0} \tanh k_{m,0} h_m. \tag{16}$$

Here, it is assumed that there is no structure over the first and last shelves, i.e., $d_1 = 0$ and $d_M = 0$.

Now, the problem is well-defined, and the estimation of the reflection and transmission coefficients are presented in the next subsection.

2.2. Eigenfunction Matching Method

To construct the complete solution using eigenfunctions, a complete set of wavenumbers is required. When there is no structure on a shelf ($d_m = 0$), in addition to the progressive wavenumber $k_{m,0}$ in Equation (16) the evanescent wavenumbers $k_{m,n}$ ($n = 1, 2, 3, \ldots$) are defined by

$$k_{m,n} = \mathrm{i}\kappa_{m,n}, \tag{17}$$

where $\kappa_{m,n}$ is the n-th smallest positive root of the dispersion relation

$$\frac{\sigma^2}{g} = -\kappa_{m,n} \tan \kappa_{m,n} h_m. \tag{18}$$

and when there is a structure over a shelf ($d_m > 0$), the wavenumber is alternatively defined as

$$k_{m,n} = \frac{\mathrm{i}n\pi}{h_m - d_m} \tag{19}$$

for $n = 0, 1, 2, \ldots$. As the incident wave is oblique, we have to define the x-component of the wavenumber $\hat{k}_{m,n}$ by Equation (14). The complex-valued wavenumbers, $k_{m,n}$ and $\hat{k}_{m,n}$, defined in Equations (14)–(19), enable the formulation of a unified EMM for all types of situations ($d_m = 0$ or $d_m > 0$; $n = 0$ or n > 0).

Based on the linear wave theory, the complete solution of the velocity potential for the m-th shelf may be expressed as

$$\phi_m(x, y, z) = \sum_{n=0}^{N} \left(A_{m,n}\xi_{m,n}^{(1)}(x) + B_{m,n}\xi_{m,n}^{(2)}(x)\right)\zeta_{m,n}(z)e^{\mathrm{i}k_y y} \tag{20}$$

for $m = 1, 2, 3, \ldots, M$, where $A_{m,n}$ and $B_{m,n}$ are unknown coefficients to be determined. Upon applying the conditions in Equations (2), (5)–(7) and by employing the method of the separation of variables, the eigenfunctions, $\zeta_{m,n}(z)$, $\xi^{(1)}_{m,n}(x)$, and $\xi^{(2)}_{m,n}(x)$, can be obtained and expressed as

$$\zeta_{m,n}(z) = \cosh k_{m,n}(h_m + z), \tag{21}$$

$$\xi^{(1)}_{m,n}(x) = \begin{cases} e^{\mathrm{i}\hat{k}_{m,n}(x-\overline{x}_{m-1})} & \hat{k}_{m,n} \neq 0 \\ 1 & \hat{k}_{m,n} = 0, \end{cases} \tag{22}$$

and

$$\xi^{(2)}_{m,n}(x) = \begin{cases} e^{-\mathrm{i}\hat{k}_{m,n}(x-\overline{x}_{m})} & \hat{k}_{m,n} \neq 0 \\ x & \hat{k}_{m,n} = 0 \end{cases} \tag{23}$$

with

$$\begin{cases} \overline{x}_m = x_m \text{ for } m = 1, 2, \ldots, M-1 \\ \overline{x}_0 = \overline{x}_M = 0. \end{cases} \tag{24}$$

By observing at Equation (21), we have $\zeta_{m,n} = 1$ for $k_{m,n} = 0$. According to the Sturm–Liouville theory [51], the following orthogonal relation is used for solving the problem

$$\left\langle \zeta_{m,n} \middle| \zeta_{m,l} \right\rangle = \Lambda_{m,n}\delta_{nl} \tag{25}$$

where n and l is a mode index varying from 0 to N, δ_{nl} is the Kronecker delta function, and Λ_n is a function of h_m and $k_{m,n}$, written as

$$\Lambda_{m,n} = \frac{2k_{m,n}(h_m - d_m) + \sinh 2k_{m,n}(h_m - d_m)}{4k_{m,n}} \tag{26}$$

For convenience, we define the inner product of two depth eigenfunctions as follows.

$$\langle F|G \rangle = \int_{-\lambda_2}^{-\lambda_1} F(z)G(z)dz, \tag{27}$$

where F and G are the depth eigenfunctions of $\zeta_{m,n}$ with arbitrary m and n; and λ_1 and λ_2 represent the structure submergence and water depths, respectively, which correspond to the first depth eigenfunction F.

It should be noted that the eigenfunction definitions of $\zeta_{m,n}(z)$, $\xi^{(1)}_{m,n}(x)$, and $\xi^{(2)}_{m,n}(x)$ are valid for all cases ($d_m = 0$ or $d_m > 0$; $n = 0$ or $n > 0$) if the complex-valued wavenumbers, $k_{m,n}$ and $\hat{k}_{m,n}$, are defined by Equations (14)–(19).

Based on the far-field conditions (Equations (12) and (13)) and the dynamic free-surface boundary condition (Equation (4)), the far-field solutions of the velocity potential can be expressed as

$$\phi_1 = -\frac{\mathrm{i}ag}{\sigma}\frac{\cosh k_{1,0}(h_1 + z)}{\cosh k_{1,0}h_1}\left(e^{\mathrm{i}\hat{k}_{1,0}x} + K_R e^{\mathrm{i}\theta_R} e^{-\mathrm{i}\hat{k}_{1,0}x}\right)e^{\mathrm{i}k_y y} \text{ as } x \to -\infty \tag{28}$$

and

$$\phi_M = -\frac{\mathrm{i}ag}{\sigma}\frac{\cosh k_{M,0}(h_M + z)}{\cosh k_{M,0}h_M}\left(K_T e^{\mathrm{i}\theta_T} e^{\mathrm{i}\hat{k}_{M,0}x}\right)e^{\mathrm{i}k_y y} \text{ as } x \to \infty. \tag{29}$$

Comparing Equations (20), (28), and (29), the following equations can be obtained as

$$A_{1,0} = -\frac{\mathrm{i}ag}{\sigma}\frac{1}{\cosh k_{1,0}h_1}, \tag{30}$$

$$B_{1,0}e^{\mathrm{i}\hat{k}_{m,n}\overline{x}} = -\frac{\mathrm{i}aK_R e^{\mathrm{i}\theta_R} g}{\sigma}\frac{1}{\cosh k_{1,0}h_1}, \tag{31}$$

$$A_{M,0}e^{-\mathrm{i}\hat{k}_{M,0}\bar{x}_{M-1}} = -\frac{\mathrm{i}aK_T e^{\mathrm{i}\theta_T}g}{\sigma}\frac{1}{\cosh k_{M,0}h_M}, \tag{32}$$

$$A_{1,n} = 0 \text{ for } n = 1,2,\ldots,N, \tag{33}$$

and

$$B_{M,n} = 0 \text{ for } n = 0,1,\ldots,N. \tag{34}$$

The other coefficients $A_{m,n}$ and $B_{m,n}$ in Equation (20) can be determined using the matching conditions, Equations (8)–(11), at two adjacent shelves. The conservation of momentum, stemming from Equation (8), gives

$$\left.\left\langle \zeta^{\text{inner}}_{m,l} \middle| \phi_m \right\rangle\right|_{x=x_m} = \left.\left\langle \zeta^{\text{inner}}_{m+1,l} \middle| \phi_{m+1} \right\rangle\right|_{x=x_m}, \tag{35}$$

where $\zeta^{\text{inner}}_{m,l}(z)$ is the inner depth eigenfunction corresponding to $d_m^{\max}$ and $h_m^{\min}$. For clarity, $\zeta^{\text{inner}}_{m,l}(z)$ is defined by Equation (21) with the wavenumbers $k_{m,n}$ obtained from Equations (16)–(19) as d_m and h_m replaced by $d_m^{\max}$ and $h_m^{\min}$, respectively. Similarly, the conservation of mass, comes from Equations (9)–(11), yields the following equation

$$\left.\left\langle \frac{\partial \phi_m}{\partial x} \middle| \zeta^{\text{outer}}_{m,l} \right\rangle\right|_{x=x_m} = \left.\left\langle \frac{\partial \phi_{m+1}}{\partial x} \middle| \zeta^{\text{outer}}_{m,l} \right\rangle\right|_{x=x_m}, \tag{36}$$

where $\zeta^{\text{outer}}_{m,l}(z)$ is the outer depth function corresponding to $d_m^{\min}$ and $h_m^{\max}$. In Equations (35) and (36), the subscripted indices go as $l = 0,1,\ldots,N$ and $m = 1,2,\ldots,M-1$. Additionally, it should be noted that Equations (35) and (36) are valid for all eight cases, as shown in Figure 2.

By using Equation (20), Equations (35) and (36) can be rewritten in the following forms

$$\begin{aligned}&\sum_{n=0}^{N}\left(A_{m,n}\xi^{(1)}_{m,n}(x_m) + B_{m,n}\xi^{(2)}_{m,n}(x_m)\right)\left\langle \zeta^{\text{inner}}_{m,l} \middle| \zeta_{m,n} \right\rangle\\ &= \sum_{n=0}^{N}\left(A_{m+1,n}\xi^{(1)}_{m+1,n}(x_m) + B_{m+1,n}\xi^{(2)}_{m+1,n}(x_m)\right)\left\langle \zeta^{\text{inner}}_{m,l} \middle| \zeta_{m+1,n} \right\rangle\end{aligned} \tag{37}$$

and

$$\begin{aligned}&\sum_{n=0}^{N}\left(A_{m,n}\frac{d\xi^{(1)}_{m,n}}{dx}(x_m) + B_{m,n}\frac{d\xi^{(2)}_{m,n}}{dx}(x_m)\right)\left\langle \zeta_{m,n} \middle| \zeta^{\text{outer}}_{m,l} \right\rangle\\ &= \sum_{n=0}^{N}\left(A_{m+1,n}\frac{d\xi^{(1)}_{m+1,n}}{dx}(x_m) + B_{m+1,n}\frac{d\xi^{(2)}_{m+1,n}}{dx}(x_m)\right)\left\langle \zeta_{m+1,n} \middle| \zeta^{\text{outer}}_{m,l} \right\rangle.\end{aligned} \tag{38}$$

Subsequently, it can be found that Equations (30), (33), (34), (37), and (38) are $2M(N+1)$ linear equations, that can be used to solve the $2M(N+1)$ unknowns, $A_{m,n}$ and $B_{m,n}$. Furthermore, Equations (37) and (38) can be reduced to the original equations of EMM for normal water-wave-scattering without structures where $\alpha = 0$ and $d_m = 0$ [28,35]. In this study, the SuperLU is used to solve the resulting sparse system of linear equations [52]. After the unknowns, $A_{m,n}$ and $B_{m,n}$, are solved, the reflection and transmission coefficients can be obtained by Equations (31) and (32), respectively. This completes the solution procedure for the EMM.

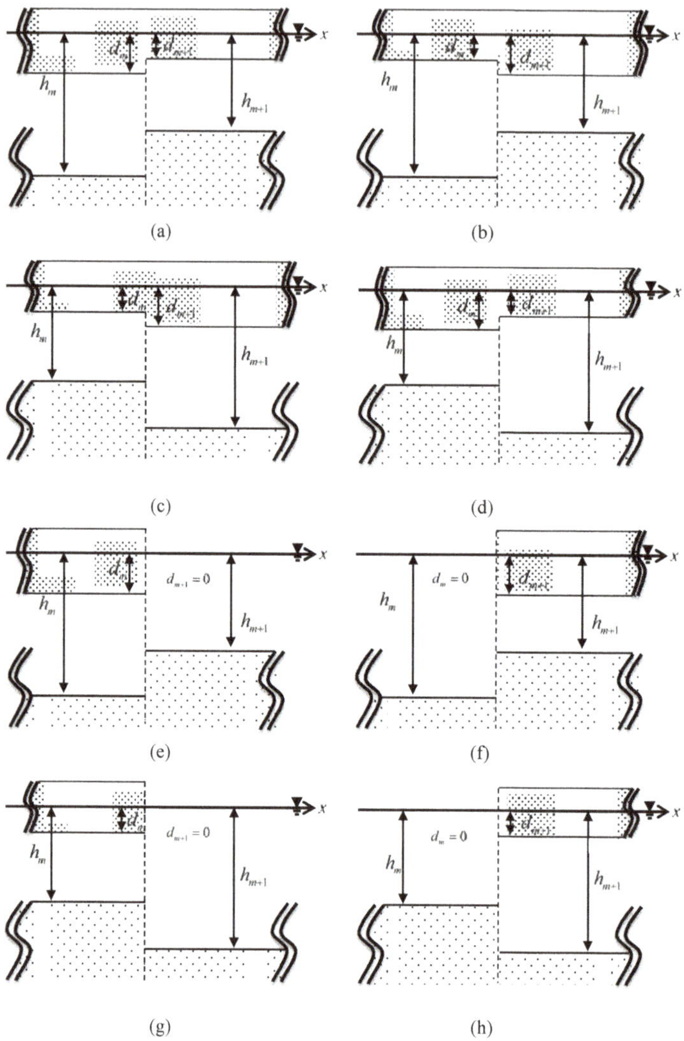

Figure 2. Schematics for eight different situations of shelves separated by abrupt connections.

3. Results

The present model is validated using three numerical examples.

3.1. Rectangular Surface-Piercing Structure over a Flat Bottom

First, we consider monochromatic wave trains with incidence angles $\alpha = 45^{\circ}$ and $\alpha = 75^{\circ}$, which propagate towards a rectangular surface-piercing structure over a flat bottom with water depth $h = 1$ m, as depicted in Figure 3. The breadth and depth of the rectangular barrier are set to $2a = 0.6$ m and $d = 0.2$ m, respectively. Figure 4 shows a comparison of the reflection and transmission coefficients obtained by the proposed EMM with the results obtained by Lebreton and Margnac [53], Bai [43], Söylemez and Gören [45]. In the figure, the convergence with respect to the increasing numbers of evanescent modes, N, is obvious. Both the reflection and transmission coefficients evaluated by the present model with $N = 5$ are in good agreement with those in the literature for the whole frequency range. This validates the present model for solving problems of oblique wave scattering by a rectangular surface-piercing structure over a flat bottom.

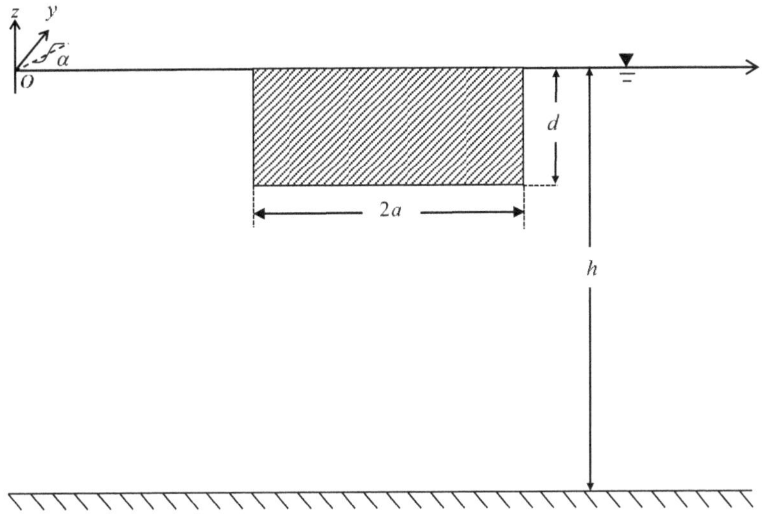

Figure 3. Schematic representation of the water wave scattering by a rectangular surface-piercing structure over flat bottom.

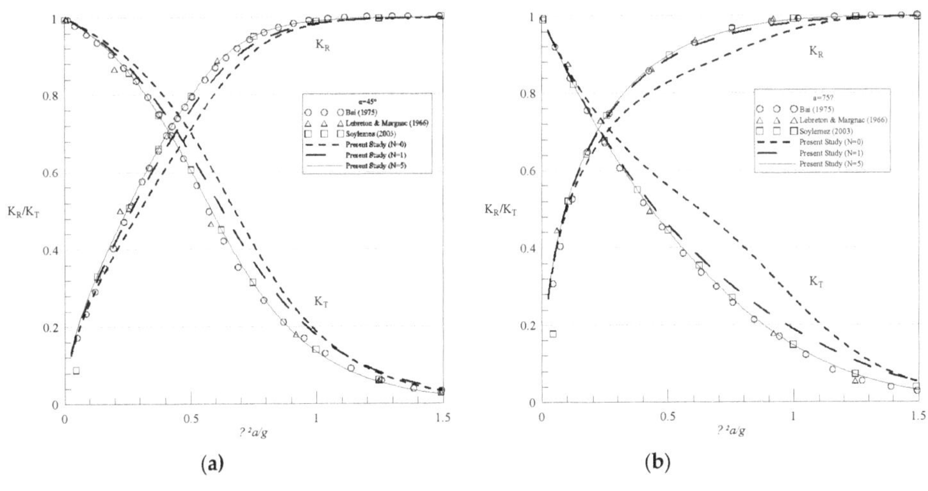

Figure 4. Comparison of the reflection and transmission coefficients from the present study with the results in the literatures for water wave scattering by a rectangular surface-piercing structure over flat bottom with incidence angles (**a**) $\alpha = 45^{\circ}$ and (**b**) $\alpha = 75^{\circ}$.

3.2. Rectangular Surface-Piercing Structure behind Parabolic Breakwater

We now consider oblique monochromatic wave trains that propagate towards a rectangular surface-piercing structure behind a parabolic breakwater defined by $z = -h(x)$ for $|x| \leq c$ as

$$h(x) = (h - h_b)\left(1 + \frac{h_b x^2}{(h - h_b)c^2}\right). \tag{39}$$

As shown in Figure 5, the other parameters are set as $h_1 = 30$ m, $h_b = 15$ m, $d = 7.5$ m, $2c = 200$ m, $w = 20$ m, and $2a = 30$ m, which are exactly the same values as those of Manisha et al. [47]. Furthermore, 40 shelves are used to approximate the parabolic breakwater after performing a preliminary convergence analysis as shown in Figure 6.

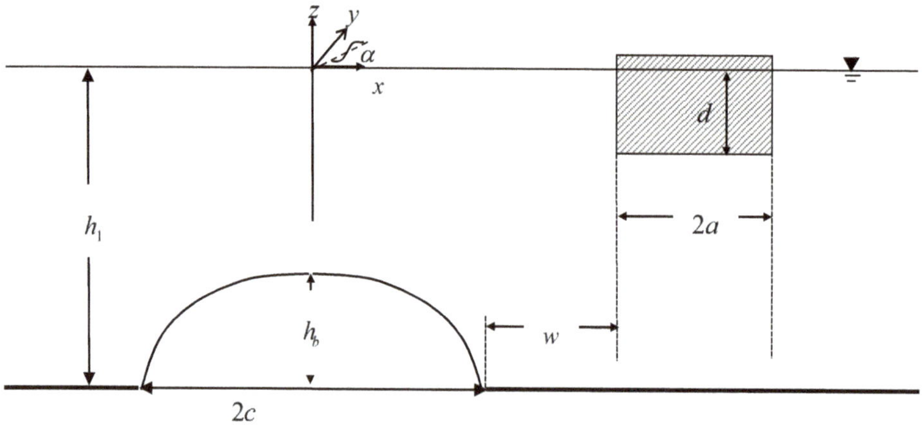

Figure 5. Schematic diagram of the water-wave-scattering by a rectangular surface-piercing structure behind a parabolic breakwater.

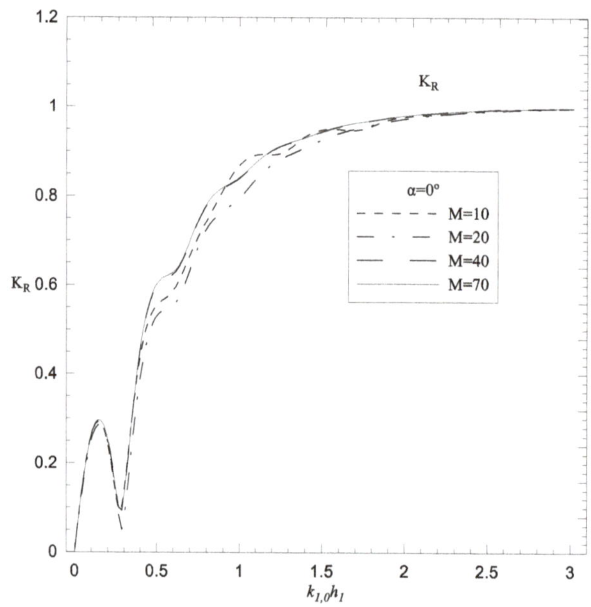

Figure 6. Convergence analysis of M for the water wave scattering by a rectangular structure behind a parabolic breakwater.

Figure 7 shows the comparison of reflection coefficients obtained by the present method and those from Manisha et al. [47]. In the figure, the convergence with respect to the increasing numbers of evanescent modes, N, can also be observed. The convergent results of the present model are in good agreement with those in Manisha et al. [47]. This validates the proposed EMM for solving oblique wave scattering by a rectangular surface-piercing structure over uneven bottoms.

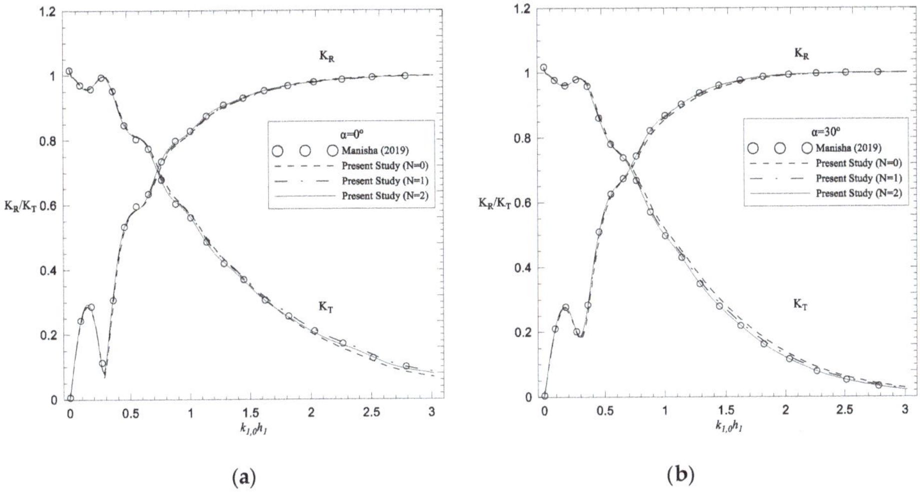

Figure 7. Comparison of the reflection coefficients from the present study with the results from the literature for the (**a**) normal and (**b**) oblique water-wave-scattering by a rectangular structure behind a parabolic breakwater.

3.3. Bragg Reflections by Periodic Surface-Piercing Structures over Flat Bottom

As the final example of validation, we consider normal monochromatic wave trains, which propagate towards a series of periodic rectangular and triangular surface-piercing structures over flat bottoms with water depth $h = 1$ m. As depicted in Figure 8, the other parameters are set as $\alpha = 0$, $L = 3$, $d/h = 0.25$, $a/h = 0.25$, and $S/h = 3$, which are exactly the same values as those of Ding et al. [10]. Typically, 10 shelves are adopted to approximate each triangular structure in this example, as shown in Figure 9b.

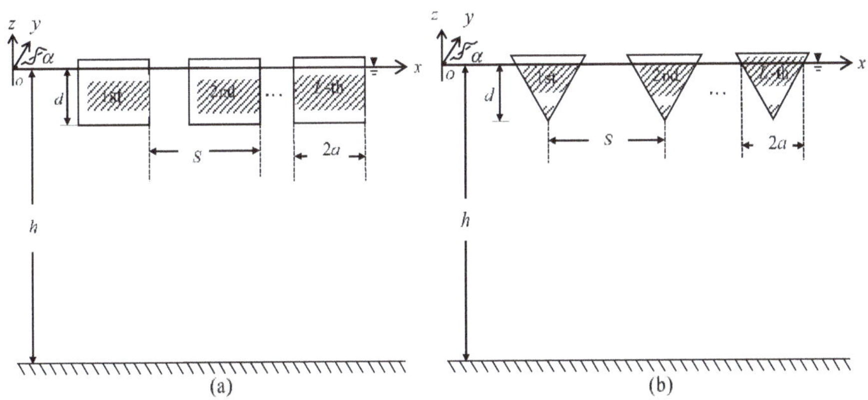

Figure 8. Schematic diagram of Bragg reflections by periodic (**a**) rectangular and (**b**) triangular surface-piercing structures over flat bottom.

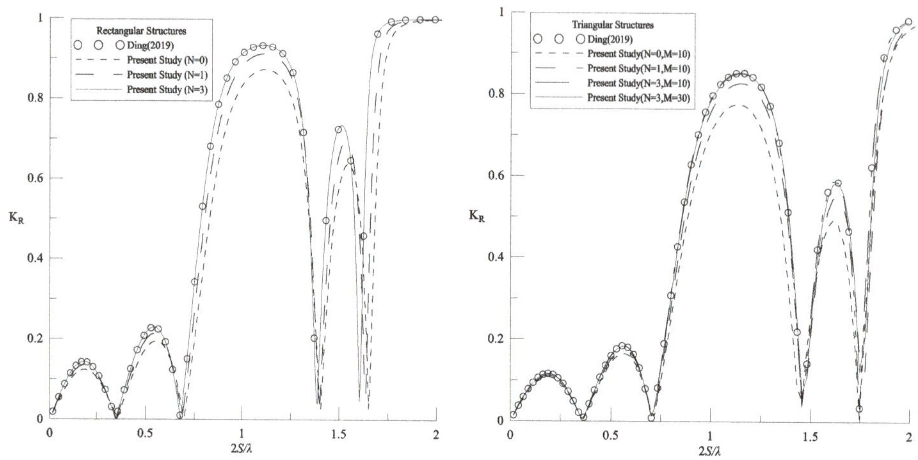

(a) (b)

Figure 9. Comparison of the reflection coefficients from the present study with the results in the literature for Bragg reflections by periodic (**a**) rectangular and (**b**) triangular surface-piercing structures over flat bottom.

Figure 9 shows the comparison of the reflection coefficients obtained by the present method and Ding et al. [10]. In the figure, the convergence with respect to the increasing numbers of evanescent modes, N, is also obvious, and the results by the present model with $N = 3$ are in good agreement with those obtained by Ding et al. [10]. In addition, Bragg's law confirms that intensive reflections occur for $2S/\lambda$, being positive integers. This validates the present model for solving Bragg reflections of normal water waves by periodic surface-piercing structures of arbitrary shapes over flat bottoms.

4. Discussion

After the model is validated, the proposed EMM model is applied to solve Bragg reflections of oblique water waves by periodic rectangular and triangular surface-piercing structures over periodic parabolic breakwaters as depicted in Figure 10. The parabolic breakwaters are defined by Equation (39) with $h_1 = 1$ m and $2c = 0.5$ m and the surface-piercing structures are either rectangular or triangular with $2a = 0.5$ m and $d = 0.25$ m, as shown in Figure 8. In addition, $S = 3$ m is set such that the problem configuration is reduced to the previous example if $h_b = 0$. The incidence angle α, structure or breakwater number L, and breakwater height h_b are the parameters studied in the following.

As depicted in Figure 11, we extend the previous example by considering waves of different incidence angles $\alpha = 15^o$ and $\alpha = 30^o$. In other words, we set $N = 3$ and $h_b = 0$, and used 10 shelves to approximate each triangular structure. In the figure, Bragg's law is observed that intensive reflections occur for $2S(\cos\alpha)/\lambda$ being equal to positive integers. Moreover, it is interesting to find that the case with a larger incidence angle results in a more intensive Bragg reflection. In addition, it is noticeable that the secondary resonance $2S(\cos\alpha)/\lambda = 2$ is stronger than the primary resonance $2S(\cos\alpha)/\lambda = 1$. In addition, total Bragg reflections occur in the secondary resonances.

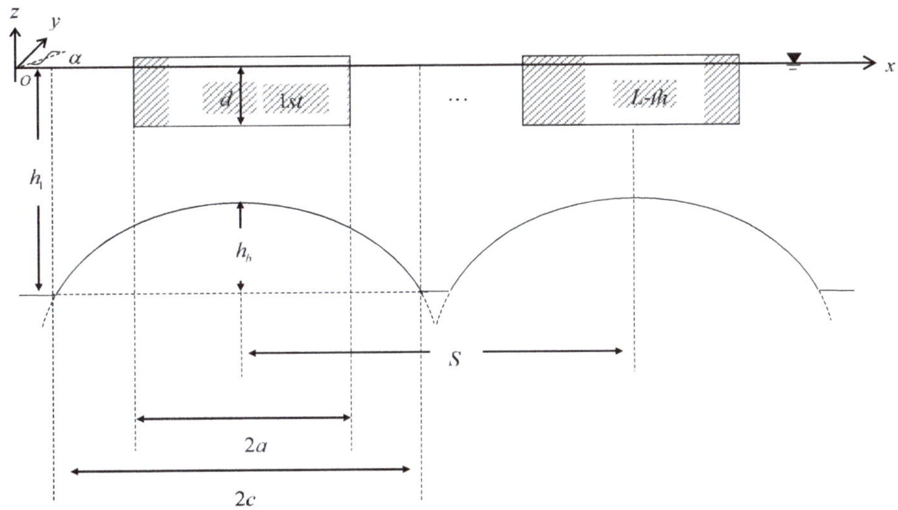

Figure 10. Schematic diagram of Bragg reflections by periodic rectangular surface-piercing structures over periodic parabolic breakwaters.

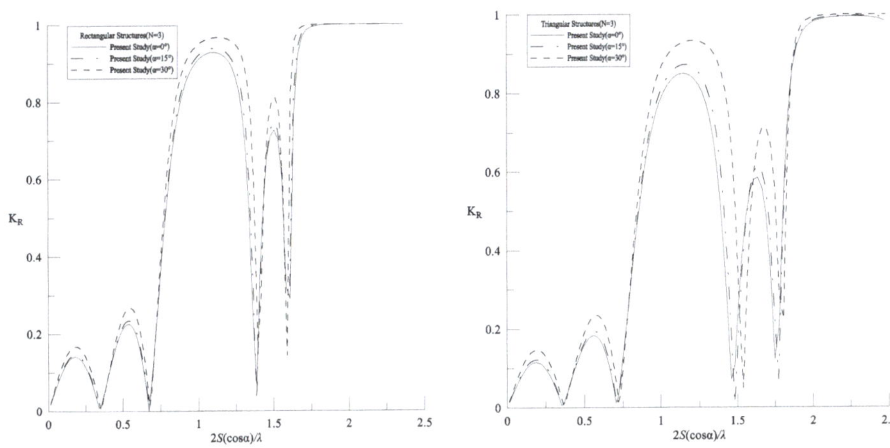

(**a**) (**b**)

Figure 11. The reflection coefficients with different incidence angles for Bragg reflections by periodic (**a**) rectangular and (**b**) triangular surface-piercing structures over flat bottoms.

Before studying the combined Bragg reflections of periodic surface-piercing and submerged breakwaters, problems of Bragg reflections solely by three periodic parabolic breakwaters of different heights are considered. Here, we use 40 shelves to approximate each parabolic breakwater as before and in the following. As depicted in Figure 12a, the case of higher breakwaters results in a more intensive Bragg reflection, as expected. In addition, Figure 12b shows the Bragg reflections for different incidence angles, which also confirm Bragg's law when $2S(\cos\alpha)/\lambda$ is equal to a positive integer. However, it is interesting to observe that a larger incidence angle results in a less intensive Bragg reflection and the primary resonance $2S(\cos\alpha)/\lambda = 1$ is stronger than the secondary resonance $2S(\cos\alpha)/\lambda = 2$, which are totally opposite to the situations of Bragg reflections caused solely by periodic surface-piercing structures.

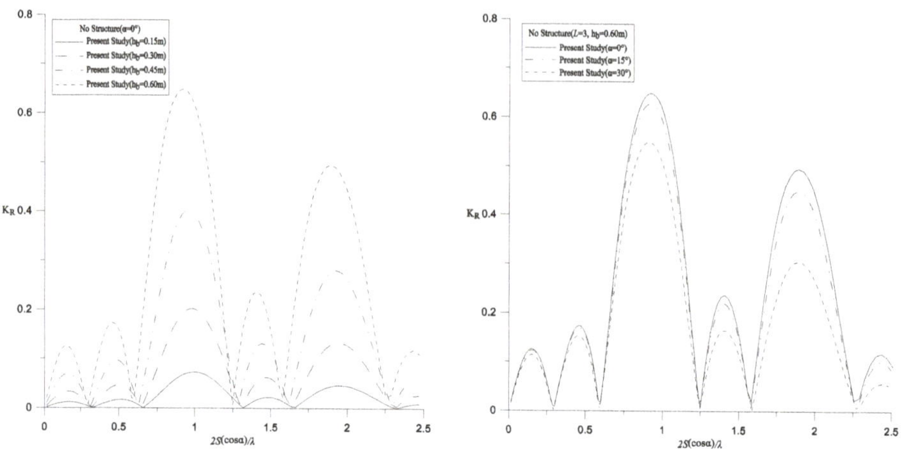

(**a**) (**b**)

Figure 12. The reflection coefficients for Bragg reflections over periodic parabolic breakwaters of (**a**) different heights and (**b**) different incidence angles.

Then the combined Bragg reflections of periodic surface-piercing and submerged breakwaters are studied by setting different heights of the periodic parabolic breakwaters. As depicted in Figure 13, the EMM is applied to solve the problems of Bragg reflections of normal incident waves by three periodic rectangular surface-piercing structures over three periodic parabolic breakwaters. In the figure, it is clear that the case of higher breakwaters results in a stronger Bragg reflection, as expected. In addition, if the periodic breakwaters are high, total Bragg reflections occur for both the primary and secondary resonances.

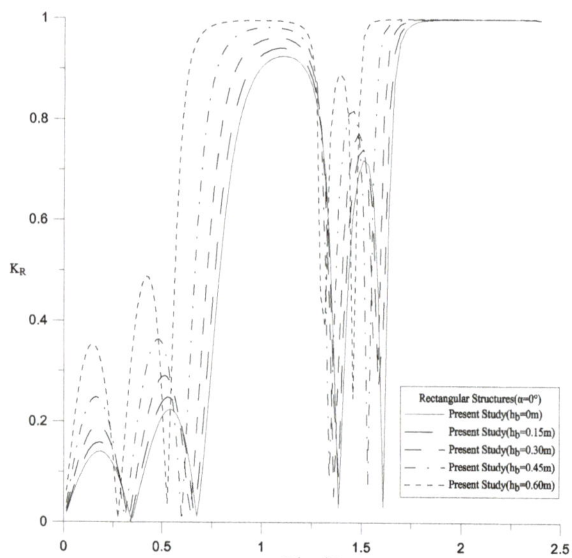

Figure 13. The reflection coefficients for Bragg reflections by periodic rectangular surface-piercing structures over periodic parabolic breakwaters of different heights.

Then, the study with $h_b = 0.6$ m in Figure 13 is extended to oblique incidence angles, as depicted in Figure 14a, which also numerically confirms Bragg's law of oblique waves. As depicted in Figure 14b, the results are similar if the study is extended to five periodic rectangular surface-piercing structures over five periodic parabolic breakwaters.

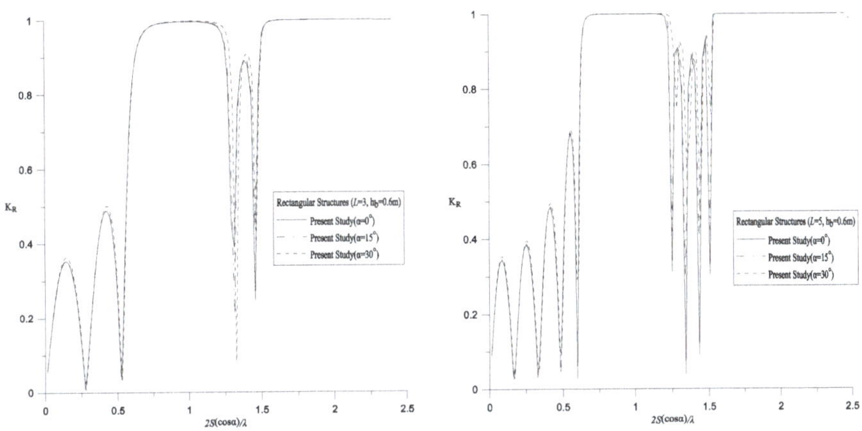

(a) (b)

Figure 14. The reflection coefficients for Bragg reflections by periodic rectangular surface-piercing structures of different numbers (**a**) $L = 3$ and (**b**) $L = 5$ over periodic parabolic breakwaters.

Finally, if the surface-piercing structures are treated as floating rather than fixed, previous studies applied the EMM to problems involving a rectangular structure with surge [54,55], heave [56], and free motions [57]. Combing the proposed step approximation with these studies, the EMM can be applied to solve water-wave-scattering by a floating structure of different shapes over uneven bottoms. This is currently under investigation.

5. Conclusions

In this study, the combined Bragg reflections of oblique water waves by periodic surface-piercing and submerged breakwaters were solved using the eigenfunction matching method (EMM). In the solution procedure, the surface-piercing structures and bottom profiles are sliced into a number of shelves separated by abrupt steps. The solution on each shelf is composed of eigenfunctions with unknown coefficients. Then, a system of linear equations is obtained by applying the conservations of mass and momentum. The present method was validated under three problems of oblique wave scattering by a rectangular surface-piercing structure over a flat bottom, behind a parabolic breakwater, and Bragg reflection by periodic rectangular/triangular surface-piercing structures over a flat bottom. Then, the proposed method was applied to solve the Bragg reflections of oblique water waves by periodic surface-piercing structures over periodic bottoms. For oblique Bragg reflections by periodic surface-piercing structures, numerical results indicated that the case with a larger incidence angle results in a more intensive Bragg reflection and the secondary resonance is stronger than the primary resonance. However, the results are completely opposite for oblique Bragg reflections by periodic submerged breakwaters. In addition, Bragg's law of oblique waves was applied in the numerical experiments. Some theoretical and physical studies on the resonance intensity and incidence angle are required in future investigation.

Author Contributions: Conceptualization, I.-F.T. and C.-C.T.; methodology, C.-C.T.; software, C.-C.T.; validation, C.-S.Y. and C.-C.T.; formal analysis, C.-S.Y.; investigation, C.-S.Y.; resources, I.-F.T. and C.-C.T.; data curation, C.-S.Y.; writing—original draft preparation, I.-F.T. and C.-S.Y.; writing—review and editing, I.-F.T. and C.-C.T.;

visualization, C.-S.Y.; supervision, I.-F.T. and C.-C.T.; project administration, I.-F.T. and C.-C.T.; funding acquisition, I.-F.T. and C.-C.T. All authors have read and agreed to the published version of the manuscript.

Funding: Ministry of Science and Technology, Taiwan: 107-2221-E-992-045-MY2.

Acknowledgments: The Ministry of Science and Technology of Taiwan is gratefully acknowledged for providing financial support to carry out the present work under the Grant No. MOST 107-2221-E-992-045-MY2.

Conflicts of Interest: The authors declare no conflicts of interest exist.

References

1. Belibassakis, K.; Athanassoulis, G. A coupled-mode technique for weakly nonlinear wave interaction with large floating structures lying over variable bathymetry regions. *Appl. Ocean Res.* **2006**, *28*, 59–76. [CrossRef]
2. Shih, R.-S. Experimental study on the performance characteristics of porous perpendicular pipe breakwaters. *Ocean Eng.* **2012**, *50*, 53–62. [CrossRef]
3. Chung, J.S. Motion of a floating structure in water of uniform depth. *J. Hydronautics* **1976**, *10*, 65–73. [CrossRef]
4. Kagemoto, H.; Fujino, M.; Murai, M. Theoretical and experimental predictions of the hydroelastic response of a very large floating structure in waves. *Appl. Ocean Res.* **1998**, *20*, 135–144. [CrossRef]
5. Mei, C.C.; Hara, T.; Naciri, M. Note on Bragg scattering of water waves by parallel bars on the seabed. *J. Fluid Mech.* **1988**, *186*, 147–162. [CrossRef]
6. Bailard, J.A.; DeVries, J.W.; Kirby, J.T. Considerations in using Bragg reflection for storm erosion protection. *J. Waterw. Port Coast. Ocean Eng.* **1992**, *118*, 62–74. [CrossRef]
7. Tsai, L.-H.; Wen, C.-C. Design of a series of submerged breakwaters for coastal protection against waves. *China Ocean Eng.* **2010**, *24*, 553–564.
8. Hsu, T.-W.; Lin, J.-F.; Hsiao, S.-C.; Ou, S.-H.; Babanin, A.V.; Wu, Y.-T. Wave reflection and vortex evolution in Bragg scattering in real fluids. *Ocean Eng.* **2014**, *88*, 508–519. [CrossRef]
9. Ouyang, H.-T.; Chen, K.-H.; Tsai, C.-M. Investigation on bragg reflection of surface water waves induced by a train of fixed floating pontoon breakwaters. *Int. J. Nav. Archit. Ocean Eng.* **2015**, *7*, 951–963. [CrossRef]
10. Ding, W.-W.; Zou, Z.-J.; Wu, J.-P.; Huang, B.-G. Investigation of surface-piercing fixed structures with different shapes for Bragg reflection of water waves. *Int. J. Nav. Archit. Ocean Eng.* **2019**, *11*, 819–827. [CrossRef]
11. Roseau, M. *Asymptotic Wave Theory*; North-Holland Pub. Co.: Amsterdam, NY, USA, 1976.
12. Berkhoff, J.C.W. Computation of combined refraction-diffraction. In Proceedings of the 13th International Conference on Coastal Engineering, Vancouver, BC, Canada, 10–14 July 1972; pp. 471–490.
13. Kirby, J.T. A general wave equation for waves over rippled beds. *J. Fluid Mech.* **1986**, *162*, 171–186. [CrossRef]
14. Chamberlain, P.G.; Porter, D. The modified mild-slope equation. *J. Fluid Mech.* **1995**, *291*, 393–407. [CrossRef]
15. Massel, S.R. Extended refraction-diffraction equation for surface waves. *Coast. Eng.* **1993**, *19*, 97–126. [CrossRef]
16. Porter, D.; Staziker, D.J. Extensions of the mild-slope equation. *J. Fluid Mech.* **1995**, *300*, 367–382. [CrossRef]
17. Athanassoulis, G.; Belibassakis, K. A consistent coupled-mode theory for the propagation of small-amplitude water waves over variable bathymetry regions. *J. Fluid Mech.* **1999**, *389*, 275–301. [CrossRef]
18. Belibassakis, K.; Athanassoulis, G. Extension of second-order Stokes theory to variable bathymetry. *J. Fluid Mech.* **2002**, *464*, 35–80. [CrossRef]
19. Belibassakis, K.A.; Athanassoulis, G.A.; Gerostathis, T.P. A Coupled-Mode Model for the Refraction–Diffraction of Linear Waves over Steep Three-Dimensional Bathymetry. *Appl. Ocean Res.* **2001**, *23*, 319–336. [CrossRef]
20. Belibassakis, K. A coupled-mode model for the scattering of water waves by shearing currents in variable bathymetry. *J. Fluid Mech.* **2007**, *578*, 413–434. [CrossRef]
21. Belibassakis, K.; Touboul, J. A nonlinear coupled-mode model for waves propagating in vertically sheared currents in variable bathymetry—Collinear waves and currents. *Fluids* **2019**, *4*, 61. [CrossRef]
22. Toledo, Y.; Agnon, Y. Three dimensional application of the complementary mild-slope equation. *Coast. Eng.* **2011**, *58*, 1–8. [CrossRef]
23. Belibassakis, K.; Touboul, J.; Laffitte, E.; Rey, V. A mild-slope system for bragg scattering of water waves by sinusoidal bathymetry in the presence of vertically sheared currents. *J. Mar. Sci. Eng.* **2019**, *7*, 9. [CrossRef]

24. Suh, K.D.; Lee, C.; Park, W.S. Time-dependent equations for wave propagation on rapidly varying topography. *Coast. Eng.* **1997**, *32*, 91–117. [CrossRef]
25. Belibassakis, K.; Athanassoulis, G. A coupled-mode model for the hydroelastic analysis of large floating bodies over variable bathymetry regions. *J. Fluid Mech.* **2005**, *531*, 221–249. [CrossRef]
26. Porter, D.; Porter, R. Approximations to wave scattering by an ice sheet of variable thickness over undulating bed topography. *J. Fluid Mech.* **2004**, *509*, 145–179. [CrossRef]
27. Porter, D. The mild-slope equations: A unified theory. *J. Fluid Mech.* **2020**, *887*, A29. [CrossRef]
28. Takano, K. Effets d'un obstacle parallelepipedique sur la propagation de la houle. *La Houille Blanche* **1960**, *15*, 247–267. [CrossRef]
29. Kirby, J.T.; Dalrymple, R.A.; Seo, S.N. Propagation of obliquely incident water waves over a trench. Part 2. Currents flowing along the trench. *J. Fluid Mech.* **1987**, *176*, 95–116. [CrossRef]
30. Kirby, J.T.; Dalrymple, R.A. Propagation of obliquely incident water waves over a trench. *J. Fluid Mech.* **1983**, *133*, 47–63. [CrossRef]
31. Devillard, P.; Dunlop, F.; Souillard, B. Localization of gravity waves on a channel with a random bottom. *J. Fluid Mech.* **1988**, *186*, 521–538. [CrossRef]
32. O'Hare, T.J.; Davies, A.G. A comparison of two models for surface-wave propagation over rapidly varying topography. *Appl. Ocean Res.* **1993**, *15*, 1–11. [CrossRef]
33. O'Hare, T.J.; Davies, A.G. A new model for surface wave propagation over undulating topography. *Coast. Eng.* **1992**, *18*, 251–266. [CrossRef]
34. Tsai, C.-C.; Lin, Y.-T.; Hsu, T.-W. On step approximation of water-wave scattering over steep or undulated slope. *Int. J. Offshore Polar Eng.* **2014**, *24*, 98–105.
35. Tsai, C.C.; Hsu, T.-W.; Lin, Y.-T. On Step Approximation for Roseau's Analytical Solution of Water Waves. *Math. Probl. Eng.* **2011**, *2011*. [CrossRef]
36. Tsai, C.-C.; Lin, Y.-T.; Chang, J.-Y.; Hsu, T.-W. A coupled-mode study on weakly viscous Bragg scattering of surface gravity waves. *Ocean Eng.* **2016**, *122*, 136–144. [CrossRef]
37. Tsai, C.C.; Lin, Y.-T.; Hsu, T.-W. On the weak viscous effect of the reflection and transmission over an arbitrary topography. *Phys. Fluids* **2013**, *25*, 043103–043121. [CrossRef]
38. Tsai, C.-C.; Lin, Y.-T.; Hsu, T.-W. Propagating of obliquely incident, weakly viscous waves over variable bathymetry. *J. Coast. Res.* **2016**, *32*, 974–982. [CrossRef]
39. Tsai, C.-C.; Tai, W.; Hsu, T.-W.; Hsiao, S.-C. Step approximation of water wave scattering caused by tension-leg structures over uneven bottoms. *Ocean Eng.* **2018**, *166*, 208–225. [CrossRef]
40. Liu, X.; Wang, X.; Xu, S. A DMM-EMM-RSM hybrid technique on two-dimensional frequency-domain hydroelasticity of floating structures over variable bathymetry. *Ocean Eng.* **2020**, *201*, 107135. [CrossRef]
41. Tsai, C.-C.; Chou, W.-R. Comparison between consistent coupled-mode system and eigenfunction matching method for solving water wave scattering. *J. Mar. Sci. Tech. Taiw.* **2015**, *23*, 870–881. [CrossRef]
42. Katō, J.; Hagino, S.; Uekita, Y. Damping effect of floating breakwater to which anti-rolling system is applied. In Proceedings of the Tenth Conference Coastal Engineering, Tokyo, Japan, September 1966; pp. 1068–1078.
43. Bai, K.J. Diffraction of oblique waves by an infinite cylinder. *J. Fluid Mech.* **1975**, *68*, 513–535. [CrossRef]
44. Kanoria, M.; Dolai, D.P.; Mandal, B.N. Water-wave Scattering by Thick Vertical Barriers. *J. Eng. Math.* **1999**, *35*, 361–384. [CrossRef]
45. Söylemez, M.; Gören, Ö. Diffraction of oblique waves by thick rectangular barriers. *Appl. Ocean Res.* **2003**, *25*, 345–353. [CrossRef]
46. Garrison, C.J. Interaction of oblique waves with an infinite cylinder. *Appl. Ocean Res.* **1984**, *6*, 4–15. [CrossRef]
47. Manisha; Kaligatla, R.B.; Sahoo, T. Effect of bottom undulation for mitigating wave-induced forces on a floating bridge. *Wave Motion* **2019**, *89*, 166–184. [CrossRef]
48. Bragg, W.H.; Bragg, W.L. The reflection of X-rays by crystals. *Proc. R. Soc. Lond. Ser. A* **1913**, *88*, 428–438. [CrossRef]
49. Mei, C.C. Resonant reflection of surface water waves by periodic sandbars. *J. Fluid Mech.* **1985**, *152*, 315–335. [CrossRef]
50. Dalrymple, R.A.; Kirby, J.T. Water waves over ripples. *J. Waterw. Port Coast. Ocean Eng.* **1986**, *112*, 309–319. [CrossRef]
51. Zettl, A. *Sturm-Liouville Theory*; American Mathematical Society: Rhode Island, RI, USA, 2005.

52. Li, X.S. An overview of SuperLU: Algorithms, implementation, and user interface. *ACM Trans. Math. Softw.* **2005**, *31*, 302–325. [CrossRef]
53. Lebreton, J.; Margnac, M. Traitement sur ordinateur de quelques problèmes concernant l'action de la houle sur les corps flottants en théorie bidimensionelle. In *Bulletin du Centre de Recherches et D'Essais de Chatou*; Chatou: Yvelines, France, 1966; pp. 3–43.
54. Lee, C.-P.; Lee, J.-F. Wave-induced surge motion of a tension leg structure. *Ocean Eng.* **1993**, *20*, 171–186. [CrossRef]
55. Lee, C.-P. Dragged surge motion of a tension leg structure. *Ocean Eng.* **1994**, *21*, 311–328. [CrossRef]
56. Lee, J.-F. On the heave radiation of a rectangular structure. *Ocean Eng.* **1995**, *22*, 19–34. [CrossRef]
57. Zheng, Y.H.; Shen, Y.M.; You, Y.G.; Wu, B.J.; Jie, D.S. On the radiation and diffraction of water waves by a rectangular structure with a sidewall. *Ocean Eng.* **2004**, *31*, 2087–2104. [CrossRef]

Journal of
Marine Science and Engineering

MDPI

Article

Structural Safety Analysis for an Oscillating Water Column Wave Power Conversion System Installed in Caisson Structure

Hsien Hua Lee [1,*], Thung-Yeh Wu [1], Chung-You Lin [1] and Yung-Fang Chiu [2]

[1] Department of Marine Environment and Engineering, National Sun Yat-sen University, Kaohsiung 804 Taiwan; m015040023@student.nsysu.edu.tw (T.-Y.W.); m015040019@student.nsysu.edu.tw (C.-Y.L.)

[2] Harbor and Marine Technology Center, IOT of MOCT, Taichung 435, Taiwan; yfchiu@mail.ihmt.gov.tw

* Correspondence: hhlee@mail.nsysu.edu.tw

Received: 22 June 2020; Accepted: 7 July 2020; Published: 9 July 2020

Abstract: In this study, an alternative way, a so called caisson based type of oscillating water column (OWC) wave energy converting system was proposed to capture and convert wave energy. Since the caisson structure is constructed to protect the coastal line or ports, it is important to know if a built-in associated OWC system will be a burden to affect the safety of the structure or it is safe enough to work appropriately. In this study, three steps of structural analysis were performed: firstly, the analysis for the structural safety of the whole caisson structure; secondly, performing the mechanic analysis for the chamber of the associated OWC system; and finally, performing the analysis for the wave induced air-pressure in the chamber under the design conditions of a local location during the wave-converting operation. For the structural safety analysis, a typical structural model associated with caisson breakwater was built and analyzed while the shape of the structure, material applied to the construction, and associated boundary conditions were all set-up according to the wave and structures. The motion and the strain distribution of the caisson structure subjected to designated waves of 50-year return period were evaluated and compared to the safety requirement by the code. For the analysis of the energy converting performance, a numerical method by using a theorem of unsteady Navier–Stokes equations in conservation form was used to analyze the proposed OWC model when the structure subjected to an incident wave of a 10-year return period.

Keywords: structural safety; breakwater design; wave energy; wave power converting system; caisson breakwater application

1. Introduction

1.1. Development of an Oscillating Water Column (OWC) Wave Energy Converter

During late 1980s and then 1990s, intensive studies for a wave-power converting system were performed. However, it was not until 1991 when a combination with breakwater system was firstly studied in Japan [1]. In that study, not only were experimental tests performed in the laboratory but also an on-site, full-scale structure was built and tested. Many valuable data were firstly acquired resulting in an encouraging way for the OWC wave power converting system. Important conclusions include the efficiency of power that can be obtained from the conversion system, the estimated direct cost for a system of comparable size and scale, and a range of dimensional ratios between the air chamber and the wavelength. The most important one was that the combination for a breakwater with the wave-power converting system would not affect the function of the breakwater. However, this conclusion related to the structural safety was only based on an overall estimation of the mass of the caisson breakwater against the wave-loading, without considering the detailed variation of the caisson with the installation of an OWC converting system, in which additional chambers with walls and slabs were built. The other difficulty was that even though the study from Goda et al. [1,2] is invaluable and pioneering, it is too hard and too costly to be repeated for an on-field experiment. It was also too confined by many local factors for a special case to apply to the caisson breakwater in other locations with different environmental conditions.

Until now, researches on the OWC wave energy converting system are still in a very active status. Some are on the improvement of the mechanism of energy harvest such as to change the shape of traditional OWC system into a U-OWC [3] or a so called BBDB-OWC by using a backward-bent duct buoy [4,5]; some are focused on the efficiency of turbine performance for outflow and inflow motions [6]; investigations on the wave-height and power taking-off damping effect are also carried out experimentally and numerically [7]; some are focused on the performance of the air-chamber [8], where the effect of neglecting the air compressibility has been studied showing that an experimental test model scaled down to 1/50 may result in an overestimation of up to 15% for the air pressure in the OWC chamber.

Since the caisson structure is constructed to protect the coastal line, it is important to know if a built-in associated OWC system will be a burden to affect the safety of the structure. Even though, studies focusing on the structural safety of an OWC wave energy converter are attractive to researchers, references related to this matter are mostly focused on the exerting forces and pressures rather than the response of structure itself [9–13]. This is indicated in a model to estimate forces acting on an OWC chamber in a caisson breakwater [9], where horizontal forces on the front (curtain) wall and the rear (in-chamber) wall are predicted. It also mentioned that unlike a conventional caisson breakwater, vertical forces acting on the caisson chamber ceiling will have considerable effect on sliding and overturning characteristics of the breakwater structure. A first OWC device embedded into a breakwater was installed in 2008 at the Port of Mutriku. Storms hit it in December 2007, March 2008, and January 2009, resulting in structural damage to both the frontal wall and roof of four chambers [14]. Fortunately, the OWC power plant was not yet under operation because the operational license had not been granted, but still it reflected that no matter how carefully the exerting forces could be calculated, the response and safety of the chamber structure was the essential matter for an OWC wave energy converting system.

Some other studies also were concerned about the safety of operation of an OWC conversion system but only very few focused on the structural safety [15]. That study considered the safety for an OWC integrated into a bottom-sitting pile structure (an OWC pile). The authors examined the wave loads on an OWC pile, which included the viscous loads and lateral loads and then safety-loading for a bottom-sitting pile design that could then be estimated. The issue of structural safety was not only essential to the operational of the system but also important to the maintenance of the system during the operation for the inspection and damage investigation that may cost a considerable fortune [16].

1.2. Objectives of the Study

In this study, by following the findings of Goda et al. and the contemporary studies [10,11], a series of studies were performed by adopting, but upgrading, a similar system, the so called the caisson based OWC type wave energy converter for green energy developments. The analysis of structural safety were performed: firstly, the analysis for the structural safety of the caisson structure and the associated built-in OWC chambers and secondly, the safety of the chamber during the wave-converting operation in the associated OWC system located at a specific water area such as Kaohsiung, Taiwan, subjected to local environmental conditions. For the structural safety analysis, usually two safety issues needed to be clarified. The first safety analysis was the overall estimation of structural resistant ability against the environmental loadings such as the one performed in Goda's study [1]. During the analysis, the whole structure was assumed to be a rigid body and therefore, the weight of the structure was the only matter to resist both the shear force and the overturning moment induced by the wave-pressure on the structure while the exertion of soil and water were both taken into account. Because this kind of analysis is more related to the weight of the structure (usually a big mass-structure like a caisson breakwater), compared to the breakwater, the weight of the associated OWC chambers, which are relatively small, will still increase the total weight of the whole structural system of the breakwater. Therefore, in this study the rigid body analysis will be skipped, and we instead focus on the second part, the structural safety estimation by using more sophisticated structural analytical method for both the whole structure and the OWC chambers. The second safety analysis was related more to a flexible type structure so that the resistance of the structure to the external loadings was not only based on the weight of the structure. The elasticity and strength of material and member related resistance must be taken into account for the structure, such as the one in this study where a traditional caisson breakwater structure was transformed with a built-in chamber of the OWC converting system. Whether the change in the structure will cause an additional safety related problem to the OWC system or the caisson breakwater must be estimated.

In this study, a typical structural model was built and analyzed while the shape of the structure, material applied to the construction and associated boundary conditions were all set-up according to the wave and structural design. The examination for the structural safety includes two parts, namely, the overall structural safety of the caisson structure with associated OWC chambers and the mechanic behavior of the chamber related structural members. In the analysis for the whole breakwater, the deformations and strains of the caisson will be examined. For the analysis of the OWC structural members, the deformations and strains of the structural members of OWC-chamber including the front curtain wall and ceiling slab built in the breakwater caisson structure subjected to designated incident waves were analyzed and compared to the safety requirement for the allowance of material deformation. The applied incident wave was based on a general design code and set as a storm of 50 years return period.

Since a more detailed study for the performance of a similar system has already been presented in a related paper [17,18], the analysis of energy conversion performance in this study will focus on the influence of the air pressure on the air-chamber induced from the heaving wave motion. A numerical method by using a theorem of unsteady Navier–Stokes equations in conservation form is used to analyze the proposed OWC model. A proposed case built-in along a breakwater structure located in a local harbor was applied for a performance study to examine the feasibility of the OWC system in the harbor. The analysis was focused on the correspondence between the wave motion and the variations of air-pressures inside the air-chamber. A comparison with Goda's result was made before carrying out the performance analysis to verify the workability of the numerical method.

It was noticed that the wave force applied to the structural safety analysis was larger than the one applied to the performance analysis. This was because the durability requirement for the structural design enforces that a more strict environmental condition be considered, while for the wave energy converting performance analysis, a normal regular wave would be enough to drive the device of an OWC converter. Therefore, a wave due to a storm of 50 years return period was applied to the structural safety analysis and on the other hand, a wave based on a storm of 10 years return period was applied for the performance analysis.

During the structural safety analysis, it was found that for the overall structural safety of the caisson structure with associated OWC chambers, the deformation of the whole structure was safe when subjected to a storm-wave of 50-year return period. However, during the structural analysis for the local behavior of the OWC air-chamber, it was found that a more strict design criteria needed to be applied to the members of the chambers if the chamber was to sustain a storm-wave of 50-year return period. During the analysis for the operational performance of the OWC converting system attached to the caisson structure, it was found that the air-pressure induced by the heaving waves in the chamber could reach as high as 31.5 MPa when the OWC was subjected to a wave of 10-year return period.

2. Analytical Model and Environmental Forces

2.1. Analytical Model of the Structure

In this study, a typical caisson structure for the breakwater and coastal protection located in water of 16 m depth was designed with the installation of accommodated OWC system. As shown in Figure 1, a schematic 3-D view of the caisson structure associated with an OWC wave energy converting system is shown, where the front side shows the square gates of air-chambers of the OWC system. Figure 3 shows the side-view of the caisson structure, where a chamber of rectangular shape for the OWC is located in the front part of the structure. Dimensions for the structure and the size of chamber are also shown in the figure.

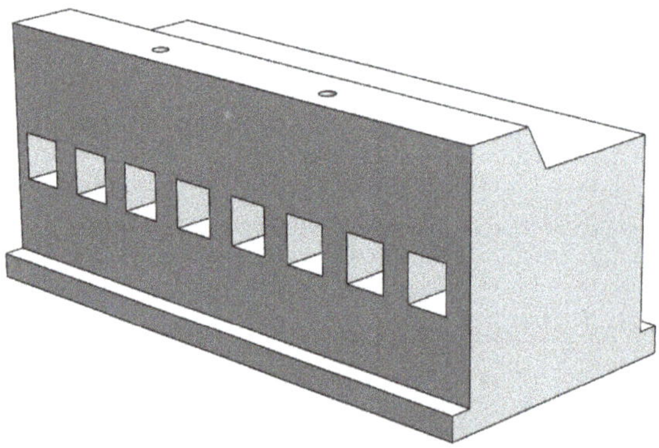

Figure 1. Schematic drawing of a caisson structure with oscillating water column (OWC) converter.

2.2. Environmental Forces

For the environmental conditions in this study, a storm of 50 years return period at the proposed location was applied to the analysis of which the wave height $H = 1.8Hd$ was 9.5 m and the corresponding period T was 10 s and Hd was the designed wave height for the breakwater of caisson structure. The incident wave was assumed to propagate in the direction normal to the front face of the caisson structure. The exerting pressure on the caisson breakwater then can be calculated based on the properties of the design-wave as prescribed.

For the analysis of structural safety, first of all, the exerting pressures from the environment including wave, water, soil, and reaction forces were all accounted and estimated. As was shown in Figure 2, it is a schematic drawing for the exerting forces around the caisson structure, which was modified based on Goda's method [19]. The calculation procedure is presented as follows:

(1) Determine the wave pressure at the still water level (SWL) as

$$P_1 = \frac{1}{2}(1+\cos\beta)\left(\alpha_1+\alpha_2 cos^2\beta\right)\omega_0 H_d \tag{1}$$

where parameter β is the angle of incident wave to the direction normal to the caisson structure, while α1 and α2 are parameters presented as

$$\alpha_1 = 0.6+\frac{1}{2}\left[\frac{\frac{4\pi h}{L}}{\sinh\frac{4\pi h}{L}}\right]^2 \tag{2}$$

and

$$\alpha_2 = min\left[\left(\frac{h_b-d}{3h_b}\right)\left(\frac{H_d}{d}\right),\left(\frac{2d}{H_d}\right)\right] \tag{3}$$

The dimension related parameters are shown in Figure 2 such as: h the water depth to the sea bottom and d the water depth to the surface of rubble-rump, while L is the designed wave length and h_b is the water depth at a distance of five times of significant wave-height away from the caisson structure.

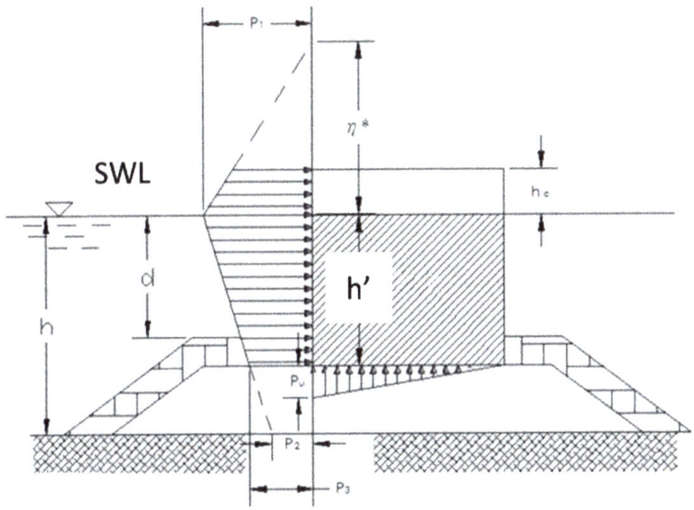

Figure 2. Schematic drawing of environmental loading on caisson structure.

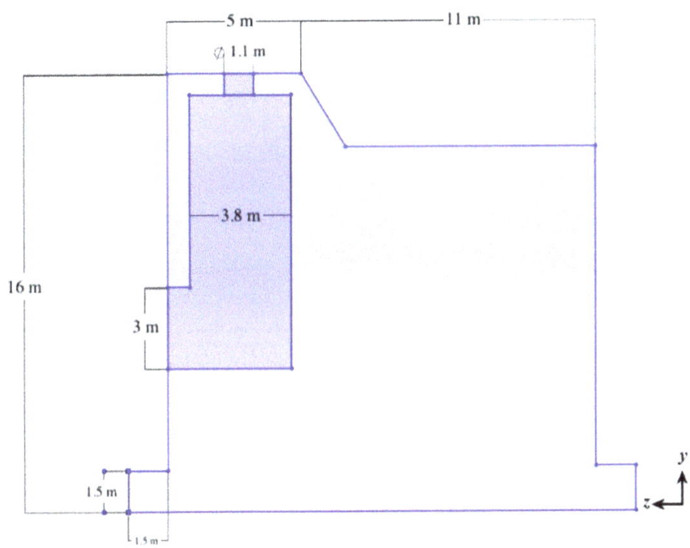

Figure 3. Side view of a caisson structure with OWC converter (redrawn after Chiu et al. [20]).

(2) Determine wave pressure at the sea bottom level as

$$P_2 = \frac{P_1}{\cosh\left(\frac{2\pi h}{L}\right)} \tag{4}$$

(3) Determine the wave pressure at the toe of the caisson as

$$P_3 = \alpha_3 P_1 \quad (5)$$

where the parameter α_3 is presented as

$$\alpha_3 = 1 - \frac{h'}{h}[1 - \frac{1}{\cosh\left(\frac{2\pi h}{L}\right)}] \quad (6)$$

The dimensional parameter h' is the submerged depth of the caisson.

3. Mechanic Analysis of Structural Safety

As mentioned in the introduction, the rigid-body analysis for the caisson structure would not be considered, that being the resistant capability from the weight of the caisson structure against the overturning moment and the sliding resistance of the structure to the horizontal wave forces. For the analysis of structural safety, the most important issue was that the stress distribution of the structure would not exceed the designed strength of the material or the strain induced from the deformation would not exceed the strain limit of the material when subjected to wave and other environmental forces. Therefore, in this study a strain analysis was performed to examine if the caisson structure was safe, along with the constructional material, when additional chambers of an OWC wave energy converter were built in, and located right at, the front side of the caisson structure facing the incident wave directly. In order to know better the structural safety, two parts of structural analysis were performed, these included the first part, a global analysis of the whole structure of caisson breakwater subjected to wave forces, and the second part, a local analysis focused on the deformation behavior of the additional chamber of an OWC wave energy conversion system.

3.1. Global Structural Analysis

For the global structural analysis, the object was to examine the global deformation of the caisson structure and find the parts that may have been subjected to severe damage or large strain. After obtaining the global deformation distribution for the whole structural system, a local analysis could be performed and a more detailed investigation carried out through the structural members of the chamber including the walls between chambers, ceiling panels on the top of chamber, and the curtain walls that normal to the direction of incident waves.

The wave forces exerting on the front face of the caisson structure was in a form of pressure distribution, as P1, P2, and P3, corresponding to various levels of depth of the caisson breakwater located in the water, which are shown in Figure 2, which can be obtained from Equations (1), (4), and (5). After obtaining the wave forces exerting on the caisson structure as was shown in Figure 4, a numerical tool for the solid mechanics of well-known ANSYS was applied and the conditions including wave forces, soil pressure, and material properties were all determined. The material for the caisson structure was reinforced concrete with material properties as the designed strength of concrete: fc = 350 (kgf/cm^2), elastic modulus of concrete: Ec = 30 (kN/mm^2), and the elastic modulus for the steel: Es = 290 (kN/mm^2) according to the ACI code for the structure exposed to an environment with chloride (marine environment) [18].

For the global structural safety analysis, a distribution of strain over the whole caisson breakwater was examined firstly and after that then a more detailed analysis on the material strength of reinforced concrete was performed for the air chamber of the OWC converter attached to the caisson breakwater. It was noticed that for the global structural analysis, since the breakwater structure is in a long and slender shape, an analytical strategy of plane strain could be applied so that only the cross section of a unit-length caisson breakwater would be analyzed and the mechanical behavior of the other sections would be assumed to be similar if the environmental conditions were similar. Even though a three-dimensional analysis could also be applied, the results would be similar to each other. Therefore, the analysis was focused on the deformations and variation of strain of the cross section.

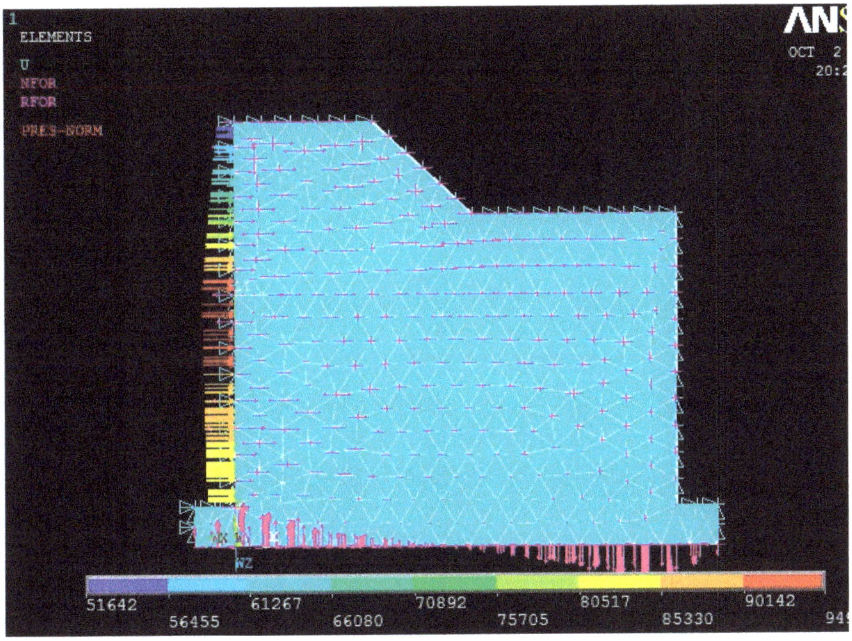

Figure 4. The distribution of the environmental loadings.

After the analysis of global structural system, Figure 5 shows the deformation of the caisson structure, which deforms along with the direction of exerting incident waves, the weak transverse direction for a breakwater. As is observed in the analytical results, the deformation varied along with the height of the structure while the largest deformation occurred near the top of the structure in the front side, which happened to be the location of the OWC chambers. In the lower part of the caisson, the deformation was quite small, even not observable because the loading was concentrated on the higher part, as shown in Figure 2, especially for the part of structure on the still water level (SWL), where the wave action was the largest, while the cross section of the caisson structure above the water level reduced gradually. This is the reason why the largest deformation occurred near the top of the structure of the front side in the global structural analysis.

The largest deformation obtained from the analysis was 0.162 mm, which was small under the loading forces calculated for this study. Compared to the corresponding dimension of the caisson breakwater, which was 16.0 m, the strain corresponding to the deformation can be calculated as 0.01×10^{-3}, which is much smaller than the strain allowance for a concrete material 0.003. However, because the location of the largest deformation was at the tip corner of the top, where the chamber of the OWC converter is built-in, a further study of the front wall of the chamber revealed that the strain corresponding to the transverse deformation was close to the allowance of the concrete material. The largest strain that occurred in the front wall was 0.00027 in terms of the thickness 0.60 m of the front wall of the chamber, as shown in the schematic drawing of Figure 3. It is one order smaller than the maximum strain (0.003) allowed for concrete material from the global structural deformation analysis. Therefore, for the global structural analysis of the caisson structure combined with an OWC wave energy converting system, it is quite safe.

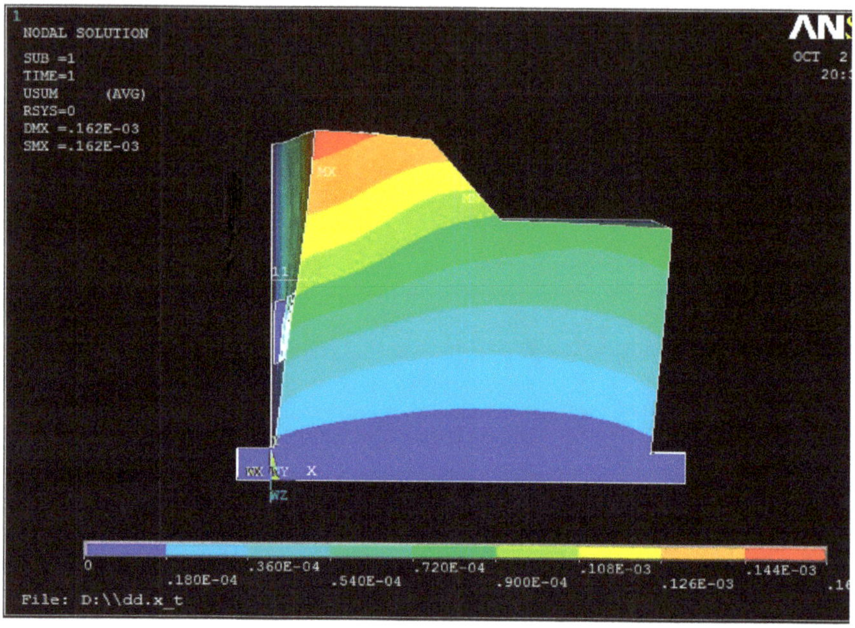

Figure 5. Deformation of caisson breakwater along the wave direction.

3.2. Structural Analysis for the Chamber of OWC Converter

As the second part for the safety analysis, a study based on the loading and deformation of the analysis of the whole structure as discussed in previous section was performed correspondingly for the air-chamber of the OWC system located in the front side of the caisson structure. Because the shape of the structure was no longer a long and slender type, a three-dimensional analysis needed to be applied and a more sophisticated solid element was utilized in the finite element numerical analysis, which included the properties of the material that must be refined and the detailed dimensions of the structural member.

On the side view of the caisson breakwater structure, the chamber of OWC system can be observed as the darker part in Figure 3, and then the OWC chamber is framed-out and redrawn as a 3-dimensional view in Figure 6. The dimension of the chamber is also shown in figures, where Figure 3 is the dimension of the chamber for the study in side-view while Figure 6 presents a detailed dimension for structural members in the chamber such as the side wall, front curtain wall, and ceiling-slab of the air-chamber. It was noticed that during the mechanic analysis for the OWC air-chamber, the orifice for in/output air was not considered for the structural safety study because the analysis focused on the response of the structure in terms of both the deformation of the structural member, and the material capacity to examine if the strain induced from extreme deformation exceeded the material capacity.

Before performing the analysis, the analytical model was divided into meshes and nodes for the structural members of the chamber. Typical meshes for the 3-D solid-element of tetrahedron type was applied and is presented in Figure 7. The arrangement of the steel reinforcement in the concrete is also presented in Figure 8. A corresponding beam element for the steel material was also applied in the structural analysis.

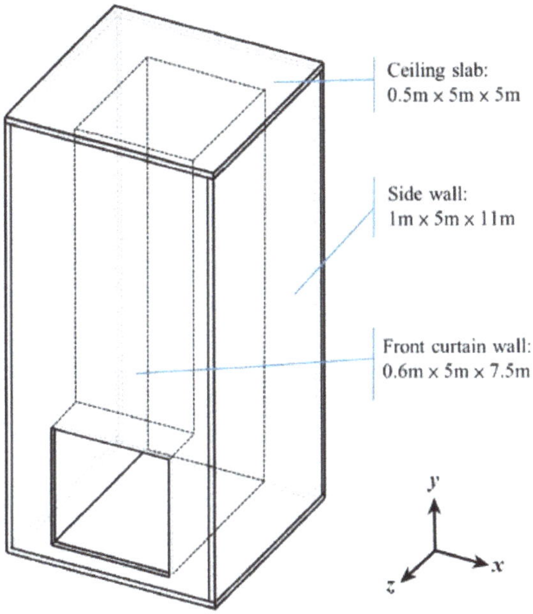

Figure 6. Dimensions of the structural members of OWC chamber in the caisson breakwater.

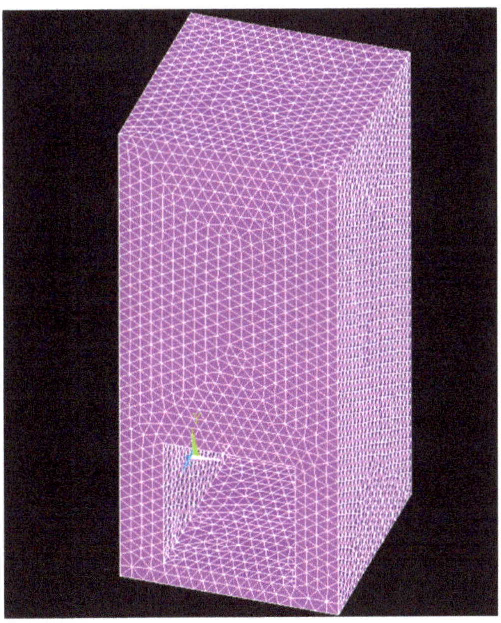

Figure 7. The grid meshes of the element for the chamber.

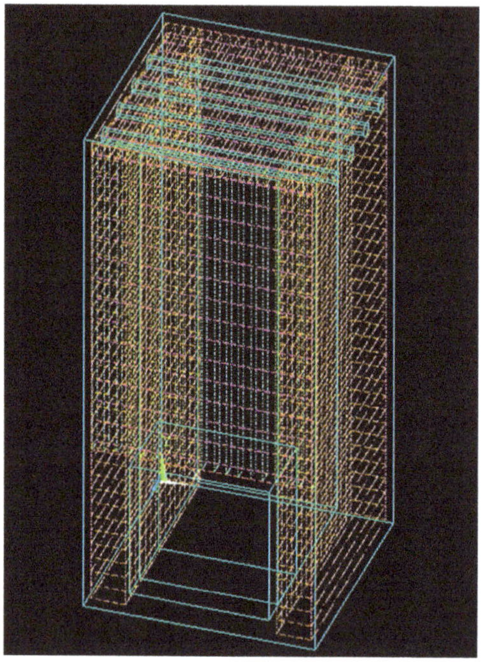

Figure 8. The arrangement of steel reinforcement in the chamber.

Results obtained from the structural analysis include the deformation distribution and corresponding strains for the chamber of OWC wave energy converter. As was shown in the global structural analysis, the deformations in the transverse direction (as the z-direction shown in Figure 6) of the caisson breakwater was the critical issue because it was related to structural safety. However, for the chambers built in an OWC converter, responses in both the horizontal transverse direction (z-direction as shown in Figure 6) and vertical transverse direction (y-direction) were essential to the structural safety of the chamber. This is because the chamber is like an empty box that was subjected to deformations all around the walls and ceiling slabs. The x-direction for the breakwater was an in-plane direction (the axis along the layout of the breakwater) that was the strong-direction subjected to smaller loads since the wave forces exerted mostly on the transverse direction (z-direction), but for the air-chamber, the x-direction was the out-of-plane direction for the side wall, of which the deformation was also critical to the structural safety.

3.2.1. Transverse Deformation for the front Curtain-Wall

As is shown in Figure 9, this is the deformation distribution in the transverse z-direction for the OWC air-chamber without an opening orifice. The color of the contour represents the seriousness of the deformation in which both the red and blue colors indicate a larger deformation corresponding to an opposite direction each other. In Figure 9, even though the largest deformation, namely, 0.003371 m, appears at the top panel of the ceiling, relative to the dimension in the z-direction 5.0 m, the corresponding strain will be 0.000742, which is much smaller than 0.003, the maximum allowed strain in concrete structure. However, the panel of the front curtain wall was normal to the transverse deformation and therefore, the strain of the curtain wall in the z-direction (out-of-plane) would be critical to the structural safety. The largest deformation of the curtain wall is about 0.0027 m and relative to the thickness 0.6 m of the wall, as shown in Figure 6, the corresponding largest strain is 0.0045, which is slightly larger than the strain capacity of a concrete material 0.003.

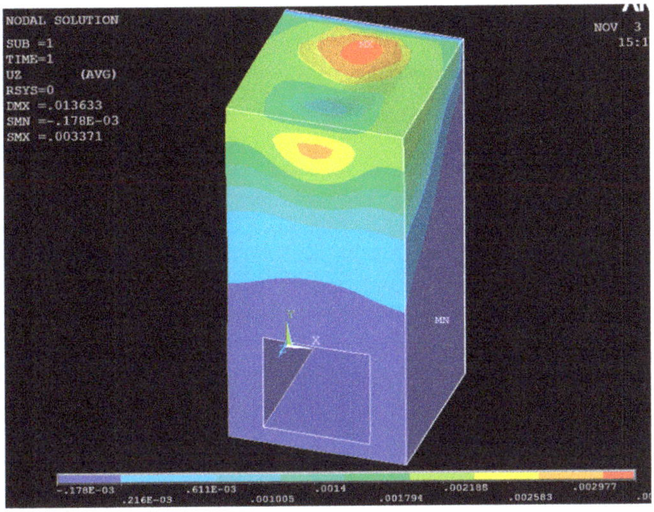

Figure 9. Deformation variation for the chamber structure in z-direction.

3.2.2. Transverse Deformation for the Top Ceiling-Slab

Shown in Figure 10 is the deformation distribution in the y-direction for the OWC air-chamber. It shows that the largest deformation in the y-direction for the chamber may occur at the lower parts of the chamber structure along the vertical side walls. Since the deformation was 0.102×10^{-5} m in-plane of the sidewall, and therefore, relative to the height of the sidewall, 11.0 m, the corresponding strain was 0.93×10^{-7}, a value much smaller than 0.003. However, y-direction was in the out-of-plane direction of the ceiling panel and the downward deformation was 0.013488 m. Relative to the thickness of the ceiling slab, 0.5 m, the corresponding strain in the out-of-plane direction was 0.026976, which is a value well over the allowed capacity for the concrete strain.

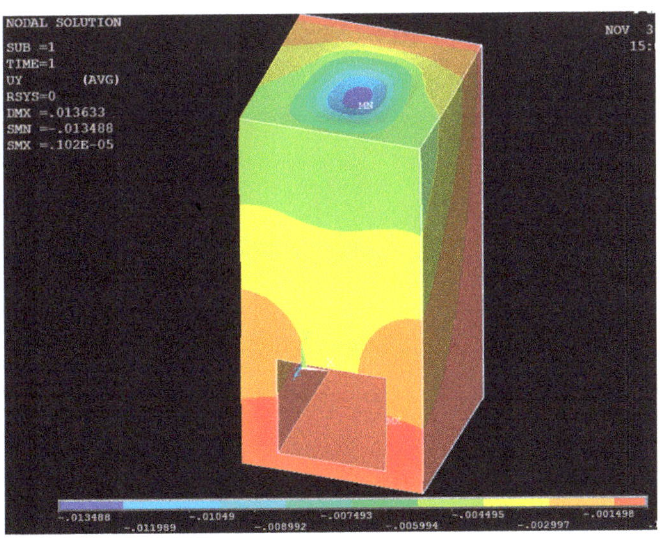

Figure 10. Deformation variation for the chamber structure in y-direction.

3.2.3. Transverse Deformation for the Side-Wall

Presented in Figure 11 is the deformation distribution in the x-direction for the OWC air-chamber. Both the maximum (positive) and the minimum (negative) exist simultaneously on the ceiling-panel. The values are similar, namely, 0.00168 m, and relative to the dimension of x-direction, the strain was 0.000336, which is much smaller than the strain capacity of concrete material. For the sidewall of the chamber deformations in the x-direction is the out-of-plane motion and critical to the structural safety. As obtained from the analysis, the largest deformation was about 0.0011 m, which based on the thickness of the sidewall, 1.0 m, the corresponding strain was 0.0011, smaller than the maximum strain of the concrete.

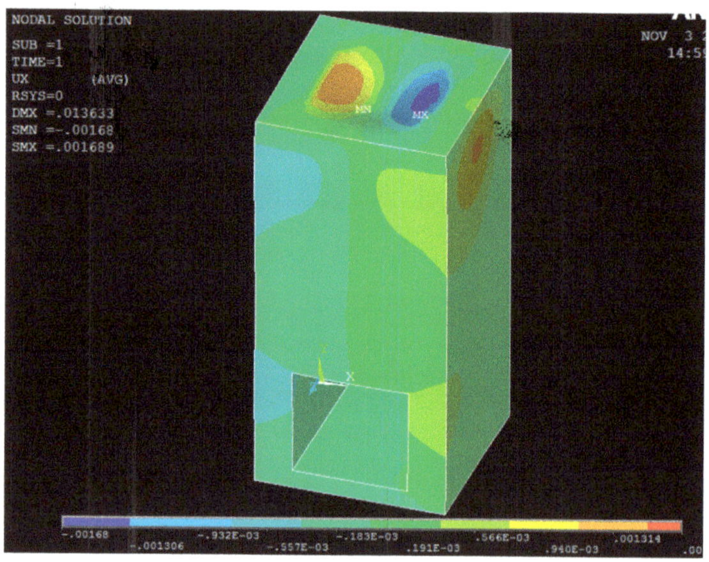

Figure 11. Deformation variation for the chamber structure in x-direction.

3.2.4. Discussions on the Deformation of Chamber Members

From the analysis presented above, we found that in terms of the structural safety, the air-chamber of an OWC conversion system attached to a caisson breakwater structure cannot satisfy the general requirement of structural safety. Both the top ceiling slab and the front curtain wall were not strong enough to sustain a design storm of 50-year return period. Therefore, some questions arise: (1) If a design-wave based on a storm with 50-year period is suitable for the design of an OWC converter based on a caisson breakwater structure? (2) If the structure can be modified to sustain the design wave based on 50-year return period? (3) Is there any other choices to improve the structural safety of the very design?

Traditionally, structures located on the coast or in the water have very large volume and mass and therefore, a rigid-body theorem is applied for the design, which means that there is no deformation in the material or the structural members. Without considering the deformation or the material properties for the structure, many kinds of design for an OWC converting system can be proposed and analyzed for their function and converting efficiency based on fluid mechanics. However, when we carefully examine the structural safety by taking account all of the factors related to the properties of material, structural dimension, and environmental conditions of the location, the analytical results remind us that structural safety is still essential.

Going back to the questions that arose after the analysis for the structural safety, a storm of 50-year return period would probably be too much for an associated structure attached to the main structure-the caisson breakwater. So long as the damage of the associated structure would not affect structural safety and operation of the main structure, a design-wave based on a storm of shorter return period can be considered. To maintain the structural safety of the chamber of the OWC system, a compromise between the converting efficiency and the structural safety probably needs to be taken into account because a larger air-chamber will take more wave forces and the associated structural members also need to be designed with stronger material. As was found in the results of structural safety analysis, the weak parts of an OWC air-chamber are the curtain wall in the gate of incident wave and the top ceiling-slab. This phenomenon was also observed during a big-storm that hit the on-field application, as in the case of the Matriku OWC Power Plant [12].

Presented in Figure 12 is the potential cracking line for the air-chamber of the OWC converter attached to the caisson breakwater. Most cracks occurred on the top ceiling of the chamber as was indicated in the analysis; the ceiling slab would suffer the most serious deformations and according to the analysis the strain in the transverse direction was well over the allowed capacity for the concrete material. This phenomenon needs to be carefully watched because when the chamber of the OWC converter is under operation, the air-pressure induced from heaving waves may further worsen the problem. Even though the cracks will occur when an extreme environmental condition such as the designed storm of 50-year return period is applied, the regular operation may also cause a fatigue damage to the ceiling slab and the curtain wall. Especially when the air in the chamber is not allowed to flow out as the case studied, the induced pressure from elevated water in the chamber could further press the walls and ceiling-slab inside the chamber and cause severe deformation and then damages for the structural members.

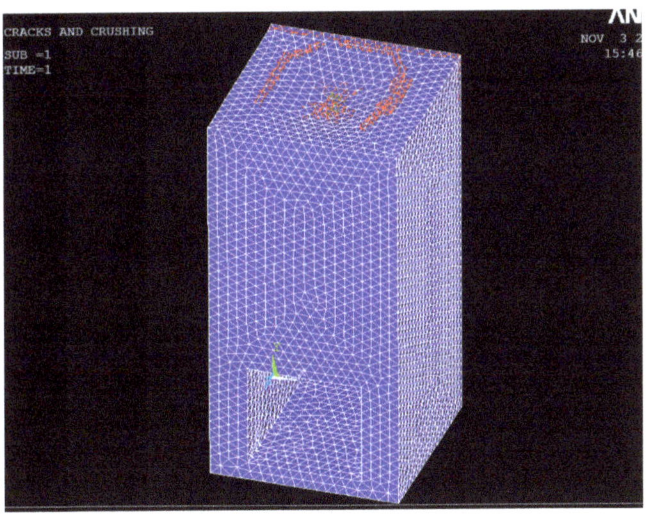

Figure 12. Potential cracking line for the chamber structure.

4. Evaluation of Air-pressure in the Chamber during the Operation of Wave Energy Conversion

For the safety analysis of the chamber during the operation of the OWC system installed in the caisson structure, the theorem of unsteady Navier–Stokes equations in conservation form including continuity equations, equation of momentums, and equation of turbulence dynamics were applied as presented in the previous section. This theorem combined with numerical tool of CFX in ANSYS was also applied to the study. Before the application of the analytical method to the study, a verification was performed by reproducing and comparing the results obtained from the on-field study of Goda's experiment [1] in Sakeda, Japan. That study is currently the sole one that may provide a full-scale on-field experimental data for a full-scale wave energy converter of which the variation of both the air pressure and water elevation in the air-chamber were shown.

4.1. Theorem of Fluid Mechanics Applied in the Study

In this study, a theorem of unsteady Navier–Stokes equations in conservation form, consisting of continuity equations, equation of momentums, and equation of turbulence dynamics, are applied and shown as follows [21,22].

Continuity equation:

$$\frac{\partial \rho}{\partial t} + \nabla\cdot(\rho U) = 0 \tag{7}$$

Equation of momentum:

$$\frac{\partial \rho U}{\partial t} + \nabla\cdot(\rho U \times U) - \nabla\cdot\left(\mu_{eff}\nabla U\right) = \nabla\cdot p' + \nabla\cdot\left(\mu_{eff}\nabla U\right)^T + B \tag{8}$$

where B is the sum of body force, μ_{eff} is the effective viscosity, and p' is the revised pressure. The effective viscosity and the revised pressure can be presented as

$$\mu_{eff}=\mu_t + \mu \tag{9}$$

$$p' = p + \frac{2}{3}\rho k \tag{10}$$

It is also noticed that μ_t is the viscosity of the turbulence, which according to the assumption of k-ε model, is related to the dynamic energy and the dissipation of the dynamic energy as presented as

$$\mu_t = C_\mu \rho \frac{k^2}{\varepsilon} \tag{11}$$

where k is obtained directly from the equation of dynamic energy and equation of energy dissipation. The equation of turbulence dynamic energy and equation of energy dissipation are presented as follows

$$\frac{\partial(\rho k)}{\partial t} + \nabla\cdot(\rho U k) = \nabla\cdot[\left(\mu + \frac{\mu_t}{\sigma_k}\right)\nabla k] + P_k - \rho\varepsilon \tag{12}$$

$$\frac{\partial(\rho \varepsilon)}{\partial t} + \nabla\cdot(\rho U \varepsilon) = \nabla\cdot\left[\left(\mu + \frac{\mu_t}{\sigma_\varepsilon}\right)\nabla\varepsilon\right] + \frac{\varepsilon}{k}(C_{\varepsilon 1}P_k - C_{\varepsilon 2}\rho\varepsilon) \tag{13}$$

where $C_{\varepsilon 1}$, $C_{\varepsilon 2}$, σ_k, and σ_ε are constant parameters to be determined empirically, while P_k is related to viscosity and floating force and can be presented as

$$P_k = \mu_t \nabla U\cdot\left(\nabla U + \nabla U^T\right) - \frac{2}{3}\nabla\cdot U(3\mu_t \nabla\cdot U + \rho k) + P_{kb} \tag{14}$$

4.2. Analysis of Wave-Induced Air Pressure in the Chamber

The verification of the numerical method applied in this study was carried out by comparing the analytical results with the on-field experimental results when the conditions were set as similar as possible [18,19]. For Goda's on-field study, a 20 minute data collection was made at Sakeda Harbor, where during the testing day, the wave-height was 2.2 m to 3.0 m (irregular waves on site). In this study a regular wave with wave-height of 2.5 m (regular wave analysis) was set and a one minute analysis was performed for the comparison of which two sections of experimental data within a 20 min were selected. It was found that even though the analytical results were slightly larger than the on-field data [23], considering the uncertainties of the on-field test such as the wave-height, which was irregular, and the data captured for the comparison, which were only small parts, the result was still encouraging.

An OWC wave converting system was proposed based on the available features for the harbor such as the depth of water in front of the breakwater, angle between wave direction and the normal direction of the caisson breakwater, and structure size of the caisson breakwater. A case is simulated for a breakwater of caisson type located in the Kaohsiung Port in Taiwan facing north-west where the water depth h = 9.5 m. According to a 10-year return period for the summer season (storm season) of local climate records, a design wave was used in simulation as: wave–height H = 1.0 m and wave-period T = 7 s. It was noticed that the wave condition applied for the case study was much smaller than the wave condition for the analysis of structural safety. This was because the design criteria applied for an offshore structure was much stricter, as a 50-year or larger return period was required, while for the case study a 10-year return period was considered suitable for regular daily performance.

Based on the shape and dimensions of the caisson structure to be modified with an OWC conversion system, a set of OWC system was designed and installed in the caisson breakwater. As shown in Figure 1, the OWC was installed at the front side of the caisson breakwater facing the incident waves so that it could capture the maximum wave energy to convert it into electricity power. It is a similar type to Goda's test but the dimension was modified based on the conditions of the location port.

The typical response of air pressure near the ceiling-slab of the chamber at various positions, namely, from 1 to 6 as referred to Figure 13 and shown in Figure 14, where it is noticed that the air chambers are connected to each other with channels while only one orifice is allowed for every set of three-chambers. It can be observed that the position with lowest positive pressure is located right at the orifice when the air pressure at the other position increases corresponding to the distance away from the orifice. The position with highest negative pressure is at position 1 just next to the orifice. According to results of the structural safety analysis for the chamber members, the ceiling-slab will suffer the largest deformation and the pressure in the chamber will also increase the deformation if the exerting pressure is in the same phase to the out-of-plane dynamic motions of the structural deformation.

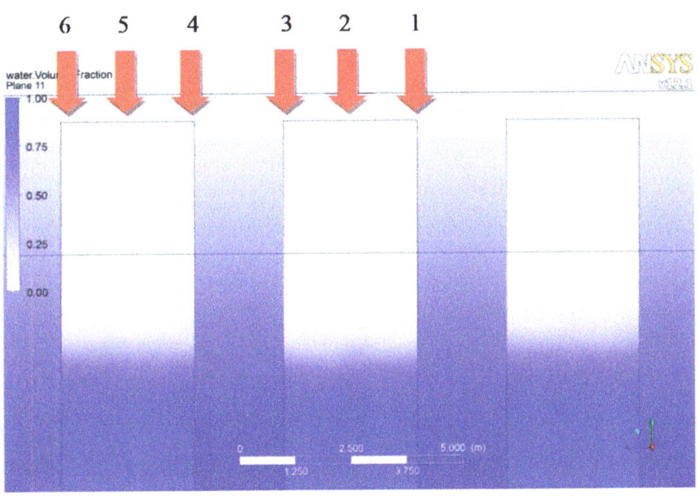

Figure 13. Schematic view of inner part of the air-chamber.

It was found that through the fluid mechanic analysis the greatest pressure exerting on position 1 of the chamber was 3210 tf/m^2 or 31.46 MPa downward (negatively) during the operation of the OWC system. It is small and about 1/10 of the designed allowed pressure that may exert on the ceiling slab. The largest pressure occurred at the corner of the ceiling-slab, where the separation-wall could also provide the tension resistance to the slab. However, even without a wall to brace the ceiling-slab, it could still hold the pressure occurs during a regular operation of OWC system subjected to a wave of

10-year return period. The concern will be more on the issue of fatigue, which was not analyzed in this study since the direction of the pressure on the ceiling slab was applied alternatively and continuously.

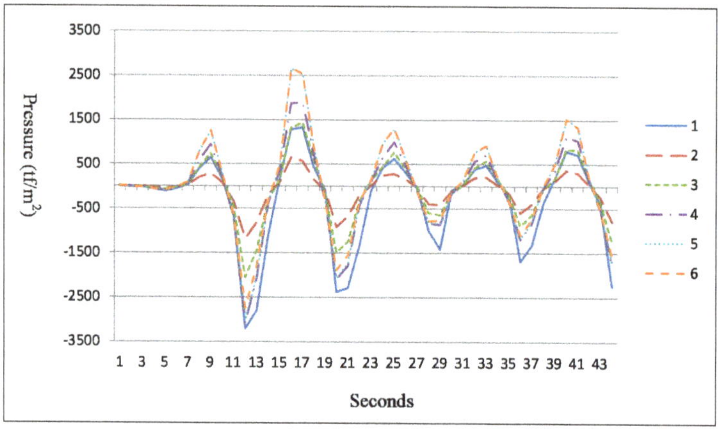

Figure 14. The air-pressure in the various location of chamber.

5. Conclusions

In this study, an OWC wave power converting system based on a caisson breakwater was proposed and analyzed for both the structural safety and the performance in the power converting efficiency. For the structural safety analysis, two parts of analysis were carried out, namely, the first part for the overall structural safety and the second part particularly for the structure of air chamber of the OWC converter. The analysis on the chamber included both the elastic structural analysis and the air pressure in the chamber built in the caisson breakwater.

(1) According to the analysis for the structural safety of this study, for the whole structural safety that included both the caisson breakwater and attached OWC system, the deformations induced from the design-wave, which was based on a storm-wave of 50-year return period, was under the capacity of the concrete material.
(2) The second part analysis of the structural safety, particularly focused on the air-chamber of the OWC converter, where the maximum deformation occurred transversely at the ceiling-slab of the air-chamber for the OWC converter, whereas the front curtain wall subjected to wave-impact directly would also deform greatly, but not so significantly. The developed strains for both the ceiling-slab and the front curtain-wall were well over the strain limit for a concrete material and therefore, when the chamber structure was subjected to a wave close to the designed storm-wave as utilized in the analysis, a major damage may occur on both the ceiling-slab and the front curtain-wall.
(3) According to the study of the air-pressure induced by the heave motion of the waves in the chamber, the pressure was small and would not significantly influence the structural safety of the OWC chamber attached to a caisson structure.
(4) It is concluded that when the OWC is attached to a traditional breakwater structure, the location will be an important factor to decide the performance of energy converting efficiency. To design an OWC structure that may sustain a storm wave of 50-year return period could be expensive. In terms of economic consideration, as long as the damage of the associated structure would not affect the structural safety and operation of the main structure, a design-wave based on a storm of shorter return period should be considered.

Author Contributions: Conceptualization, H.H.L.; Methodology, H.H.L.; Software, C.-Y.L. and T.-Y.W.; validation, H.H.L.; Writing—original draft preparation, H.H.L.; Writing—review and editing, H.H.L.; Supervision, H.H.L. and Y.-F.C.; Project administration, H.H.L. and Y.-F.C.; Funding acquisition, H.H.L. and Y.-F.C.; All authors have read and agreed to the published version of the manuscript.

Acknowledgments: This research was financially supported in part by the Ministry of Transportation and Communication (MOTC), Taiwan, under grants "MOTC IOT-103-H2DB005c". It is gratefully appreciated.

Conflicts of Interest: The authors declare no conflict of interest.

References

1. Goda, Y.; Nakada, H.; Ohneda, H.; Suzuki, M.; Takahashi, S.; Shikamori, M. Results of field experiment of a wave power extracting caisson breakwater. *Proceedings Ocean. Dev.* **1991**, *7*, 143–148. [CrossRef]
2. Goda, Y.; Shinda, T.; Chiyama, S.; Ohneda, H.; Suzuki, M.; Takahashi, S.; Shikamori, M.; Takaki, Y. Experiment of a wave power extracting caisson breakwater. *Proceedings Ocean. Dev.* **1989**, *5*, 1–6.
3. Spanos, P.D.; Strati, F.M.; Malara, G.; Arena, F. An approach for non-linear stochastic analysis of U-shaped OWC wave energy converters. *Probabilistic Eng. Mech.* **2018**, *54*, 44–52. [CrossRef]
4. Boccotti, P. Comparison between a U-OWC and a conventional OWC. *Ocean. Eng.* **2005**, *34*, 799–805. [CrossRef]
5. Sheng, W. Power performance of BBDB OWC wave energy converter. *Renew. Energy* **2019**, *132*, 709–722. [CrossRef]
6. Ansarifard, N.; Fleming, A.; Henderson, A.; Kianejad, S.S.; Orphin, J. Comparison of inflow and outflow radial air turbines in vented and bidirectional OWC wave energy converters. *Energy* **2019**, *1821*, 159–176. [CrossRef]
7. Elhanafi, A.; Kim, C.J. Experimental and numerical investigation on wave height and power take–off damping effects on the hydrodynamic performance of an offshore–stationary OWC wave energy converter. *Renew. Energy* **2018**, *125*, 518–528. [CrossRef]
8. Simonetti, I.; Cappietti, L.; Elsafti, H.; Oumeraci, H. Evaluation of air compressibility effects on the performance of fixed OWC wave energy converters using CFD modelling. *Renew. Energy* **2018**, *119*, 741–753. [CrossRef]
9. Pawitan, K.A.; Dimakopoulos, A.S.; Vicinanza, D.; Allsop, W.; Bruce, T. A loading model for an OWC caisson based upon large-scale measurements. *Coast. Eng.* **2019**, *145*, 1–20. [CrossRef]
10. Viviano, A.; Musumeci, R.E.; Vicinanza, D.; Foti, E. Pressures induced by regular waves on a large scale OWC. *Coast. Eng.* **2019**, *152*, 103528. [CrossRef]
11. Garrido, A.J.; Otaola, E.; Garrido, I.; Lekube, J.; Maseda, F.J.; Liria, P.; Mader, J. Mathematical modeling of fluid and structure interaction in ocean engineering. *Ocean. Eng.* **2015**, *2015*, 727982. [CrossRef]
12. Medina-Lopez, E.; Allsop, W.; Dimakopoulos, A.; Bruce, T. Conjectures on the failure of the OWC breakwater at Mutriku. In Proceedings of the Conference Coastal Structures, Boston, MA, USA, 9–11 September 2015.
13. Benrugeig, P.; Murphy, J. Modelling air compressibility in OWC devices with deformable air chambers. *J. Mar. Sci. Eng.* **2019**, *7*, 268. [CrossRef]
14. Kim, J.-S.; Nam, B.W.; Kim, K.-H.; Park, S.; Shin, S.H.; Hong, K. A numerical study on hydrodynamic performance of an inclined OWC wave energy converter with nonlinear turbine–chamber interaction based on 3D potential flow. *J. Mar. Sci. Eng.* **2020**, *8*, 176. [CrossRef]
15. Huang, Z.; Xu, C.; Huang, S. A CFD simulation of wave loads on a pile-type oscillating-water-column device. *J. Hydrodyn.* **2019**, *31*, 41–49. [CrossRef]
16. Ambühl, S.; Sørensen, J.D. Sensitivity of risk-based maintenance planning of offshore wind turbine farms. *Energies* **2017**, *10*, 505. [CrossRef]
17. Lee, H.H.; Chen, C.-H. Parametric study for an oscillating water column wave energy conversion system installed on a breakwater. *Energies* **2020**, *13*, 1926. [CrossRef]
18. Lee, H.H.; Chiu, Y.-F.; Lin, C.-Y.; Chen, C.-H.; Huang, M.-H. 2016, Parametric study on a caisson based OWC wave energy converting system. *World J. Eng. Technol.* **2016**, *4*, 3D.
19. Goda, Y. *Random Seas and Design of Maritime Structures*; Tokyo University Press: Tokyo, Japan, 1985.

20. ACI 357.3R-14. Guide for Design and Construction of Waterfront and Coastal Concrete Marine Structures. ACI Committee 357. 2014. Available online: https://www.concrete.org/Portals/0/Files/PDF/Previews/357_3R_14preview.pdf (accessed on 2 July 2014).
21. Sarpkaya, T.; Isaacson, M. *Mechanics of Wave forces on Offshore Structures*; Van Nostrand Reinhold Company: New York, NY, USA, 1981.
22. Milne-Thomson, L.M. *Theoretical Hydrodynamics*; The MacMillan Co.: Toronto, ON, Canada; Macmillan: New York, NY, USA, 1960.
23. Chiu, Y.-F.; Lee, H.H.; Lin, C.-Y.; Chang, F.-T. Study on a Caisson Based OWC Wave Energy Converting System. 2014. Available online: http://www.hcirp.jornal.wjet (accessed on 27 July 2015).

Journal of
Marine Science and Engineering

MDPI

Article

Multiple TLDs on Motion Reduction Control of the Offshore Wind Turbines

Po-Hung Yeh, Shao-Hua Chung and Bang-Fuh Chen *

Department of Marine Environment and Engineering, National Sun Yat-sen University, Kaohsiung 804, Taiwan; kuroshio@mail.nsysu.edu.tw (P.-H.Y.); allen82324@gmail.com (S.-H.C.)

* Correspondence: chenbf@g-mail.nsysu.edu.tw

Received: 26 May 2020; Accepted: 22 June 2020; Published: 24 June 2020

Abstract: This study explores the damping effects of tuned liquid dampers (TLDs) on a monopile offshore wind turbine (OWT). The fluid–solid coupling of ANSYS was used to simulate the damping effect of a TLD on the structures. The environmental conditions refer to the IEC-61400-3 and the Design Load Case (DLC) 1.2 for the annual average environmental conditions and DLC 6.2 for the 50-year regression period, and the extreme environmental conditions were used in the study. The turbulent wind field simulation was performed by TurbSim, and the load of wind waves on structures was generated by FAST, which were all developed by the NREL (National Renewable Energy Laboratory). In addition to wind and waves, the seismic force was also considered. The cylindrical TLD was located above the rotor nacelle assembly (RNA). A TLD has different damping effects when acting under wind, wave, and earthquake loads, respectively. The effect of the TLD regarding motion reduction on the OWT under coupled wind, wave, and seismic loads was studied. This study also designed a simple experiment to verify the correctness of the numerical simulation results. Fatigue analysis shows that multi-layer TLDs can extend the fatigue life (37%) of an OWT. In addition, under extreme environmental load conditions, multi-layer TLDs have a better vibration damping performance than single-layer TLDs. The study demonstrates that multi-layer TLDs can be considered as a vibration reduction damper for OWTs.

Keywords: motion reduction control; renewable energy; TLD; offshore wind turbine

1. Introduction

Although coal is the earliest type of fossil fuel used by human beings, it has always a serious air pollution impact. The nuclear waste treatment and the warm drainage are the major environmental impact issues. Therefore, at present, most countries have started to use renewable energy. Taiwan is an island located in the eastern part of Asia and on the west side of the Pacific Ocean. Taiwan's imported energy accounted for 98.16% of its total energy usage in 2018. It is urgent for the Taiwanese government to develop and increase its renewable energy percentage. Taiwan's northeast monsoon is very strong, especially in the Taiwan Strait. According to the 4C Offshore reported in 2014, the world's 23-year average wind speed observation found that the Taiwan Strait has accounted for 16 places of the world's 20 most windy offshore wind farms. For this reason, the Taiwanese government approved the "Thousand Sea and Land Wind Turbines". The goal is to complete the land wind farm installation with a capacity of 1200 MW. In addition to land farm, the offshore wind farm project is expected to be completed in 2025 with a capacity of 3000 MW and 1000 wind turbines will be installed in the near west coast of Taiwan.

Taiwan is located in the volcanic belt of the Pacific Rim, and it is in the active seismic zone. High seismic activity and frequent typhoon attacks are the threats to offshore structures. Therefore, the offshore structure is usually equipped with a damping device to suppress the vibration of the structure under extreme loading conditions and reduce the fatigue damage caused by the long-term cyclic loading, thereby prolonging the service life of the offshore structure.

Offshore Wind Turbine (OWT)

Offshore wind turbines are developing rapidly in recent years because the OWTs have higher wind speeds, less noise, and less land occupation than on-land WTs. The OWTs may generate large amounts of energy as the extreme wind and wave loads act on the OWTs, which usually have a slender supported tower and a monopile foundation. Therefore, the damping device is needed to reduce the dynamic response of OWTs. Structural control includes active, semi-active, and passive control, and the passive control is the simplest control method. The tuned mass damper (TMD) is a passive control that absorbs structural vibration by mass swing. For example, the control system of the top floor of Taipei 101 is a tuned mass damper. Murtagh et al. [1] applied TMD to control the vibration of the OWT, and two TMDs were placed in the cabin of OWT to reduce the vibration in the fore-aft and side-to-side directions. Sun and Jahangiri [2] designed a three-dimensional pendulum tuned mass damper (3d-PTMD) to control the multi-directional vibration of the OWT, and it has 10% more damping effect than double TMDs have. Hemmati and Oterkus [3] developed a semi-active TMD for the vibration reduction of OWTs; they used a short-time Fourier transform to adjust the damping system to the instantaneous frequency of the system to achieve a better damping effect. The results showed that the performance of a semi-active TMD with a mass ratio (the ratio of mass of TMD to mass of the structure) of 1% is better than a passive TMD with a mass ratio of 4%. Since the installation and maintenance costs of TMD are quite expensive, in order to reduce the cost, the tuned liquid damper (TLD) has been developed. The sloshing liquid counteracts with the externally applied force to achieve the shock absorption effect. The working principle of TLD is based on sloshing of the liquid to absorb a portion of the dynamic energy of the structure subjected to dynamic loadings and thus controls the structural vibration.

As early as 1950, the TLD has been used to stabilize ships. Modi and Welt [4] used the liquid sloshing in the annular tank to dissipate the sloshing energy and evaluate whether the damper can attenuate the low-frequency vibrations of the structure in aerodynamics, and that damper can effectively suppress the vibration of the structure under wind loads. Fujii et al. [5] developed multi-layer tuned liquid dampers (TLDs), which were used in the Nagasaki Airport Building (42 m high) and the Yokohama Marine Building (101 m high) under strong winds, which can reduce structural vibration by about 50%. Kareem [6] avoids calculating complex free liquid levels by converting the sloshing of the liquid into a particle spring system, and then simulated the effect of TLD installation in high-rise buildings. Sun et al. [7] established a rectangular nonlinear TLD model, using the shallow water wave theory to consider the damping effect and breaking waves of the liquid, and then using fluid–solid coupling to obtain the effect of TLD on the structure. Wakahara et al. [8] studied the optimization of TLD design on high-rise buildings, explored the parameters of changing TLD design, and developed new methods to predict the suppression of high-rise buildings by TLDs. Tamura et al. [9] set up TLDs on buildings to verify the motion reduction effect of the TLD. It turns out that the TLD can significantly improve the durability of the building.

Sakai [10] proposed another TLD-based damping method, tuned liquid column damper (TLCD), which used a U-tube as a container for liquids, and the vibration of the structure is reduced by the sway of the liquid in the tube. Colwell and Basu [11] used a TLCD to reduce the vibration of OWTs under wind and wave loads, and they applied Miner's law in their fatigue analysis. They reported that TLCD can effectively reduce the vibration generated by strong winds and waves, and it can also reduce the fatigue damage caused by wind waves. However, because the OWT is a long and narrow structure, the performance of the TLCD depends on the width of the bottom and the total length proportion. The greater the proportion of the bottom width, the better the shock absorption performance, but the width of the OWT is limited (cannot be large), and the performance of the container is limited as well. On the other hand, the TLCD can only absorb vibration in one direction, but the direction of the wind and waves on the sea does not necessarily coincide, so the load action on the OWT is not only in one direction. Colwell suggests that it can install multiple sets of TLCDs in different directions or use TLD to achieve multi-directional damping. Jin et al. [12] used a cylindrical TLD to control the vibration of the offshore platform, and they used the lumped mass method to simulate the effect of the TLD on the structure. The results are consistent with the experiment, and they concluded that the TLD could effectively reduce the structure motion during earthquakes. However, the earthquakes were the only loads considered in their study.

In the previous studies, not all the environmental loads, such as wind, waves, and earthquakes, were considered. Therefore, this study proposed setting multiple TLDs on the top of OWTs to reduce the motion of OWTs under various environmental load conditions including wind, waves, and earthquakes. The most recent reported study of Chen and Yang [13] indicated once the natural frequency (ω_l) of a TLD is equal to the exciting frequency (ω_x), the TLD may have the best motion reduction effect. As a result, the natural frequency of the TLD may be tuned to the natural frequency (ω_s) of the OWT, and good motion reduction of the OWT can be expected. The study mainly refers to the International Electrotechnical Commission (IEC) and the internationally renowned certification unit DNV (Det Norske Veritas) [14] for the environmental load simulation and analysis. The supported tower, monopile foundation, and wind turbine models refer to the 5 MW WTs designed by Jonkaman et al. [15] and Jonkman and Musial [16]. Wind and wave loads are generated by the FAST program, which was developed by the National Renewable Energy Laboratory (NREL). The FAST program can perform the wind analysis of the wind turbine during operation, startup, parked, shutdown, and standby states. The turbulent wind field simulation was made by the TurbSim program, which was also developed by NREL. The wind and wave conditions refer to the results of the feasibility study of Taipower's second phase offshore wind power project. The wave conditions are the appropriate wave height and period at the water depth of 20 m in the Changbin offshore project wind field in Taiwan, and the corresponding wind speed is at 90 m elevation. The required wind field and wave conditions were determined according to IEC 61400-3 design load (Design load case) regulations [17].

Turbsim simulates the wind field in the frequency domain using single-point power spectral density, spatial correlation function, and Taylor turbulence hypothesis. The FAST uses the blade element momentum theory to calculate the force of the wind acting on the blade. In this study, the hydrodynamic load evaluation of the supporting structure is referred to the above relevant specifications, and it considered the regular wave and the irregular wave models. After obtaining the dynamic characteristics of the wave particles, the hydrodynamic load can be calculated by using the Morison equation.

The TLDs were set at the top of the tower, and the two-way fluid-solid coupling module of ANSYS was used to simulate the interaction between the TLD and the structure. This study also designed simple experiments to verify the accuracy of numerical simulation. Then, this study explored the effect of TLD on offshore wind turbines under various wind and wave conditions. We also used the rainflow-counting method and the Miner's rule to do the fatigue analysis. Section 2 introduces the methods used in this study and they include FAST and Turbsim, as well as ANSYS-Fluent and ANSYS-Mechanical modules. The environmental loads according to IEC 61400-3 were also described in the section. Section 3 reports the simulation results and experimental measurements and numerical results validation. The effects of multiple TLDs on structural motion control were investigated in the section. The fatigue analysis of OWT with TLDs was also reported in this section. The final section lists the concluding remarks found in this study.

2. Methods Used

FAST is a computer-aided engineering (CAE) software developed by NREL, which is mainly used to simulate the dynamic responses of a wind turbine. FAST incorporates aerodynamic models (AeroDyn), hydrodynamic models of offshore structures (HydroDyn), control and motor system dynamic models (ServoDyn), and structural dynamic model (ElastoDyn and SubDyn). FAST can simultaneously couple air, fluid, motors, and structures in the same time domain.

2.1. Environmental Loads

The environmental conditions included wind, waves, and current. A full field wind speed above mean sea level (MSL) was established based on the field data measured by the Central Weather Bureau, Taiwan. Regarding hydrodynamic loading, the ocean conditions considered in this study include waves, ocean currents, and water levels.

2.1.1. Wind Model

This study selects the wind speed and wind field model of the wind field according to the IEC 61400-3 specification, and then it uses the TurbSim to simulate the wind field. TurbSim is a random, global, turbulent wind field simulator. Simulating the time series of three wind speed vectors on two-dimensional orthogonal grid points, TurbSim generates the wind spectrum frequency by wind speed, turbulence model, wind field model, etc. Then obtains the time history of the wind field by inverse Fourier transform. The data generated by TurbSim can be used by inputting AeroDyn in FAST. Then, it inputs TurbSim's two-dimensional wind field data in AeroDyn, converts it into a three-dimensional wind field by Taylor's frozen turbulence hypothesis, linearly interpolates the wind speed on the node, and converts it into an external force on the node. The force on the blade is theoretically analyzed by blade element momentum theory, and the blade tip loss correction factor is also considered. The FAST simulation procedure is shown in Figure 1.

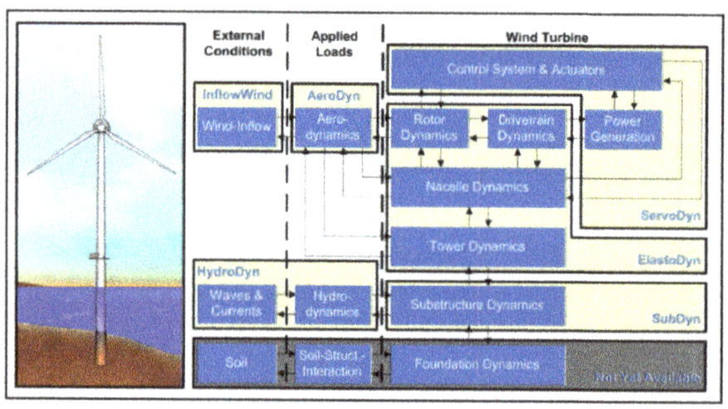

Figure 1. The simulation procedure of FAST (National Renewable Energy Laboratory, or NREL).

The wind load of the rotor nacelle assembly (RNA) and the structure (the tower and substructure are included) were evaluated separately. The wind load of the support structure is based on DNV-RP-C205 (DNV 2014) [18], while FAST analysis provides the wind load of RNA. According to the IEC 61400-1 specification (IEC 2005) [19], the relevant parameters of the standard wind-turbine (WT) class are shown in the Table 1.

Table 1. Wind related parameters of the wind-turbine (WT) classes.

WT class	I	II	III
V_{ref} (m/s)	50	42.5	37.5
Turbulence class	A	B	C
I_{ref}	16%	14%	12%

V_{ref} is the 10-min average wind speed at RNA, and I_{ref} is the turbulence intensity within 10 min.

According to IEC 61400-1, the WT class I-A was adopted, and the reference wind speed V_{ref} and turbulence intensity I_{ref} were set as 50 m/s and 0.16, respectively. The longitudinal turbulence scale parameter Λ_1 is shown in Equation (1) as

$$\Lambda_1 = \begin{cases} 0.7z, \ z_{ref} \leq 60\ m \\ 42, \ z_{ref} > 60\ m \end{cases} \tag{1}$$

where z is the height above the still water level, and z_{ref} is the height at the RNA center.

The longitudinal standard deviation of the normal turbulence model is the standard deviation of 90% of the average wind speed at the height of the hub (nacelle), and it can be estimated by

$$\sigma_1 = I_{ref}(0.75V_{ref} + b), \ \mathrm{b} = 5.6\ \mathrm{m/s}. \tag{2}$$

The turbulent extreme wind speed model has a 10-min average wind speed of V_{50} and V_1, respectively, in the 50-year and 1-year regression periods:

$$V_{50} = V_{ref}(z/z_{ref})^{0.11} \tag{3}$$

$$V_1 = 0.8V_{50}. \tag{4}$$

The standard deviation of the longitudinal turbulence is:

$$\sigma_1 = 0.11V_{ref}. \tag{5}$$

The annual data for buoys and tidal observations provided by the Central Weather Bureau, Taiwan were used in this paper, and monthly wind speed statistics and hourly wave records were obtained from the Hsinchu Buoy Station and monthly tidal statistics were obtained from the Waipu Tide Station in the Taiwan Strait.

2.1.2. Wave Model

According to the design load case of IEC 61400-3 (IEC 2009) [17], regular waves and irregular waves can be used. Regular waves include linear and nonlinear waves, and the linear wave is applicable when the wave height is small. When the wave height increases or the water depth becomes shallow, the wave peaks become sharp, and the troughs become flat. At this time, the linear wave theory is not enough to describe the waveform and the movement of water particles, so the nonlinear wave should be used, and one can use the streamline function theory to describe nonlinear waves in the most water depths. Once the water depth, period, and wave height were determined, the required wave theory can be determined accordingly (DNV OSJ101, 2014) [18]. Irregular waves usually describe in terms of their spectra, and the JONSWAP and Pierson–Moskowitz spectra are the most commonly used spectra. In this study, the water depth of 20 m was assumed, the stream wave function was used in the regular wave model, and the JONSWAP spectrum was applied to the stochastic irregular wave model.

2.1.3. Current Model

Ocean currents always vary in space and time, but they are usually defined as a uniform horizontal flow field with a direction and flow rate that varies only with water depth. According to IEC 61400-3, the current model is divided into three types: sub-surface currents, near surface currents, and surf underflow currents. In this study, the sub-surface current model was used and they can be expressed as:

$$U_{ss}(z) = U_{ss}(0)\left(\frac{z+d}{d}\right)^{1/7}. \tag{6}$$

The free surface current speed U_{ss} (0) was measured at the OWT site. After selecting the wave theory and current model, the speed and acceleration of water particles can be obtained, and Morison's equation was used to calculate the forces acting on the structure by the waves and currents.

2.2. ANSYS Fluent

ANSYS Fluent is a CFD (Computational Fluid Dynamics) software developed by ANSYS. To study TLDs on the motion reduction for OWT, we combined the fluent and structural analysis modules of ANSYS to simulate the fluid–structure interaction between TLDs and OWTs. We adopted Reynolds-Average Navier–Stokes Equation Models (RANS) and a standard k-ε model to solve the flow passing marine turbines. The detailed description of the model is as follows:

The RANS momentum equations can be written as

$$\rho\left(\frac{\partial \overline{u}_i}{\partial t} + U_k\frac{\partial \overline{u}_i}{\partial x_k}\right) = -\frac{\partial p}{\partial x_i} + \frac{\partial}{\partial x_j}\left(\mu\frac{\partial \overline{u}_i}{\partial x_j}\right) + \frac{\partial R_{ij}}{\partial x_j} \tag{7}$$

where ρ is fluid density, μ is dynamic viscosity, p is pressure, u_i is velocity component, x_i is coordinate, t is time, and R_{ij} is the Reynolds stress. The eddy viscosity model was used, and the Reynold shear stress can be written as

$$R_{ij} = -\rho\overline{u'_i u'_j} = \mu_t\left(\frac{\partial \overline{u}_i}{\partial x_j} + \frac{\partial \overline{u}_j}{\partial x_i}\right) - \frac{2}{3}\mu_t\frac{\partial \overline{u}_k}{\partial x_k}\delta_{ij} - \frac{2}{3}\rho k\delta_{ij} \tag{8}$$

δ_{ij} = 1 when $i = j$ and 0 when $i \neq j$. Lai et al. (2016) used three turbulence models to investigate the optimal turbine blades, and they are the "Realizable k-ε two layer", "standard k-ε", and "k-ε SST". They found that the results of three different models are about the same. Then, we chose the "standard k-ε" model. Then, we have μ_t = f($\rho k^2/\varepsilon$), where k and ε are the turbulent kinetic energy and turbulent kinetic energy dissipation rate, and they can be defined as

$$k \equiv \overline{u'_i u'_j}/2 \text{ and } \varepsilon \equiv \overline{\nu\partial u'_i/\partial x_j(\partial u'_i/\partial x_j + \partial u'_j/\partial x_x)}. \tag{9}$$

Turbulent energy equation of the standard k-ε model equations:

$$\frac{D(\rho\varepsilon)}{Dt} = \frac{\partial}{\partial x_j}\left[\left(\mu + \frac{\mu_t}{\sigma_\varepsilon}\right)\frac{\partial \varepsilon}{\partial x_j}\right] + C_{\varepsilon 1}\frac{\varepsilon}{k}G_k - \rho C_{\varepsilon 2}\frac{\varepsilon^2}{k} \tag{10}$$

where G_k can be defined as

$$G_k = \rho\nu_t\frac{\partial \overline{u}_i}{\partial x_j}\left(\frac{\partial \overline{u}_i}{\partial x_j} + \frac{\partial \overline{u}_j}{\partial x_i}\right). \tag{11}$$

The values of all the constants are empirically as C_μ = 0.09, $C_{\varepsilon 1}$ = 1.44, $C_{\varepsilon 2}$ = 1.92, and σ_k = 1.0.

ANSYS Mechanical

ANSYS Mechanical is a CAE (Computer Aided Engineering) software that simulates structural responses by the finite element method. The simulation analysis includes structural strength, stress, vibration, etc., and it can be combined with other ANSYS software for coupling analysis. In some cases, the fluid flow will cause the displacement of the solid, and the displaced solid will change the flow field of the fluid. Fluid–structure coupling is divided into one-way (1-way) and two-way (2-way) fluid–solid coupling. One-way fluid–solid coupling is to treat an already calculated result as another boundary condition to be calculated. Bidirectional coupling is divided into fully coupled, implicit, and explicit coupling. Full coupling calculates the simultaneous equations of fluid and solid as a matrix, which is often difficult to be solved. Implicit coupling is generally used by iteration until the convergence target is reached. Explicit coupling requires very a small time step and is very time consuming; therefore, it is not recommended. Figure 2 shows the FSI (Fluid-Structure interaction) module of ANSYS, and the implicit method was chosen in this study to perform the fluid–structure coupling analysis.

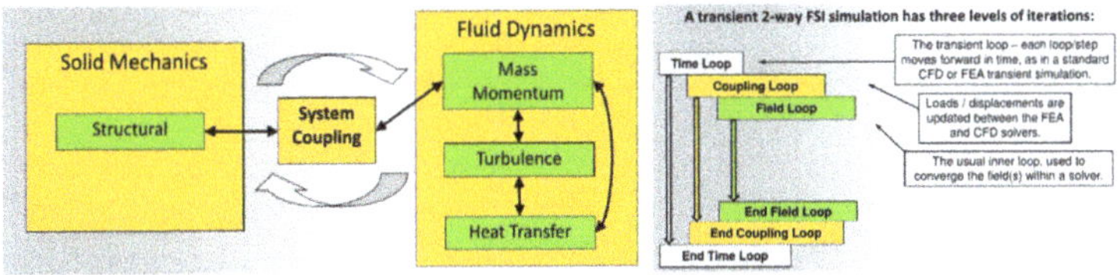

Figure 2. The system coupling and the two-way FSI simulation in ANSYS workbench.

2.3. Monopile OWT

This paper refers to the 5 MW offshore wind turbine developed by NREL. The radius of the three blades is 61.5 m, and the hub height is 90 m above the water surface. The water depth is assumed to be 20 m. The top of the tower has a diameter of 3.87 m and a thickness of 0.019 m. The bottom diameter is 6 m, and the thickness is 0.027 m. The total length of the tower is 77.6 m. The diameter and the thickness of the pile foundation are 6 m and 0.060 m respectively, and the length of the pile foundation is 30 m. The material of the supporting structures is steel, and the properties are as follows: the Young's modulus is 210 GPa, the shear modulus is 80.8 GPa, and the density is 7850 kg/m^3. In this study, the rotor, wind turbine blade, and nacelle are assembled as a lump mass (RNA). Figure 3 shows the 5 MW OWT and the simplified OWT model for the simulation.

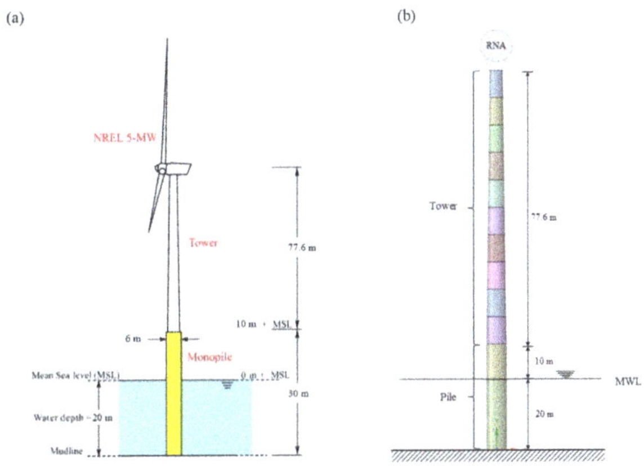

Figure 3. (**a**) The 5 MW offshore wind turbine (OWT) developed by NREL; (**b**) a simplified OWT model.

To reduce the dynamic response of the OWT, a TLD is installed at the top of the OWT. Figure 4 shows the simulation chart. To perform the simulation, the environmental loads were determined first. Since the turbulent wind field was considered, the TurbSim was used to generate a turbulent wind field, and then we input the obtained data to FAST. Then, the force data generated by FAST was input into ANSYS Transient Structural. The TLD geometry was determined based on the tuned frequency of TLD. We used ANSYS Fluent to simulate the sloshing flow in TLD, and then we used system coupling to complete the fluid–structure coupling simulation.

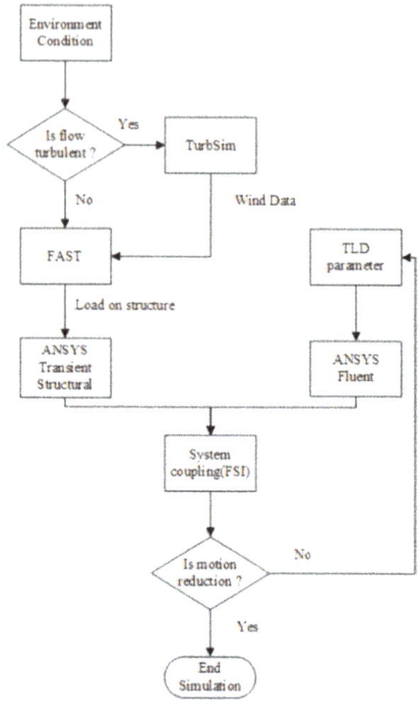

Figure 4. The flow chart of the simuation performed in this study.

One TLD was assumed first, and the simulated responses were compared to those without TLD, and the motion reduction ratio was calculated to evaluate the motion control effect. Different load conditions were assumed, such as harmonic ground motion, real earthquake excitation, coupled wind and wave loads, etc. Once the motion reduction was insignificant, more TLDs (2-TLD and 3-TLD) were then added to evaluate the possible increasing on motion reduction control. For fluid–solid interaction simulation, the SpaceClaim model was imported into Fluent, and the area of the structure was deactivated except for the TLD part. After the meshing was generated, the fluid–solid coupling interface was defined. The VOF calculation and the multi-phase flow were used in Fluent analysis. The turbulence mode was a standard $k - \epsilon$ model.

3. Results and Discussion

3.1. Model Validation

Before presenting the numerical simulations made in this study, the experimental measurements were performed to validate the simulation results obtained in this study. A simple experimental model (shown in Figure 5) was set up. The supported tower was a 3-meter height PVC pipe with a diameter of 12 cm. The RNA (rotor nacelle assemble) was replaced by a lump weight of 1 kg. The cylindrical TLD was set at the top of the RNA. The density, Young's modulus, and Poisson's ratio of the PVC pipe are, respectively, 1532 kg/m^3, 3070 MPa, and 0.4. As illustrated in Figure 6, the OWT model was set on a shaking table that can be moved back and forth with an AC motor. The maximum moving distance (r) of the shaking table is ±30 mm, and the highest revolutions of the motor is 2000 r.p.m. The AC motor can reciprocate according to the programmed displacement path and frequency. The motion trajectory of the instrument is based on the sine wave. With a maximum amplitude of ±5 mm reciprocating movement, a circular platform is attached to the square platform on which the OWT was installed. A Keyence laser displacement meter is set at the proper elevation to measure the time history of displacement of the RNA.

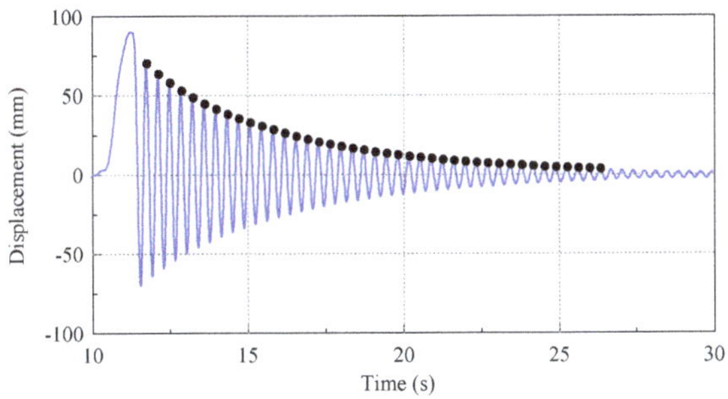

Figure 5. The dynamic displacement decay history of free vibration of a scaled model OWT.

The free vibration experiment was firstly made to evaluate the damping coefficient of the system. Figure 5 shows the decay history of the dynamic displacement, and the damping ratio of the experimental set up can be calculated according to the following equation

$$\ln\left(\frac{x_1}{x_j}\right) = \frac{2\pi j\zeta}{\sqrt{1-\zeta^2}}. \tag{12}$$

According to the decay history, the damping ratio $\zeta = 0.0119$ was determined.

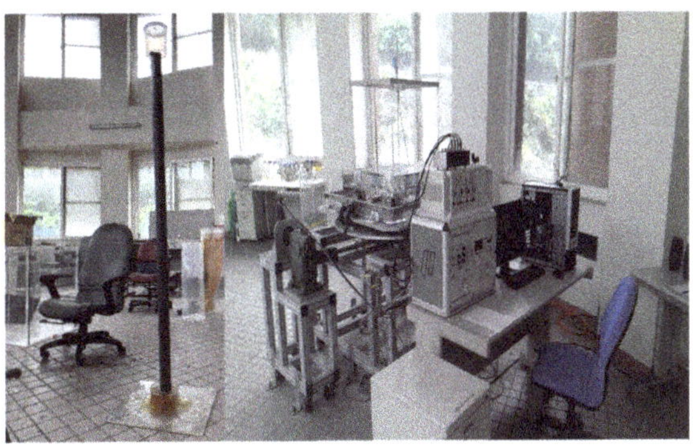

Figure 6. The simple experimental set up of OWT with a tuned liquid damper (TLD) on top.

3.1.1. Simulated Natural Frequency and Dynamic Response Validation

To confirm the accuracy of experimental measurements, the fundamental frequency of the scaled OWT model was checked. The base platform was excited by a harmonic force, $F(t) = A\omega_x \sin(\omega_x t)$, where A is the amplitude of the excitation and ω_x is the excitation frequency. ANSYS was used to calculate the corresponding dynamic displacement response. The ANSYS FEM-Modal set up was based on the experimental model, and the natural frequency of the model was calculated. The calculated natural frequency (2.751 Hz) is very close to the experimental measurement (2.75 Hz). Figure 7 clearly shows the agreement of the simulation and experimental results, and the peak dimensionless displacement occurred when the exciting frequency (ω_x) is equal to the natural frequency (ω_s) of the model, i.e., $\omega_x/\omega_s = 1$. Figure 8 shows the comparison of the dynamic displacement obtained by ANSYS simulation and those of the experimental measurements and the agreement is also very good. The dynamic displacements measured in this section were also used as a reference to evaluate the effects of TLD on the response reduction of the OWT.

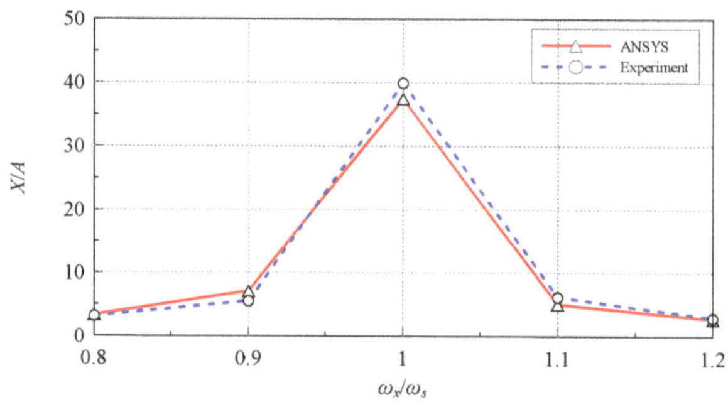

Figure 7. The comparison of ANSYS simulated results and the experimental measurements; X/A: the dimensionless displacement of OWT; $\omega_x/\omega_s = 1$: excitation frequency (ω_x) is equal to the natural frequency of OWT (ω_s).

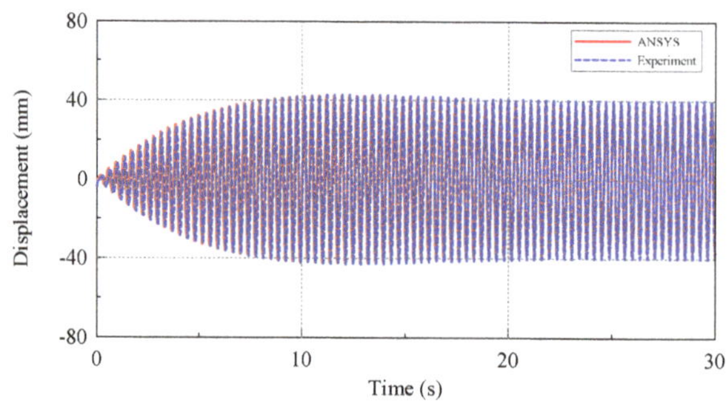

Figure 8. The history of the displacement at the rotor nacelle assembly (RNA) of OWT under excitation wih exciting frequency = natural frequency of the OWT.

3.1.2. Fluent TLD Simulation Validation

In the ANYSY fluid–solid coupling, the accuracy of the hydrodynamic force calculation is important, and Fluent may correctly transmit the force of the fluid acting on the tank wall to the structure. Figure 9 illustrates the comparison of the Fluent-simulated force with the experimental measurements of Krabbenhøft (2011) [20], and the agreement is very good.

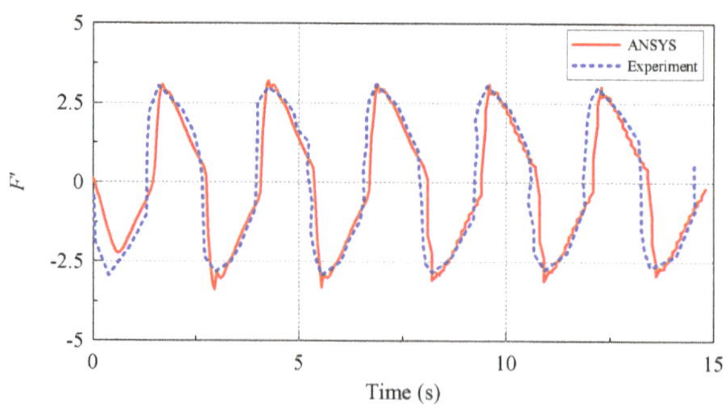

Figure 9. The comparison of the forces acting on the tank wall; solid line: ANSYS results; dashed line: experimental measurements (Krabbenhøft, 2011) [20]. Tank length = 0.59 m, water depth = 0.02 m, the exciting amplitude = 0.02 m, and the exciting frequency = 2.36 rad/s.

3.1.3. ANSYS Mechanical Model (Fluid–Structure Interaction) Validation

The comparison made in the previous sections validated the accuracies of simulations made by the ANSYS Fluent and ANSYS structure models. In this subsection, we further validated the accuracy of the fluent and structure coupling model. The TLD was added on the top of the scaled OWT model, and an experiment was performed to investigate the motion reduction of the OWT.

The exciting force is a harmonic motion, the displacement amplitude is set as 0.001 m, and the exciting frequency is set to equal to the natural frequency of the OWT. ANSYS-Transient-Structural and ANSYS-Fluent are connected to System Coupling for the data transmission setting and time control. Figure 10 illustrates the TLD mesh model and the fluid–structure coupling interface. Figure 11 shows the schematic of the projects included in the simulation of the interaction between the TLD and OWT.

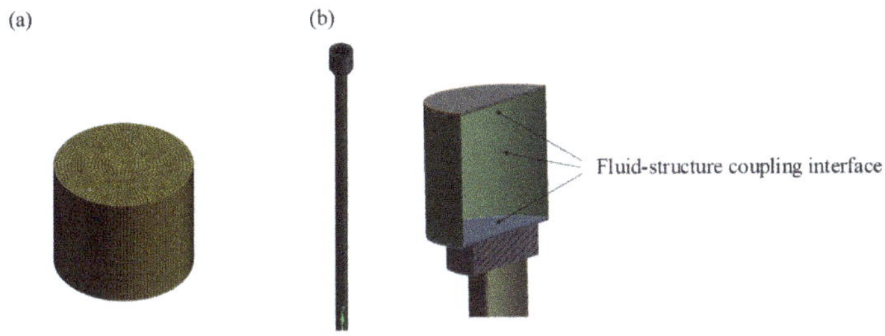

Figure 10. (**a**) The mesh arrangements of fluid in TLD and (**b**) the fluid–structure coupling interface.

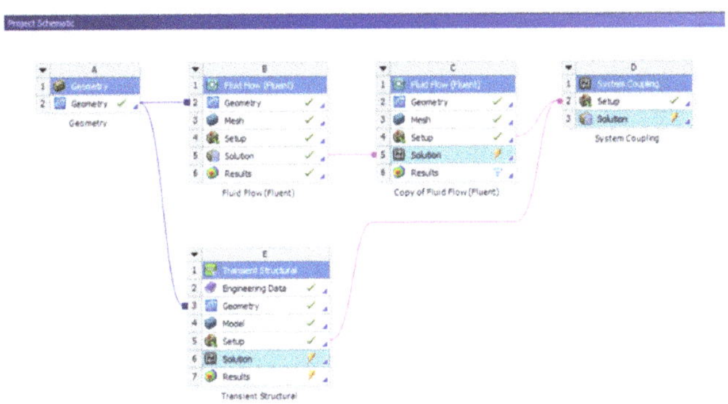

Figure 11. The schematic projects included in the TLD + OWT interaction simulation.

Figure 12 further compares the dynamic displacements of RNA of OWT with and without TLD, and the motion of RNA of OWT is remarkably reduced when the TLD is installed at the top of the RNA of OWT. Figure 13 also shows that the results obtained by ANSYS simulation and the experimental measurements agree very well. The amplitude of the OWT with the TLD is much smaller than that without the TLD; therefore, we use different vertical axes to clarify the difference between the experiment and simulation results. The response of the OWT with the TLD is much more complicated than that without TLD, and more difference between experiment and simulation can be expected, but the deviation is still below 10%. The accuracy of the ANSYS system coupling was confirmed and can be used to study the motion reduction effect of the TLD on the OWT.

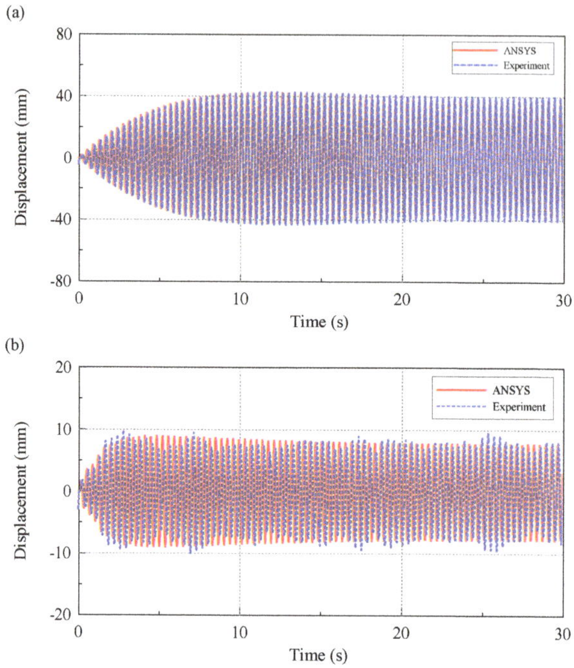

Figure 12. Dynamic displacement of RNA of OWT: (**a**) without TLD; (**b**) with TLD.

3.2. TLD on Motion Reduction of OWT

As shown in Figure 12, the TLD may have a very good motion reduction effect on OWT when it is under a harmonica ground excitation. In the following sections, we will investigate the damping effect of the TLD on the OWT when it is under real environmental load conditions. The design standard IEC 61400-3 defines the load conditions DLC (Design Load Case) for the structural design of offshore wind turbines, including all operating conditions of OWTs, such as startup, normal operation, shutdown, etc. The structural design of the offshore wind turbine often considers the extreme 50-year regression period, while the fatigue analysis is based on the general sea conditions. The environmental loading used in this study refers to the feasibility study of the Taiwan Power Company's offshore wind power generation second phase plan. Table 2 lists the estimated wind and wave conditions and extreme wind and waves at the Zhangbin Industrial Zone project.

Table 2. The wind and wave conditions at the Zhangbin Industrial Zone project. DLC: Design Load Case.

Wind Condition (Elevation 95 m)		Wave Condition (Water Depth 20 m)	
Annual mean wind V_{ave} (m/s) (DLC 1.2)	9.2	H_s (m)	1.4
		T_p (s)	6.1
50 years, V_{ref} (m/s) (DLC 6.2)	54.16	H_{s50} (m)	8.24
		T_{p50} (s)	12.1

3.2.1. Wind Field Simulation

As mentioned in the previous sections, TurbSim was used to simulate the turbulent wind field of the whole domain, and the output binary file was directly used by AeroDyn in FAST. The input parameters of TurnSim are shown in Table 3.

Table 3. Input file description (Turbsim) (DLC 1.2).

Parameter	Description	Value
WrADHH	Output format for AeroDyn	TRUE
NumGrid_Z	Vertical mesh number	31
NumGrid_Y	Horizontal mesh number	31
TimeStep	Time-step (s)	0.05
AnalysisTime	Analysis time (s)	630
UsableTime	Usable time (s)	100
HubHt	Hub height (m)	90
GridHeight	Vertical domain height (m)	145
GridGridWidth	Width of domain (m)	145
TurbModel	Turbulence model	IECKAI
IECstandard	IEC version, IEC-61400-3	3
IEC_WindType	IEC wind type	NTM
RefHt	Reference height (m)	95
URef	Wind-speed at reference height (m/s)	9.2

For the wind field condition of DLC 6.2, the U_{ref} is changed to 54.16 m/s, and the generated file was submitted to AeroDyn to expand into a global wind field. In the AeroDyn, the direction of the wind can be set. In the case of the annual average wind speed of the wind field conditions of the DLC 1.2, the wind turbine directly faces the windward direction, so the wind direction angle is set to 0°. In the extreme load DLC 6.2, the side winds of the 10-year average extreme wind speed and the 50-year regression period are blown toward the blades, so the wind direction angle is set to 90°. Figure 13 shows the wind speed in three directions with an annual average wind speed at 90 m elevation. Since the annual average wind speed is blown toward the wind turbines, the wind speed in the x direction is greater than the other two. Figure 14 shows the corresponding 10-min average extreme wind speed in a 50-year regression period. The wind field is designed to have a crosswind effect on the wind turbines, so the wind speed in the y direction is the largest. The fatigue load in DLC 1.2 is in normal operation, and the turbine blades rotate normally, but in the DLC 6.2 limit load condition, the wind turbine is parked because of excessive wind, and the blade is turned parallel to the wind to reduce the force on the blades. Meanwhile, in the FAST setting, the blades are parked, and the blade angle is set to 90°. Therefore, we may expect that RNA has larger force in the x direction, whereas at extreme wind speeds, RNA has larger force in the y direction, since the wind blows from the y direction.

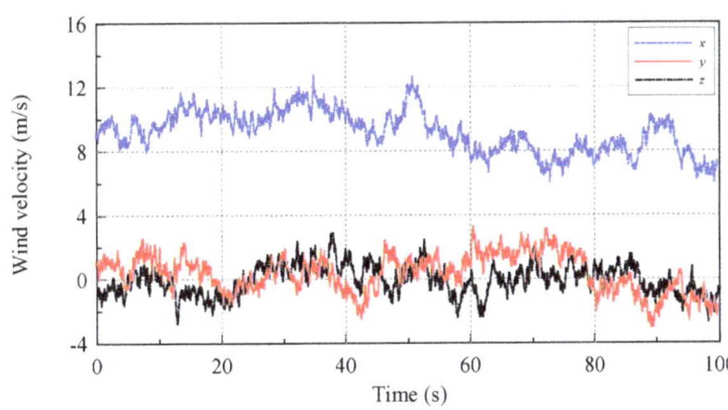

Figure 13. Annual wind speed V_{ave} = 9.2 m/s at 90 m elevation.

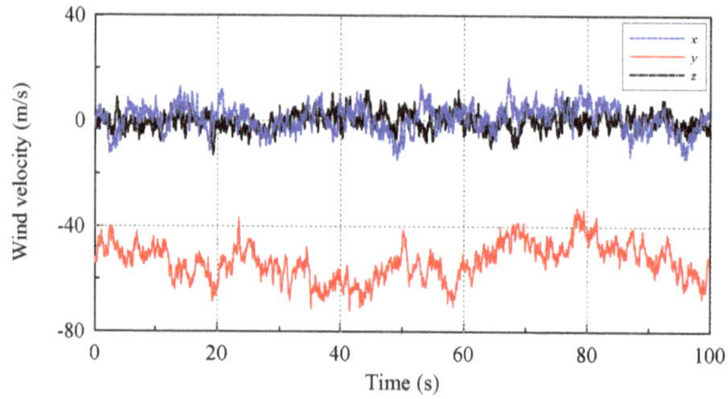

Figure 14. 50-year, 10 min average extreme wind speed V_{ref} = 54.16 (m/s) at 90 m elevation.

3.2.2. Wave Simulation

In FAST, HydroDyn is used to calculate hydrodynamic loads. The regular wave and irregular wave can be selected. In the wave condition of fatigue load, an irregular wave is selected. The PM wave spectrum was selected for the normal sea state, whereas the JONSWAP spectrum is selected for the ultimate load condition, because the JONSWAP spectrum can show the characteristics of extreme sea conditions. Figure 15 shows the wave histories of normal and extreme sea states.

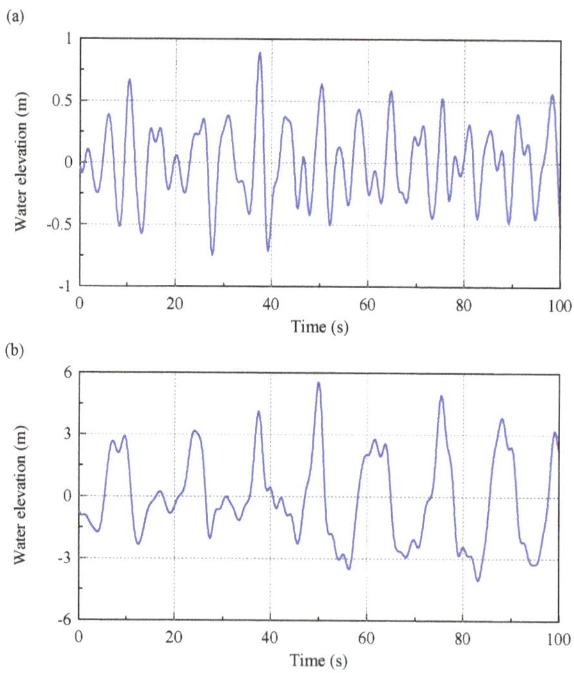

Figure 15. (**a**) Normal wave condition, H_s = 1.4 m, T_p = 6.1 s; (**b**) 50-year extreme wave condition, H_s = 8.24 m, T_p = 12.01 s.

3.2.3. Structural Load

The forces exerted on the OWT can be further calculated by FAST once the wind field and wave fields were established. The OWT includes a rotor, a nacelle, a tower, and a pile. The focus of the study is on the tower, which does not consider the pile part and the deformation of the blade and the rotational speed of the blades. Therefore, the rotor, the blade, and the nacelle are combined here (RNA) as a lump mass point.

3.2.4. The Loads on RNA

Figure 16 shows the forces of the RAN acting on the top of the tower in DLC 1.2 and DLC 6.2 conditions, respectively. As shown in the figure, RNA has larger force in the x direction for the average wind speed condition, whereas at extreme wind speeds, RNA has larger force in the y direction, since the wind blows from the y direction.

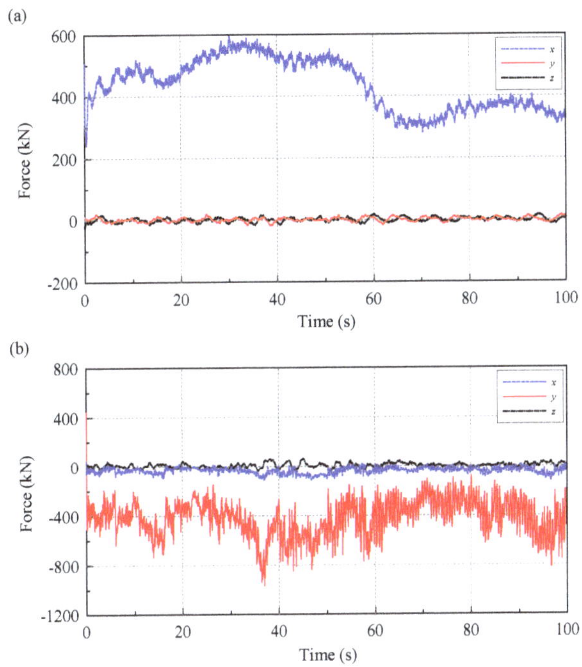

Figure 16. The forces on RNA: (**a**) normal wind condition; (**b**) extreme wind condition.

3.2.5. The Loads on Wind Tower

It can be seen from Figure 6 that the tower with a total length of 77.6 m in FAST is divided into 10 sections, and the length of each section is 7.76 m. The force of the tower is calculated by AeroDyn; Figure 17 depicts the force acting on each tower section at the average wind speed condition, and the force in the x direction is obviously greater than in the other directions. Figure 18 depicts the corresponding forces under extreme wind speed, and the force in the y direction is the largest among all directions.

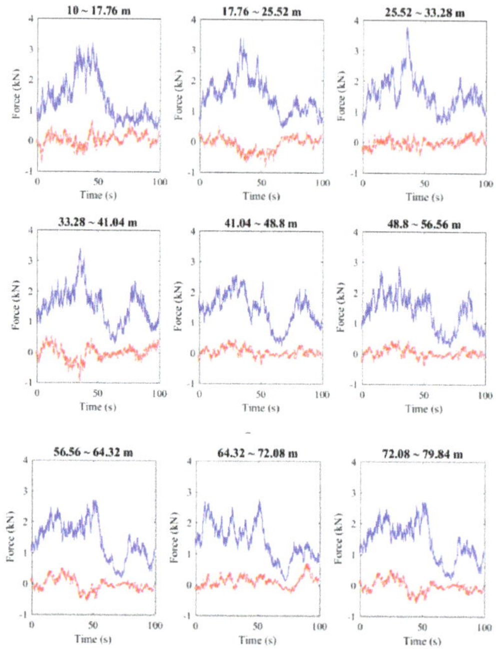

Figure 17. The force on each section of the tower under normal wind conditions (blue line: x direction; red line: y direction).

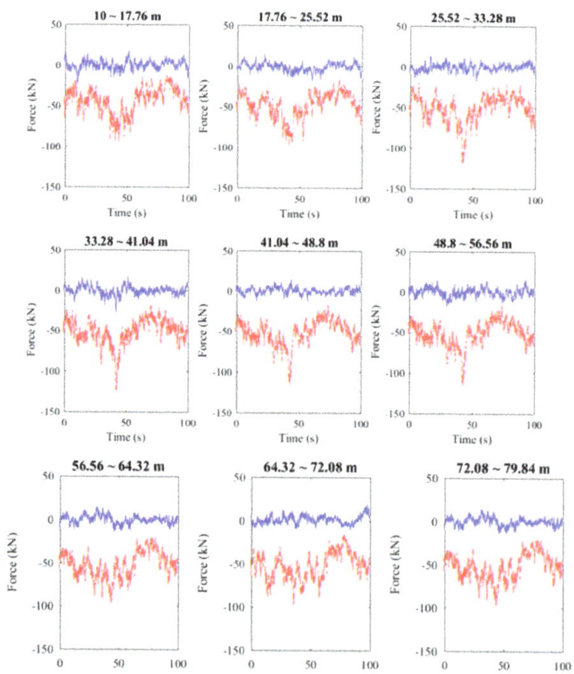

Figure 18. The force of each section of the tower under extreme wind conditions (blue line: x direction; red line: y direction).

3.2.6. The Loads on Pile of the OWT

The length of the supported pile of the OWT is 30 m and it is fixed on the sea bottom, and 20 m and 10 m of it are below and above the sea surface, respectively. HydroDyn is used to calculate the force of the wave on the pile. Figure 19 shows the corresponding forces applied on the pile.

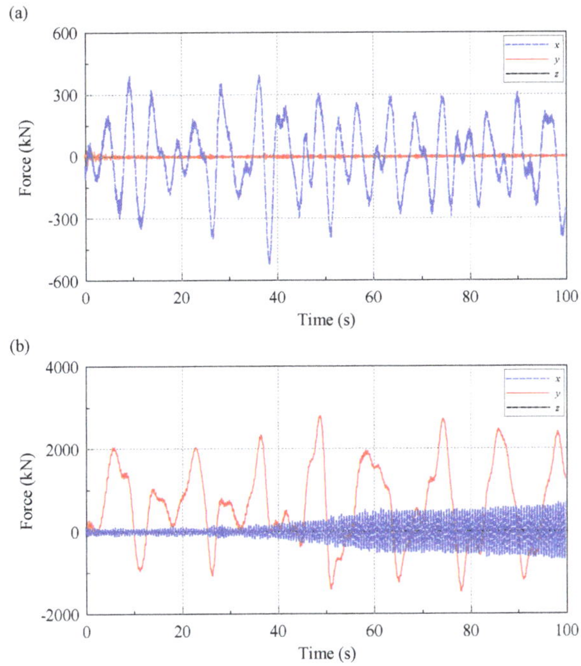

Figure 19. Force applied on the pile of the OWT: (**a**) normal wave condition; (**b**) extreme wave condition.

3.2.7. Tuned Frequency of TLD

As reported in Chen and Hunag (2015) [21] and Chen and Yang (2018) [13], the best motion reduction effect of TLD on the structure may occur when the natural frequency of the TLD is tuned to the frequency of the exciting forces. The effect of waves on the tower top displacement is much less than that of wind, so it is useless to tune the frequency of the TLD to the frequency of waves. In addition, because turbulent wind is a random variable, we did a spectrum analysis of the wind and found that no specific peak exists, so it is not feasible to tune the TLD frequency to the frequency of the wind. However, tuning the frequency of the TLD to the frequency of the structure also has a good damping effect (Chen and Huang, [21]; Frandsen, [22]). Therefore, the natural frequency of the TLD used in this study was tuned to the fundamental frequency of the supported tower of the OWT.

3.2.8. The Displacement of Tower Top

The calculated forces in the previous section were applied to the finite element method (FEM) model of ANSYS. The dynamic response of offshore wind turbines, including the deformations and the stresses of piles and towers, were calculated in this section. Figure 20 shows the displacements of the top of the tower under DLC 1.2 and DLC 6.2 conditions.

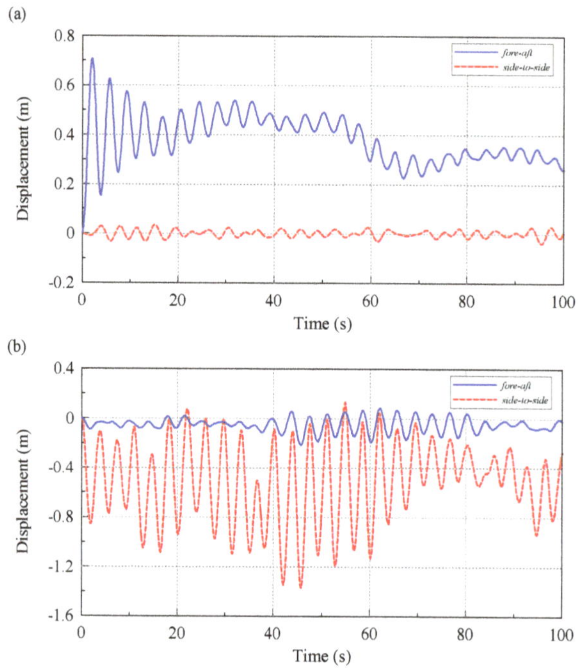

Figure 20. The displacement of the tower top in (**a**): DLC 1.2 condition; (**b**) DLC 6.2 condition.

3.3. TLD Application on Motion Reduction of OWT

The monopile offshore wind turbine is subject to wind and wave loads, which may cause dynamic responses of the structure. This study tried to install the cylindrical TLD on the top of the offshore wind turbine, as shown in Figure 21, to suppress those responses. The diameter of the TLD is 3.846 m, which is about the same size of the inner diameter of the tower top. The natural frequency of the liquid in the cylindrical tank can be calculated by $\omega^2 = \frac{\lambda g}{R}\tanh\frac{\lambda h}{R}$. The water depth of TLD can be determined when we tune the natural frequency of TLD to be equal to the natural frequency of the structure; then, the water depth of TLD = 0.35 m was calculated.

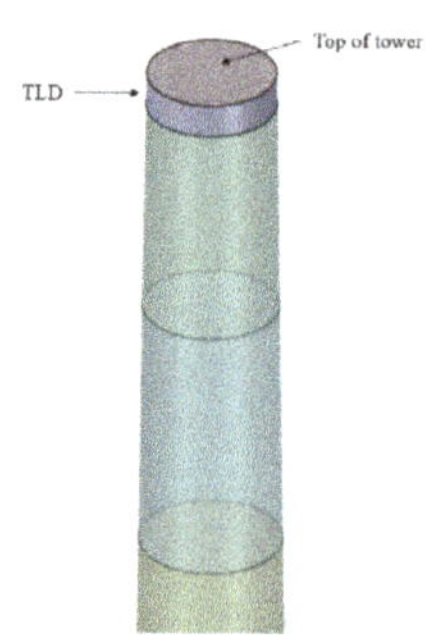

Figure 21. The conceptual sketch of the TLD on the top of the tower.

The mechanical properties of the OWT are shown in Figure 22. The OWT model is drawn by ANSYS SpaceClaim and TLD was also added, which were all input to the Transient Structural. In the Transient Structural, the parameters of the material such as density, Young's modulus, and damping coefficient, etc. were set. Then, the FAST-calculated wind and wave loads were input. The fluid–solid coupling interface was set as shown in Figure 23.

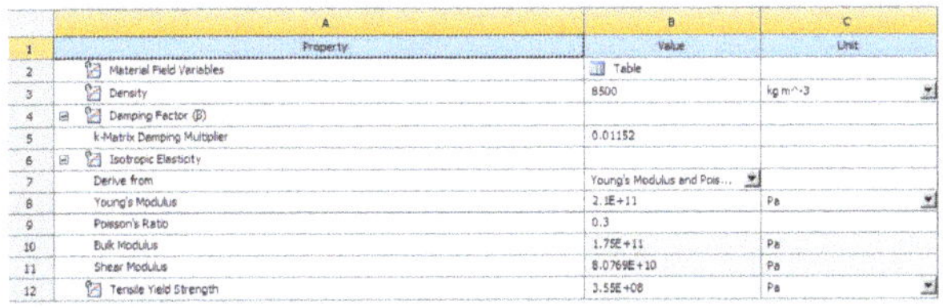

	A	B	C
1	Property	Value	Unit
2	Material Field Variables	Table	
3	Density	8500	kg m^-3
4	Damping Factor (β)		
5	k-Matrix Damping Multiplier	0.01152	
6	Isotropic Elasticity		
7	Derive from	Young's Modulus and Pois...	
8	Young's Modulus	2.1E+11	Pa
9	Poisson's Ratio	0.3	
10	Bulk Modulus	1.75E+11	Pa
11	Shear Modulus	8.0769E+10	Pa
12	Tensile Yield Strength	3.55E+08	Pa

Figure 22. Transient Structural (the parameters setting).

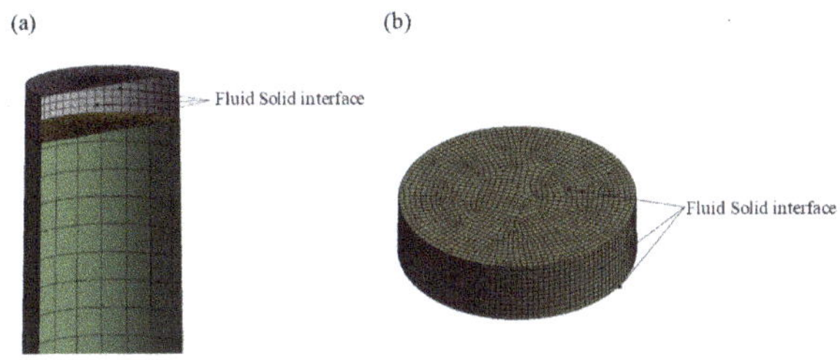

Figure 23. Fluid–solid interface.

Then, the SpaceClaim model was imported into Fluent, and the area of the structure was deactivated except the TLD part. After the meshing was generated, the fluid–solid coupling interface was defined. The VOF calculation was started in Fluent to simulate the multi-phase flow, and the solid boundary was set as the wall. The turbulence mode was a standard k-ε model. In general multi-phase flow, air and water are often set as incompressible fluids to simplify the model, but in the fluid–structure interaction calculation, the boundary of the fluid changes with the solid and the volume of the fluid changes with time, which is very prone to problems in the calculation of Fluent, so it is necessary to set the air as an ideal compressible gas to avoid problems in calculation.

3.3.1. Convergence Test

In order to save calculation time, a hexahedral structural mesh is used here. Three difference mesh sizes (1.4 m, 1.0 m, and 0.7 m) were used to do the convergence tests, and the DLC 6.2 conditions were used as the external force conditions. Table 4 lists the comparison of the root square of displacements of RNA of various mesh sizes used, and all the results were about the same; the mesh size = 1.4 m was selected in the later simulation. Table 5 lists the convergence tests for mesh selection for TLD simulation, and the results of 0.09 m and 0.06 m are about the same and therefore, the mesh size = 0.09 m was selected in the later simulation. The time steps = 0.005 s and 0.0025 s were tested, and the results are about the same, and the time step = 0.005 s was used in the later simulations.

Table 4. The convergence test for tower FEM mesh selection.

Mesh Size (m)	Mesh Number	Root Square of Displacement (m)	Relative Difference with 0.7 m (%)
1.4	3956	0.86877	0.1
1.0	7502	0.86827	0.05
0.7	14,950	0.86784	0

Table 5. The convergence test for TLD mesh selection.

Mesh Size (m)	Mesh Number	Root Square of Displacement (m)	Relative Difference with 0.06 m (%)
0.3	940	0.63282	2
0.09	19,572	0.62020	0.035
0.06	61,778	0.62041	0

3.3.2. TLD on Motion Reduction of OWTs

As shown in Figure 14, the TLD has a significant motion reduction effect when OWT is under a harmonic excitation. While the OWTs are mostly under wind and wave loads, in this section, we investigated the motion reduction effects of TLD on OWT when it is under DLC 1.2 and DLC 6.2 load conditions. Figure 24 illustrates the time histories of fore-aft and side-to-side displacement of RNA of OWT with and without TLD. Although the results presented in the Figure 25 do not show obvious damping effects of TLD on OWT motion reduction, Figure 26 shows the FFT of the displacement responses and indicates that the TLD may reduce response peak intensity by 44% and 24% in fore-aft and side-to-side displacements, respectively. When the environmental load condition DLC 6.2 was applied, the results shown in Figure 27 clearly demonstrate the effects of TLD on the motion reduction of OWTs.

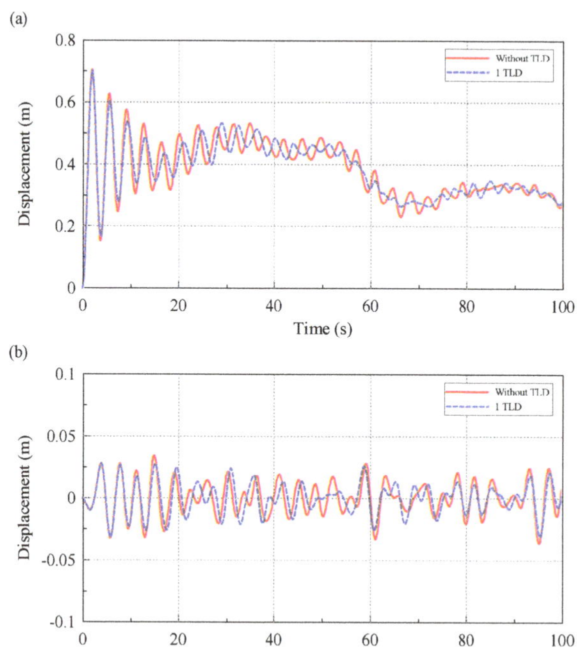

Figure 24. Displacement of RNA for an OWT under the DLC 1.2 condition: (**a**) fore-aft movement; (**b**) side-to-side movement.

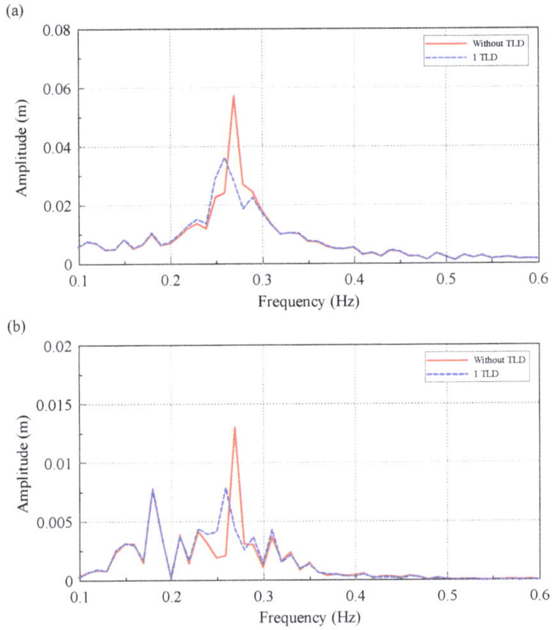

Figure 25. Amplitude spectrum of RNA displacement for an OWT under the DLC 1.2 condition: (**a**) fore-aft movement; (**b**) side-to-side movement.

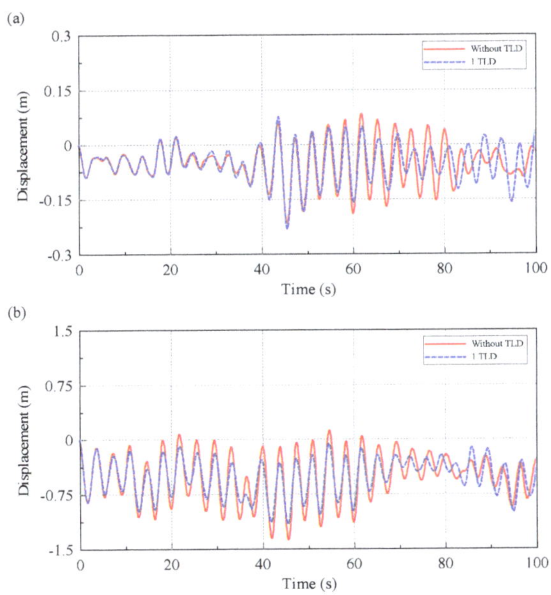

Figure 26. Displacement of RNA for an OWT under the DLC 6.2 condition: (**a**) fore-aft movement; (**b**) side-to-side movement.

3.3.3. Multiple TLDs on Motion Reduction of OWTs

In the previous section, one TLD presents mild motion reduction effects on OWT when it is under DLC 1.2 and DLC 6.2 load conditions. As reported in Chen and Yang's study, the best damping effect of

TLD on the structure might occur when the natural frequency of TLD is tuned to the natural frequency of the structure. Since the OWT is a slender structure and the natural frequency is small, therefore, the water depth of the liquid in TLD is also small. The damping effect of a single TLD might be limited. Then, we increase the number of TLDs, which all have same natural frequency as the natural frequency of the structure. Figure 27 shows the motion reduction effect of multiple TLDs on OWTs when it is under harmonic ground excitation, and the 3-TLD has the best motion reduction effect, which is nearly 100% better than that of the 1-TLD.

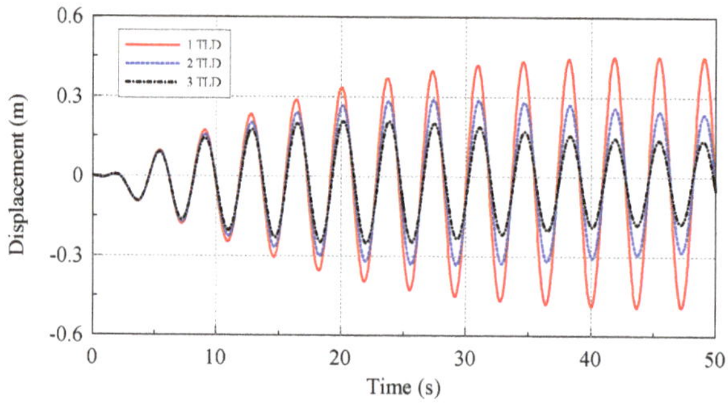

Figure 27. Multiple TLDs on OWT motion reduction and harmonic ground excitation.

As mentioned by Jin et al. (2007) and the results obtained in this study, the TLD did have an excellent motion reduction effect on OWT when it is under harmonic ground excitation. Meanwhile, the environmental loads that OWT may experience include wind, waves, and real earthquakes. Then, we further investigated the motion reduction effects of the multiple TLDs on OWT when it is under real earthquake, wind, and wave loads. Figure 28 depicts the comparison of the damping effects of various TLDs on an OWT under DLC 1.2 and DLC 6.2 load conditions. The motion reduction effects of various TLDs are about the same when the OWT is under DLC 1.2 load conditions. Figure 28 even shows larger peak performance for the 3-TLD, whereas the narrower band can be found for the 3-TLD, and the root mean square of the FFT spectrum of three cases (1-TLD, 2-TLD, and 3-TLD) are about the same in the DLC1.2 condition. The response of OWT under DLC 1.2 is virtually small, and the difference among various TLDs is also insignificant. The force of the DLC 6.2 condition is much larger in the y-direction, and thus, the side-to-side displacements were shown in Figure 28 when the OWT was under the DLC 6.2 condition. The much larger displacement occurred when the OWT was under extreme loading condition (DLC 6.2), and the effect of TLD on motion reduction control became more obvious; more TLDs also enhance the reduction effects. The corresponding comparison of OWT under real earthquakes is shown in Figure 29, and 3-TLD also has the best motion reduction effects among others.

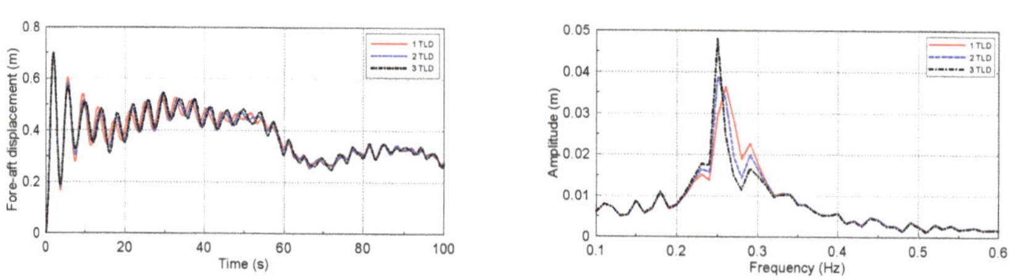

(a) (b)

Figure 28. *Cont.*

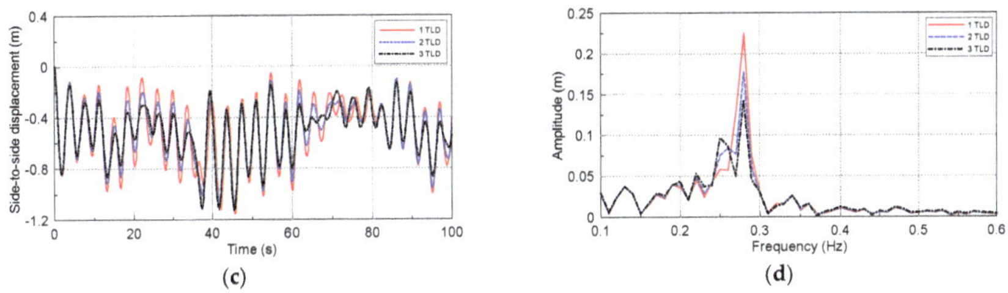

Figure 28. (**a**) and (**b**): The fore-aft displacement and corresponding amplitude spectrum of OWT with various TLDs under the DLC 1.2 condition; (**c**) and (**d**): The side-by-side displacement and corresponding amplitude spectrum of OWT with various TLDs under the DLC 6.2 condition.

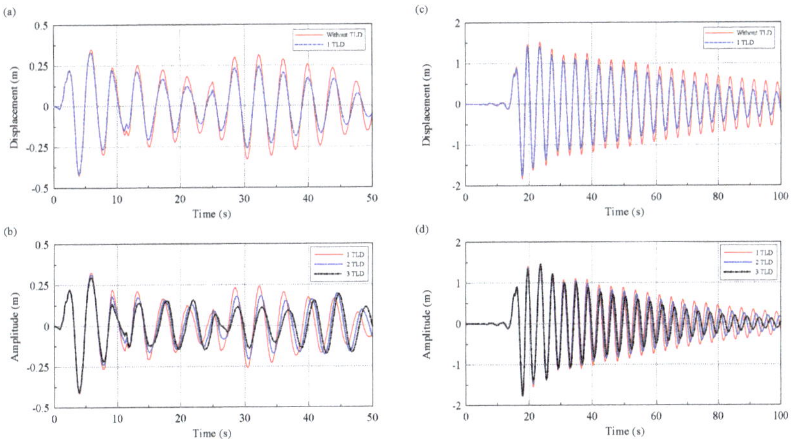

Figure 29. The side-to-side displacement of the OWT with various TLDs: (**a**) and (**b**), El-Centro earthquake; (**c**) and (**d**), Chi-chi earthquake.

3.4. Fatigue Analysis

Figure 30 shows the location where the maximum stress occurs in the absence of TLD according to ANSYS simulation. This position is the intersection of the tower and the pile. Then, we made the fatigue analysis of the stress at the junction of the tower and the supported pile. Figure 31 shows the time series of the maximum stress at the intersection of the tower and the pile.

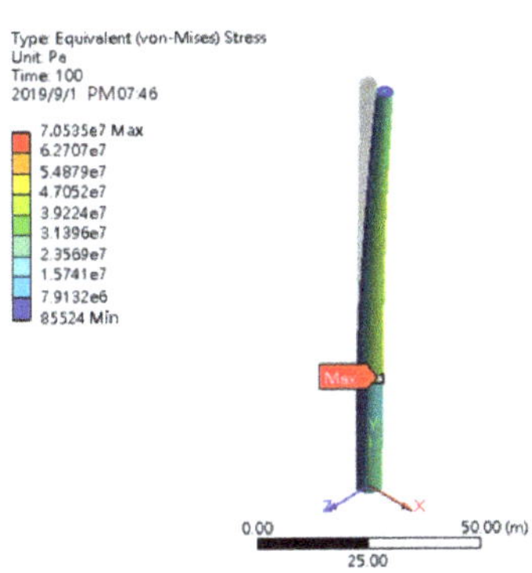

Figure 30. The location of the maximum stress occurred in the OWT.

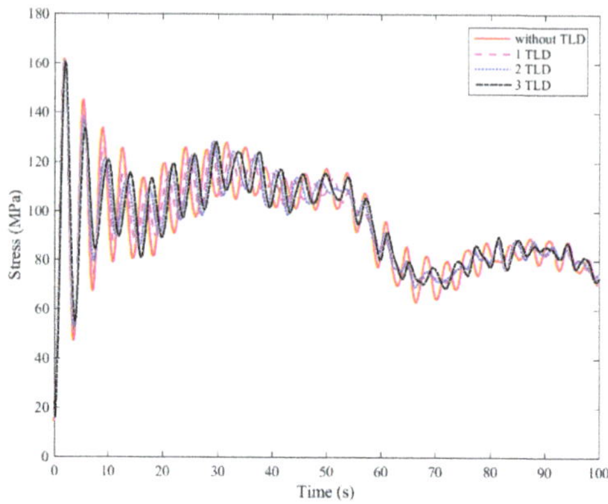

Figure 31. The comparison of the history of the stress at the interaction of the tower and pile of the OWT with various numbers of TLDs.

For a typical offshore structure, the fatigue load history spans a period of 20 years corresponding to about 10^8 wave load cycles (an average wave load period of 6 s). We used the S-N curve to determine the fatigue life at the interaction of tower and pile of the OWT (Ju et al. [23]). In this study, the S-N curve (structural detail class E) of DNVGL-RP-C203 [18] was used to assess the fatigue damage at the interaction of the tower and pile of the OWT. The S-N curve can be expressed as follows

$$\log N = \log \bar{a} - m \log \Delta\sigma \tag{13}$$

where N = predicted number of cycles to failure for stress range; m = the negative inverse slope of the S-N curve; and $\log \bar{a}$ = the intercept of log N-axis by the S-N curve. Table 6 lists the details of the parameters of the S-N curve.

Table 6. S-N curves for tower and pile (Structural detail class E).

Environment	m_1	$\log \bar{a}_1$	m_2	$\log \bar{a}_2$
Air	$N \leq 10^6$ cycles		$N \geq 10^6$ cycles	
	3.0	11.61	5.0	15.35

The rainfall counting was used to obtain the stress counting. The Miner cumulative damage theory was used to estimate the degree of damage to the structure. The Miner's rule states that if there are k different stress levels (with linear damage hypothesis) and the average number of cycles to failure at the ith stress, S_i, is N_i, then the damage fraction, C can be expressed as

$$\sum_{i}^{k} \frac{n_i}{N_i} = C \tag{14}$$

where n_i is the number of cycles accumulated at stress S_i and C is the fraction of life consumed by exposure to the cycles at the different stress levels. Usually, fatigue damage occurs when $C > 1$.

Figure 32 shows the histogram of stress rainflow counting of the uncontrolled OWT and controlled OWT with difference numbers of TLDs. Finally, we apply the Miner cumulative damage theory and S-N curve to infer the fatigue damage value of the OWT. Table 7 lists the fatigue life evaluated based on Miner's rule of uncontrolled OWT and OWT with different numbers of TLDs, and an OWT with 3 TLDs may increase the fatigue life for 20 years. The OWT with a single TLD can also increase the fatigue life for 13 years. The 3-TLD can not only reduce the motion response but also increase the fatigue life (37% more) of the OWT.

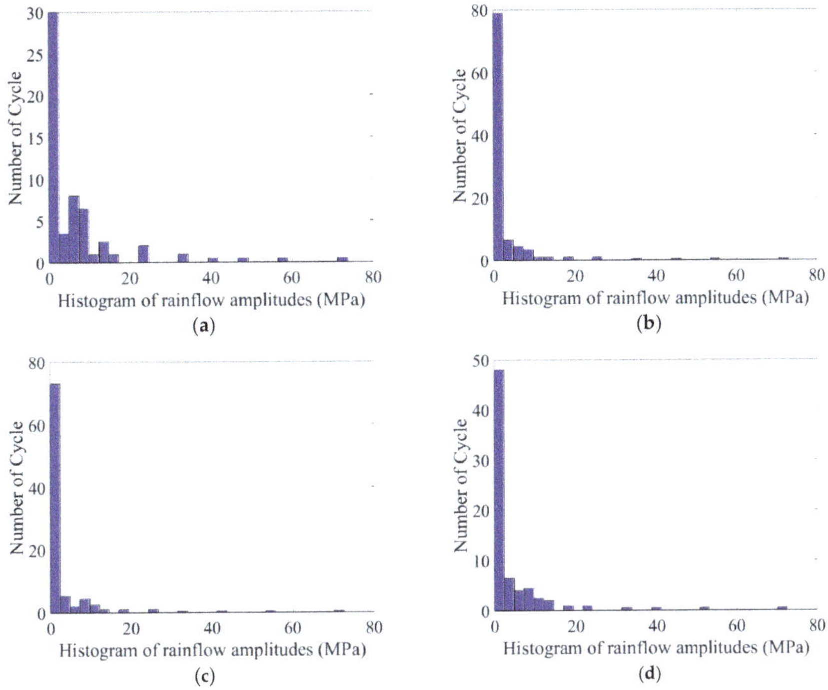

Figure 32. The histogram of stress rainflow counting of uncontrolled OWT and controlled OWT with difference numbers of TLDs.

Table 7. Fatigue life of using various TLDs.

Number of TLD	Fatigue (year)
NO TLD	54.03
1-TLD	67.21
2-TLD	71.25
3-TLD	74.47

4. Conclusions

This study used NREL developed FAST and Turbsim software to generate the environmental loads, which were based on the regulation code IEC 61400-1 and 3. ANSYS-Fluent and ANSYS-Mechanical modules were also used to perform the fluid–structure interaction between OWT and TLDs. The convergence tests and numerical simulation validation were made to confirm the accuracy of the simulation tools used in the study. Extensive simulation cases were made, and the following conclusion were found.

(1). A sample simulation was made to valid the readiness of using FAST, and a simple experimental model was set to perform the numerical validation of accuracy of the ANSYS simulation.

(2). The experimental measurements confirm the significant structural motion control effects of TLD on OWT under a harmonic ground excitation.

(3). The motion reduction effect of a single TLD was studied first. The FFT of the displacement responses of OWT under DLC 1.2 loads indicates that the TLD may reduce response peak intensity by 44% and 24% in fore-aft and side-to-side displacements, respectively. Similar damping effects of TLD can be found in DLC 6.2 load conditions.

(4). The simulation of the multiple TLDs setting was also made. When the OWT is under harmonic ground acceleration, the natural frequency is easily tuned to equal to the ground exciting frequency, and the simulation results show that better motion reduction control (almost 100% better) can be achieved when the number of the TLD is increased from 1 to 3. Meanwhile, for real earthquake excitation, the frequency contents are various, and the motion reduction effects of multiple TLDs on an OWT under real earthquake conditions are not as significant as those of an OWT under a harmonic excitation.

(5). The motion reduction effects of various TLDs are about the same when an OWT is under DLC 1.2 load conditions, whereas the reduction effect of a 3-TLD is the best among others when the OWT is under DLC 6.2 loading conditions. The corresponding comparison of an OWT under real earthquakes also indicates that 3-TLD has the best motion reduction effects.

(6). The fatigue analysis shows that the fatigue life of an OWT may increase 37% when a 3-TLD was installed. The multiple TLDs are recommended and may be applied on the structural motion control of OWTs.

(7). As a result of the time-consuming nature of the calculation, the environmental conditions of this study only picked DLC 1.2 and DLC 6.2 for simulations. In IEC 61400-3, there are many different DLCs, of which DLC 6.2 also has many different wind and wave angles. More simulations can be made, and the effectiveness of the proposed direction-free TLD application may be further confirmed.

Author Contributions: Conceptualization, B.-F.C.; methodology, B.-F.C. and S.-H.C.; software, S.-H.C.; validation, S.-H.C. and P.-H.Y.; writing—original draft preparation, S.-H.C. and B.-F.C.; writing—review and editing, B.-F.C. and P.-H.Y.; visualization, P.-H.Y.; supervision, B.-F.C. and P.-H.Y.; project administration, B.-F.C.; funding acquisition, B.-F.C. All authors have read and agreed to the published version of the manuscript.

Funding: This study is partially supported by a grant of Ministry of Science and Technology of ROC under a grant-number MOST-102-2221-E-110-039-MY3.

Conflicts of Interest: The authors declare no conflict of interest.

References

1. Murtagh, P.J.; Ghosh, A.; Basu, B.; Broderick, B.M. Passive control of wind turbine vibrations including blade/tower interaction and rotationally sampled turbulence. *Wind Energy Int. J. Prog. Appl. Wind Power Convers. Technol.* **2008**, *11*, 305–317. [CrossRef]
2. Sun, C.; Jahangiri, V. Bi-directional vibration control of offshore wind turbines using a 3D pendulum tuned mass damper. *Mech. Syst. Signal Process.* **2018**, *105*, 338–360. [CrossRef]
3. Hemmati, A.; Oterkus, E. Semi-active structural control of offshore wind turbines considering damage development. *J. Mar. Sci. Eng.* **2018**, *6*, 102. [CrossRef]
4. Modi, V.J.; Welt, F. Damping of wind induced oscillations through liquid sloshing. *J. Wind Eng. Ind. Aerodyn.* **1988**, *30*, 85–94. [CrossRef]
5. Fujii, K.; Tamura, Y.; Sato, T.; Wakahara, T. Wind-induced vibration of tower and practical applications of tuned sloshing damper. *J. Wind Eng. Ind. Aerodyn.* **1990**, *33*, 263–272. [CrossRef]
6. Kareem, A. Reduction of wind induced motion utilizing a tuned sloshing damper. *J. Wind Eng. Ind. Aerodyn.* **1990**, *36*, 725–737. [CrossRef]
7. Sun, L.M.; Fujino, Y.; Pacheco, B.M.; Chaiseri, P. Modelling of tuned liquid damper (TLD). *J. Wind Eng. Ind. Aerodyn.* **1992**, *43*, 1883–1894. [CrossRef]
8. Wakahara, T.; Ohyama, T.; Fujii, K. Suppression of wind-induced vibration of a tall building using tuned liquid damper. *J. Wind Eng. Ind. Aerodyn.* **1992**, *43*, 1895–1906. [CrossRef]
9. Tamura, Y.; Fujii, K.; Ohtsuki, T.; Wakahara, T.; Kohsaka, R. Effectiveness of tuned liquid dampers under wind excitation. *Eng. Struct.* **1995**, *17*, 609–621. [CrossRef]

10. Sakai, F. Tuned liquid column damper-new type device for suppression of building vibration. In Proceedings of the 1st International Conference on High-Rise Buildings, Najin, China, 1989; pp. 926–931. Available online: https://ci.nii.ac.jp/naid/10007252264 (accessed on 15 March 2020).
11. Colwell, S.; Basu, B. Tuned liquid column dampers in offshore wind turbines for structural control. *Eng. Struct.* **2009**, *31*, 358–368. [CrossRef]
12. Jin, Q.; Li, X.; Sun, N.; Zhou, J.; Guan, J. Experimental and numerical study on tuned liquid dampers for controlling earthquake response of jacket offshore platform. *Mar. Struct.* **2007**, *20*, 238–254. [CrossRef]
13. Chen, B.F.; Yang, B.H. Experimental study of a hybrid TMD and TLD on structure motion reduction. *Ocean Eng.* **2018**, *165*, 538–549. [CrossRef]
14. Veritas, D.N. *Fatigue Design of Offshore Steel Structures*; Det Nor Ske Veritas: Oslo, Norway, 2010.
15. Jonkman, J.; Butterfield, S.; Musial, W.; Scott, G. *Definition of a 5-MW Reference Wind Turbine for Offshore System Development*; National Renewable Energy Lab. (NREL): Golden, CO, USA, 2009.
16. Jonkman, J.; Musial, W. *Offshore Code Comparison Collaboration (OC3) for IEA Wind Task 23 Offshore Wind Technology and Deployment*; National Renewable Energy Lab. (NREL): Golden, CO, USA, 2010.
17. International Electrotechnical Commission. *Wind Turbines—Part 3: Design Requirements for Offshore Wind Turbines*; International Electrotechnical Commission: Geneva, Switzerland, 2009.
18. Standard, O. *Design of Offshore Wind Turbine Structures*; Det Nor Ske Veritas: Oslo, Norway, 2014.
19. International Electrotechnical Commission. *Wind Turbines—Part 1: Design Requirements*; International Electrotechnical Commission: Geneva, Switzerland, 2005.
20. Krabbenhoft, J. Shallow Water Tuned Liquid Dampers. Ph.D. Thesis, Department of Civil Engineering, Technical University of Denmark, Copenhagen, Denmark, 2011.
21. Chen, B.F.; Huang, S.M. A numerical study of liquid tank and structure interaction. *J. Mar. Sci. Technol.* **2015**, *23*, 781–791.
22. Frandsen, J.B. Sloshing motions in excited tanks. *J. Comput. Phys.* **2004**, *196*, 53–87. [CrossRef]
23. Ju, S.H.; Feng-Chien, S.; Yi-Pei, K.; Min-Hsuan, X. Fatigue design of offshore wind turbine jacket-type structures using a parallel scheme. *Renew. Energy* **2018**. [CrossRef]

Journal of
Marine Science and Engineering

MDPI

Article

Propagation of Solitary Waves over a Submerged Slotted Barrier

Yun-Ta Wu and Shih-Chun Hsiao *

Department of Hydraulic and Ocean Engineering, National Cheng Kung University, Tainan 701, Taiwan; ytwu@mail.ncku.edu.tw
* Correspondence: schsiao@mail.ncku.edu.tw; Tel.: +886-6-275-7575 (ext. 63262)

Received: 24 May 2020; Accepted: 6 June 2020; Published: 9 June 2020

Abstract: In this article, the interaction of solitary waves and a submerged slotted barrier is investigated in which the slotted barrier consists of three impermeable elements and its porosity can be determined by the distance between the two neighboring elements. A new experiment is conducted to measure free surface elevation, velocity, and turbulent kinetic energy. Numerical simulation is performed using a two-dimensional model based on the Reynolds-Averaged Navier-Stokes equations and the non-linear k-ε turbulence model. A detailed flow pattern is illustrated by a flow visualization technique. A laboratory observation indicates that flow separations occur at each element of the slotted barrier and the vortex shedding process is then triggered due to the complicated interaction of those induced vortices that further create a complex flow pattern. During the vortex shedding process, seeding particles that are initially accumulated near the seafloor are suspended by an upward jet formed by vortices interacting. Model-data comparisons are carried out to examine the accuracy of the model. Overall model-data comparisons are in satisfactory agreement, but modeled results sometimes fail to predict the positions of the induced vortices. Since the measured data is unique in terms of velocity and turbulence, the dataset can be used for further improvement of numerical modeling.

Keywords: solitary wave; submerged breakwater; slotted barrier; experiment; PIV; RANS model

1. Introduction

Coastal structures are typically employed to reduce wave energy so as to mitigate coastal hazards for protecting the local residence [1] and coastal species [2]. On designing the structure, not only providing strong protection for the shore but also involving environmentally-friendly consideration should be balanced. In recent years, submerged-type structures have been extensively considered as alternative choices [3,4] to enhance water exchange and retain natural coastal landscape for a recreational purpose. On the other hand, coastal structures may have permeable parts, which leads to attenuate additional wave energy through viscous dissipation within the porous media [5,6]. A typical permeable structure mostly consisted of rubble-mounted elements. However, permeable objects can be built using several impermeable parts with slots to vary the porosity, which is known as screen-type barriers [7,8]. The classic type of barrier feature is thin, rigid, vertical, perforated, and surface-piercing, which is beneficial to account for economic and environmental concerns.

As reviewed in Huang et al. [9], most available studies in the literature have focused on evaluating the hydraulic performance in terms of wave reflection (R), transmission (T), and dissipation (D) coefficients, where surface-piercing-type barriers received more attention than those of submerged-type ones. Wu and Hsiao [7] numerically investigated solitary waves over a submerged dual-slotted-barrier system using a well-validated wave model based on the Reynolds-Averaged Navier-Stokes equations (RANS) by providing a simple empirical formula for estimating RTD coefficients, where wave conditions and porosities of each barrier are considered as the primary parameters for the estimations. However, the flow fields of wave interactions with slotted barriers were studied sparsely. Although several numerical studies have provided simulated flow fields around slotted barriers [8,10], flow separation is one of the complicated phenomena in fluid mechanics and may not be able to be resolved accurately using numerical models unless the model has been rigorously validated through detailed model-data comparisons [11]. Using the particle image velocimetry (PIV), Liu and Al-Banaa [12] studied non-breaking solitary waves runup on a vertical surface-piercing barrier, and Wu et al. [13] investigated breaking solitary waves over a submerged bottom-mounted barrier. However, the flow fields due to the interaction of solitary waves and a submerged slotted barrier were not understood.

In practical applications, the elements of slotted barrier can be installed either horizontally or vertically, and, thus, the problem to be solved results in two-dimensional and three-dimensional setup for horizontal and vertical slotted barriers, respectively. Thomson [14] stated that the orientations of slotted barriers had an influence on transmitted waves based on experimental observation, where the horizontal slotted barrier appeared to be more effective in reducing wave transmission. In addition, choosing the shape of slotted barriers is one of the factors affecting the hydraulic performance. Krishnakumar et al. [15] stated that the slotted barrier with sharp edge elements such as square, rectangle, and triangle result in lower wave transmission but higher wave reflection than those consisting of circular shape elements. Additionally, Huang et al. [9] indicated that the rectangular element of perforated barriers may help generate more energy dissipation due to flow separation around the sharp edge elements of slotted barriers. Therefore, based on statements mentioned in available literature, the horizontal slotted barrier with rectangular elements may be the optimized setup as effective coastal structures, which can be considered a two-dimensional (2D) problem.

In this study, the primary aim is to investigate and understand the flow fields of solitary waves interacting with a submerged slotted barrier experimentally and numerically. A new experiment is performed in a laboratory-scale wave flume to measure the free surface displacement time series, the ensemble-averaged flow velocities, and the turbulent kinetic energy. Numerical simulation is carried out based on the RANS equations for the mean flow fields and the non-linear k-ε turbulence closure model to approximate the Reynolds stresses [16,17]. Detailed flow fields are addressed based on laboratory observations. Model-data comparisons in terms of the free surface elevation time series, the mean velocities, and the turbulent kinetic energy are performed to examine the accuracy of the numerical model and point out the limitation of numerical simulations.

2. Research Methods

2.1. Experiment

An experiment was conducted in a 2D glass-walled and glass-bottomed wave flume, which allowed the use of optical-based and image-based measuring systems. The flume dimensions are 22.0 m long, 0.50 m wide, and 0.76 m deep, located at Tainan Hydraulics Laboratory, National Cheng Kung University, Taiwan. A computer-controlled piston-type wavemaker was installed at one end of the flume and a sloping beach with a layer of concrete units was constructed at the other end to dissipate the transmitted wave energy. The slotted barrier was designed with an overall dimension of 10 cm high and 2 cm thick, which is identical to the study of a solid barrier under solitary waves [13], and was consisted of three identical square elements with a dimension of 2 cm × 2 cm made by transparent acrylic units. Therefore, the resulting porosity of the slotted barrier is 0.40. The slotted barrier was suspended in the wave flume with 2 cm freely from the top of the structure to the free surface and from the lower end of the barrier to the seafloor. The slotted barrier was installed around the middle of the flume at the constant water depth region, where the water depth h is 14 cm. Two different wave heights H were considered in the experiment for which H/h = 0.29 and 0.18.

Figure 1 shows the experimental layout, apparatus, definitions of variables used in this study and flowchart for instrument synchronization. Four capacitance-type wave gauges were employed to measure the free surface elevation time series in which two of them were positioned in front of the slotted barrier for recording incident and reflected waves, and the other two are instrumented behind the barrier for measuring transmitted waves. All wave gauges were synchronized with the wavemaker for 60-s recording with a sampling rate of 100 Hz. The first wave gauge (WG1) was used to define the wave height and reference the time origin as the wave crest passing over this wave gauge. The origin of the coordinate system $(x, z) = (0, 0)$ was defined at the intersection of the seafloor and the leading edge of the submerged slotted barrier.

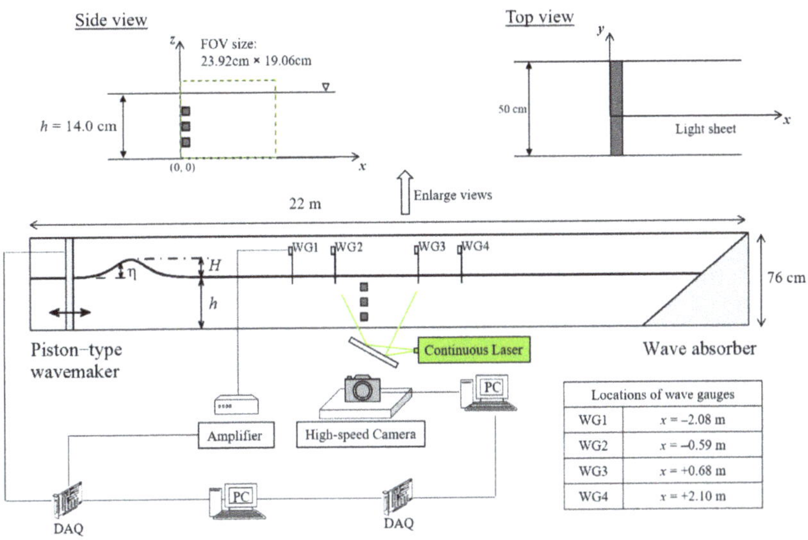

Figure 1. Overview of experimental set-up, facilities (not to scale), and definitions of the variable assigned for the submerged slotted barrier.

Flow velocities were measured by a time-resolved PIV system, which consisted of a high-speed camera and a continuous laser [18]. Instantaneous particle images were captured by an 8-bit digital CMOS camera (MS55K2, Canadian Photonic Labs Inc) with a resolution of 1280 × 1020 pixels and a maximum 1,000 framing rate per second (fps). Images were captured by using an in-house software of the camera, which was also used in Reference [19], and the camera was triggered by an external signal from the wavemaker. A single field of view (FOV) that covered an area of 239.2 mm × 190.6 mm in the vicinity of the slotted barrier (see Figure 1 for relative location) was used. A 2w continuous laser was employed as a light source to illuminate the measuring region. The frame rate of the PIV system was set to 200 fps and, thus, the temporal resolution of the velocity field was 199 fps while a 50-fps recording was set for flow visualization, which provides the path line of the flow fields to help identify the induced vortices qualitatively. Since the same PIV system with a similar setup has been used, the estimation of uncertainty for velocity determination can be referred to Reference [20]. Raw PIV images were processed using a multi-pass algorithm [21]. The analyses were starting from 128 × 128 pixels and ending with 32 × 32 pixels with a 50% overlap. Spurious velocity vectors were removed from the cross-correlated velocity fields using a dynamic mean value filter and a local median filter (3 × 3 vectors) [22]. No attempt was made to smooth or interpolate the cross-correlated velocity fields. The experiments were repeated under identical initial and boundary conditions up to 20 and 10 times for the cases of $H/h = 0.29$ and 0.18, respectively. The Reynolds-decomposition method was used to obtain the ensemble-averaged free surface elevation and the flow velocity. Due to the limited repeated runs, the turbulent kinetic energy could only be estimated for the case of $H/h = 0.29$.

2.2. Numerical Model

In recent years, the RANS-type model has been extensively used in coastal and ocean engineering applications, such as tsunami runup [23,24], wave-current interactions [25], and waves interacting with structures [26,27]. In this study, a 2D depth-resolving and phase-resolving viscous numerical wave model is used to simulate the interaction of a solitary wave and a submerged slotted barrier. The physics behind the model is based on the RANS equations to describe the mean flow fields and the non-linear k-ε closure model to approximate the Reynold stresses by means of the turbulent kinetic energy (k) and the turbulent dissipation rate (ε). The model solves the RANS equations by using the finite-difference two-step projection method [28]. The free surface displacement during wave-structure interactions is traced by the volume of fluid method [29]. The no-slip condition is implemented at the solid boundaries and the zero-stress condition is applied to the mean free surface for neglecting the air-flow effect. The surface tension is not considered in this study. Solitary waves, where the wave conditions are identical to those of experiments, are generated through the inflow boundary by giving the theoretical solutions [30] in terms of free surface displacement along with the corresponding horizontal and vertical velocities. The radiation boundary condition is utilized for allowing the wave outgoing the computational domain to eliminate significant reflection. More detailed information about the numerical implementation can refer to References [16,17]. The model used in this study has been rigorously validated against experimental measurements in terms of velocity and turbulence for coastal-related problems such as solitary wave interactions with a submerged impermeable breakwater [31], submerged permeable structure [5], surface-piercing barrier [12], and submerged bottom-mounted barrier [13]. However, the flow-field accuracy for solitary wave interactions with a slotted barrier, especially focusing on flow separations, has not been investigated yet.

The numerical setup used herein can be very similar to the studies of solitary waves interacting with a bottom-mounted barrier [13] and dual-solid-barrier [7]. However, the wave-induced flow fields are much more complicated due to flow separation that occurs by each element of the slotted barrier. Furthermore, those induced vortices are expected to interact with each other to further create a complex flow pattern like the vortex shedding process [32]. As a result, such complicated phenomena are very difficult to accurately simulate. The flow separation is closely linked and inseparable from the development of the boundary layer flow, especially near the surfaces of the bottom boundary and the object elements. Typically, to avoid tremendous computational efforts, the RANS model always employed a log-law model to simplify the boundary layer flow using a logarithmic velocity profile because the tiny thickness of the boundary layer may not be able to be directly resolved, especially for the wave condition with a high Reynolds number. However, such simplification may lead to inaccurate results in the vicinity of flow separation regions under wave actions [31]. The laminar boundary layer characteristics under solitary waves can be estimated from the formula derived by Reference [33] and later improved by Reference [34]. The estimated boundary layer thickness (BLT) is expressed as:

$$\begin{cases} \mathrm{BLT} = 2\sqrt{\nu/(KC)} \\ K = \sqrt{\frac{3H}{4h^3}} \\ C = \sqrt{g(H+h)} \end{cases} \quad (1)$$

in which C is the wave celerity of a solitary wave, KC is to estimate the duration of a solitary wave, and υ is the kinematic viscosity of the fluid. According to Equation (1) with h = 14 cm, the estimated BLT for the case of H/h = 0.29 is around 1.0 mm whereas, for the case of H/h = 0.18, the estimated BLT is around 1.1 mm. The higher the H/h, the thinner the estimated BLT. Following Reference [34] for the treatment of solid boundaries, the region within $0 \leq z/\mathrm{BLT} \leq 5$ is resolved by 20 grids whereas the region within $5 \leq z/\mathrm{BLT} \leq 10$ is resolved by 10 grids with uniform distributions of numerical meshes. By doing this, for the case of H/h = 0.29, the minimum resolution is set to 0.25 mm. After performing sensitivity analyses on the use of different grid resolutions, it is found that the modeled results for the region within $0 \leq z/\mathrm{BLT} \leq 10$ resolved by 40 grids and more grids are almost the same, so that the results presented in this study employ 40 numerical grids to resolve 10 times of the estimated BLT for all solid boundaries, including the seafloor and the surfaces in the vicinity of each element of the slotted barrier. The computational domain is designed as $-2.5\ \text{m} \leq x \leq 2.4\ \text{m}$ and $0.0\ \text{m} \leq z \leq 0.2\ \text{m}$, where the origin is defined at the weather side of the slotted barrier like the definition used in the experiments.

3. Results and Discussion

3.1. Free Surface Elevation

To verify the initial force of solitary waves, comparisons between experimental measurements and numerical simulations are performed for time histories of the free surface elevations at selected locations. Figure 2 shows the model-data comparisons for four wave gauges (see Figure 1 for relative locations) where the left and right columns demonstrate the results obtained from the cases of $H/h = 0.29$ and 0.18, respectively. Since the numerical wave tank is shorter than that of the physical experiment, numerical wave heights are decided by matching the measured wave heights at WG1, where the time origin is also referenced. For experiments, all instantaneous measurement and mean free surface elevations are plotted in the same figure for each position of wave gauges. All measured data almost overlap with each other and show the high repeatability of experiments. Quantitatively, the standard deviation for incident wave heights obtained from all 20 trials for the case of $H/h = 0.29$ is around 0.2 mm while the standard deviation for the case of $H/h = 0.18$ obtained from 10 repetitions is less than 0.1 mm. This, once again, indicates the high repeatability of the present experiments. In addition, model-data comparisons reveal that numerical calculations for both wave conditions fit the measurements well for the main waveforms in terms of incident and reflected waves recorded by WG1 and WG2 and transmitted waves recorded by WG3 and WG4. Moreover, the theoretical solutions of the solitary wave [30] are also plotted at WG1 of Figure 1 for both cases. It is evident that comparisons show satisfactory agreements between theoretical, measured, and model waveforms of solitary waves. Furthermore, the wave transmission coefficients (C_T) can be simply calculated by the wave height ratio, which is obtained from WG4 and WG1. The estimated C_T was around 0.86 and 0.83 for the cases of $H/h = 0.29$ and 0.18, respectively.

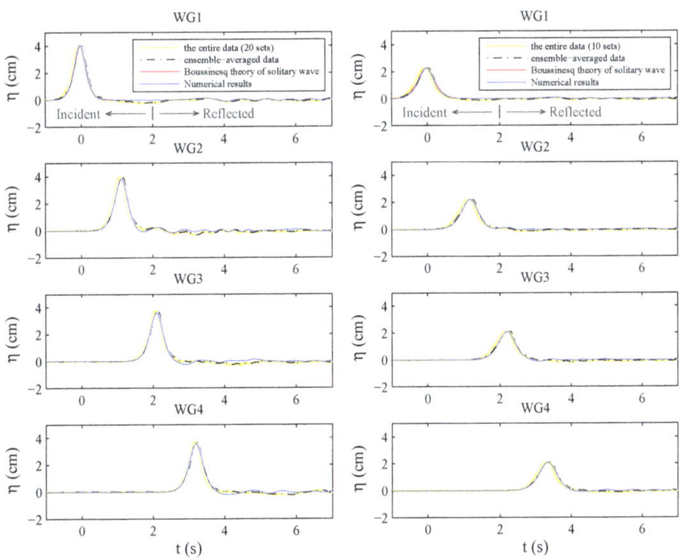

Figure 2. Model-data comparison in terms of free surface elevation time series at four different locations, i.e., WG1 to WG4, for the cases of $H/h = 0.29$ (**left column**) and $H/h = 0.18$ (**right column**), respectively.

3.2. Flow Visualization

Resolving the generation and evolution of vortices at the initial stage of flow separation relies on a very high-resolution PIV system, as the scale of those vortices are too small to be fully resolved and the velocities there are also small compared to the velocities due to pure wave actions. Some previous studies [35,36] have raised the same difficulty, so that further efforts on quantitating those tiny vortices are necessary. To provide a better understanding of the flow separation induced flow fields, a flow visualization technique was used to qualitatively define the flow characteristics. The particle tracking technique was employed for flow visualization by carefully adjusting the camera exposure to generate path lines of the induced flow pattern. The size of FOV for using the particle tracking technique is identical to that used in the PIV measurement. However, the temporal resolution is reduced to 50 fps. Although the induced flow fields for those of two wave conditions reveal different strengths and movements of induced vortices, the overall phenomena in terms of the flow pattern are nearly the same. Therefore, only the flow visualization for the case of $H/h = 0.29$ is presented herein and their differences of varying wave conditions are provided by means of a model-data comparison for velocity fields.

Flow visualization images were re-sized in order to better demonstrate those tiny vortices due to flow separation. Figure 3 shows a close view for the initial stage of flow separation while Figures 4 and 5 show the stage of the vortices' interaction. Based on laboratory observations, the phenomena of flow separation occurred at both sides of each element and, for all three elements of the slotted barrier, the vortices in the lower end of the element are induced first and the vortices in the upper side of the element are induced later. This may be partly due to the slight pressure difference between the upper and lower parts of the element. The lower part has relatively large hydrostatic pressure whereas the upper side suffers a relatively low pressure. The induced vortices are labelled in Figures 3–5 and the rule for labelling vortices is followed. The upper, middle, and lower elements are, respectively, named as A, B, and C, and the sequence of induced vortices are numbered. For example, the first vortex induced by the first element is labelled as A1.

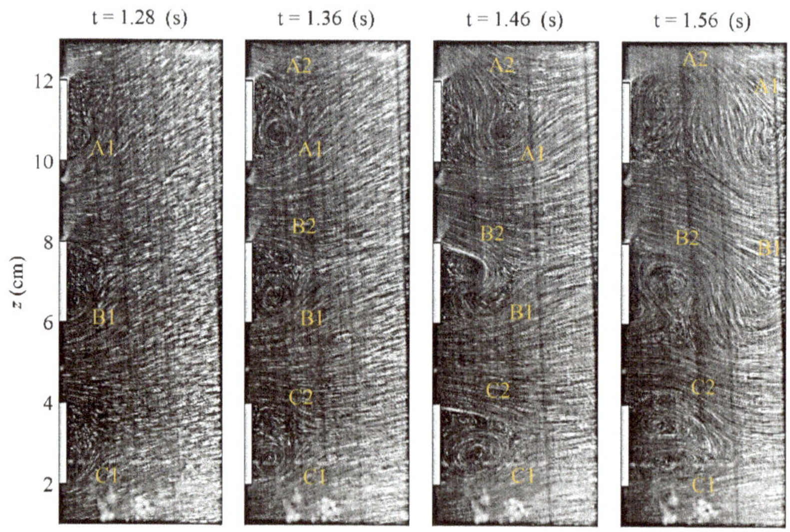

Figure 3. Flow visualization at the initial stage of flow separation at different time instants from t = 1.28–1.56 s for the case of $H/h = 0.29$.

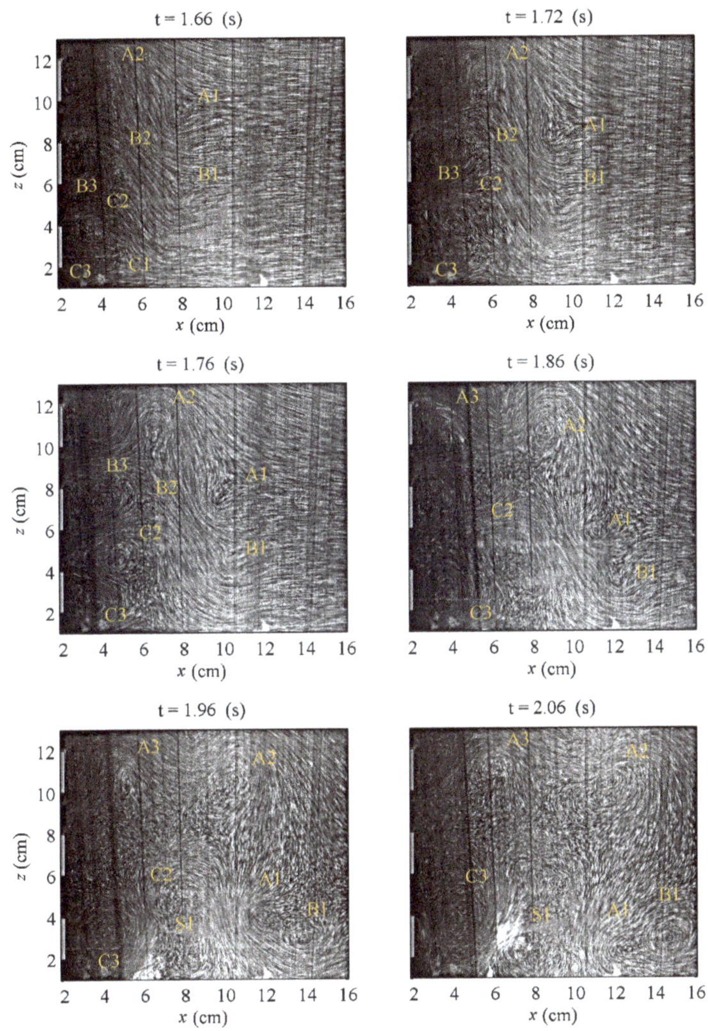

Figure 4. Flow visualization at the stage of a vortices' interaction at different time instants from t = 1.66–2.06 s for the case of $H/h = 0.29$.

As shown in Figure 3 at t = 1.28 s, distinct vortices are visible at the lower end of the elements, i.e., A1, B1, and C1. At t = 1.36 s, the size and its strength of the lower vortices gradually increase and then the vortices induced by the upper sides of the elements, i.e., A2, B2, and C2, are generated slightly later, where their sizes are smaller and strengths are weaker than those of A1, B1, and C1. At t = 1.46 s, the vortices A1 and B1 convect downstream and its pathline is cut out by the vortices induced by the upper end of the elements, i.e., A2 and B2, and new vortices are then generated from the edges of the elements due to flow separation. Such phenomena of vortices' interaction are like the vortex shedding process of uniform flow passing through an obstacle [32]. Moreover, it is found that the time instant of shedding out of the vortices induced by the lower end of the elements is different. As can be seen at t = 1.56 s, the vortex A1 sheds out of the vortex street first. This is followed by B1. The vortices induced by the lowest element form a pair of almost symmetric vortices, i.e., C1 and C2, with different rotating directions.

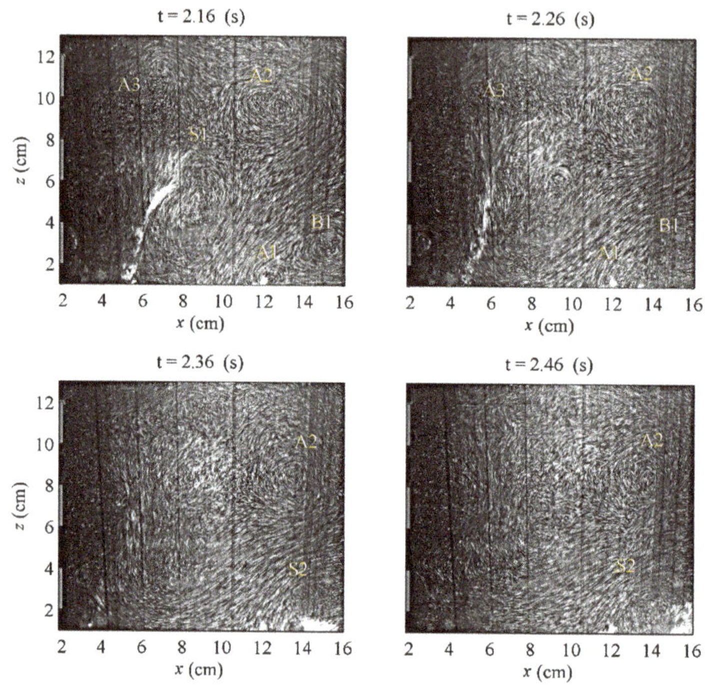

Figure 5. Flow visualization at the stage of vortices' interaction at different time instants from t = 2.16–2.46 s for the case of $H/h = 0.29$.

In Figure 4, before t = 1.56 s, the flow is dominated by the acceleration part of the solitary wave, i.e., before the arrival of the wave crest. After t = 1.56 s, the deceleration part of the solitary wave then passes over the slotted barrier. As t = 1.66 s to 1.72 s, the vortices A1 and B1 move downward due to the decrease of the free surface. Then, until t = 2.06 s, those two vortices move downward to the bottom and eventually merge together to form a distinct counterclockwise vortex due to the same rotating direction, i.e., t ≥ 2.36 s in Figure 5. In between t = 1.66 s to 1.86 s, the vortices A2, B2, and C2 are then shed out of the vortex street to further induce vortices. Vortex A2 moves upward to reach the free surface. Vortices B2 and C2 also slightly move upward and then eventually diffuse due to the complicated vortices' interaction. In addition, a distinct vortex A3 is later induced by the first element with a clockwise rotation, i.e., at t = 1.86 s, and a counterclockwise-rotating vortex C3 occurs by the lowest element. At t = 1.96 s, a clockwise-rotating vortex is generated due to the viscous effect of the bottom boundary, i.e., S1. Then, a strong upward vertical velocity is visible at t = 2.06 s due to the opposite rotating directions between C3 and S1, which indicates possible sediment suspension of the seafloor. In Figure 5, a considerable amount of seeding particles originally accumulated near the bottom is suspended upward due to the strong vertical velocity gradient at t = 2.16–2.26 s. Furthermore, the vortex S2 merged from A1 and B1 may also lead to sediment suspension at t = 2.36–2.46 s. In practical engineering using submerged breakwaters, the scour of seafloor is mostly found near the toe of the breakwater. However, in this study, possible scour by means of seeding particle suspension is found in the shoreward direction. It will be interesting and of great importance as an extended work to conduct a mobile seabed experiment for the same obstacle setup.

3.3. Velocity Fields

Given that free surface deformations are clearly illuminated by a laser and imaged by the PIV camera, the free surface displacements can be detected and used for model-data comparisons in spatial variations. Figures 6–10 present detailed model-data comparisons in terms of spatial distribution of free surface elevation (top column), velocity fields (middle column), and corresponding velocity profiles in both horizontal and vertical components (lower column) for the cases of $H/h = 0.29$ (Figures 6–9) and $H/h = 0.18$ (Figure 10). For velocity profiles, seven cross-sections are uniformly selected from $x = 0.06$ m to $x = 0.18$ m with an identical interval of 0.02 m. It is remarked in these cases that the selected locations of velocity profiles may not be the same between experiments and numerical modeling because their resolutions in space are essentially different. As such, the closest locations between measurements and simulations are selected for comparisons, and no interpolation of the results is attempted. Since the laser was illuminated from the bottom of the flume, those near each element of the slotted barrier and free surface may not be well-illuminated, so that those data have been removed in order to avoid providing inaccurate velocity information.

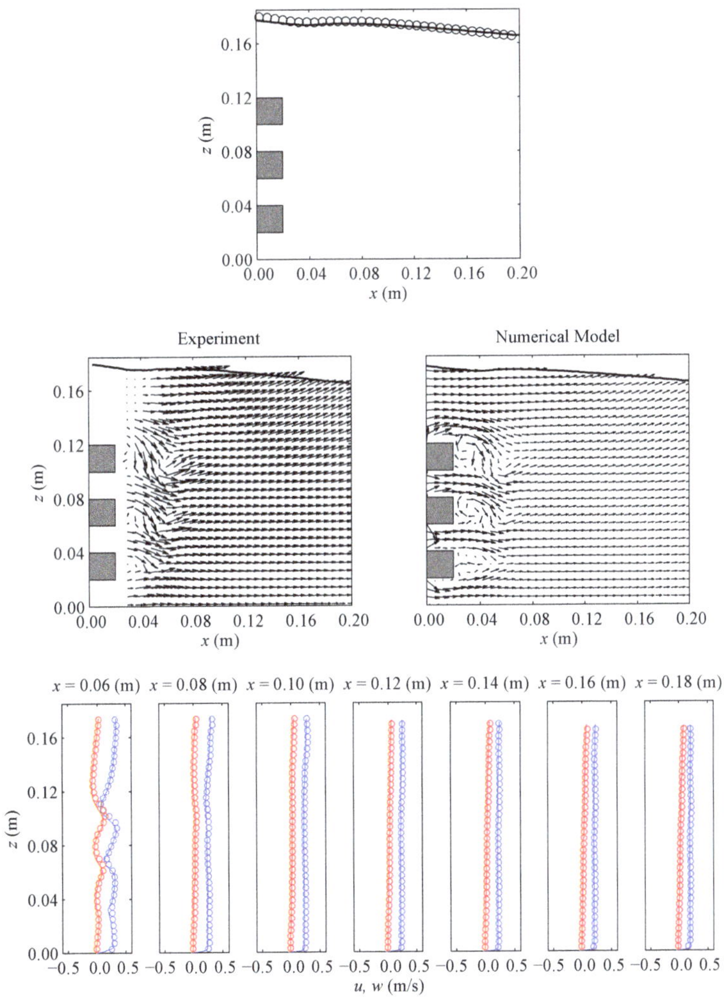

Figure 6. Comparison between measured (o) and modeled (–) results for the case of $H/h = 0.29$ at t = 1.56 s in terms of free surface elevation (**top panel**), velocity fields (**middle panel**), and corresponding velocity profiles (**lower panel**). For the lower panel, blue and red indicate horizontal and vertical velocities, respectively.

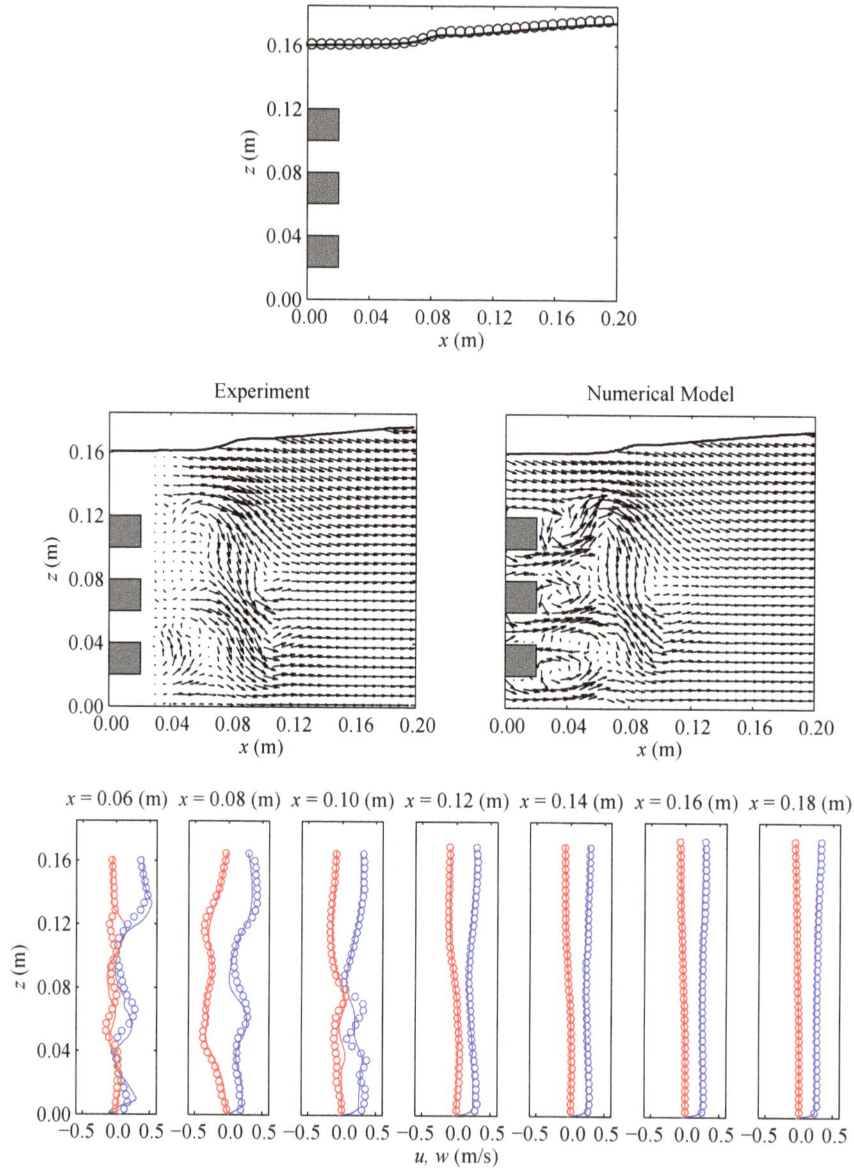

Figure 7. Comparison between measured (o) and modeled (–) results for the case of H/h = 0.29 at t = 1.76 s in terms of free surface elevation (**top panel**), velocity fields (**middle panel**), and corresponding velocity profiles (**lower panel**). For the lower panel, blue and red indicate horizontal and vertical velocities, respectively.

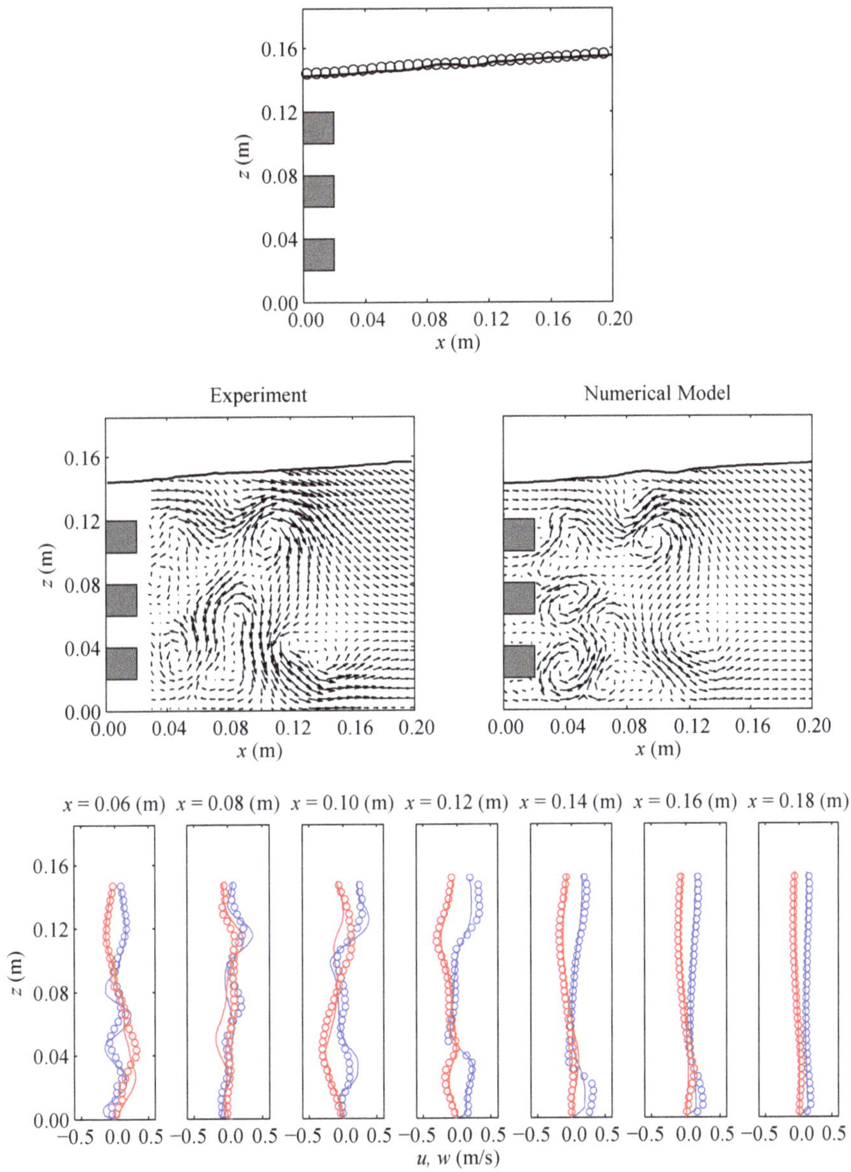

Figure 8. Comparison between measured (o) and modeled (–) results for the case of $H/h = 0.29$ at t = 1.96 s in terms of free surface elevation (**top panel**), velocity fields (**middle panel**), and corresponding velocity profiles (**lower panel**). For the lower panel, blue and red indicate horizontal and vertical velocities, respectively.

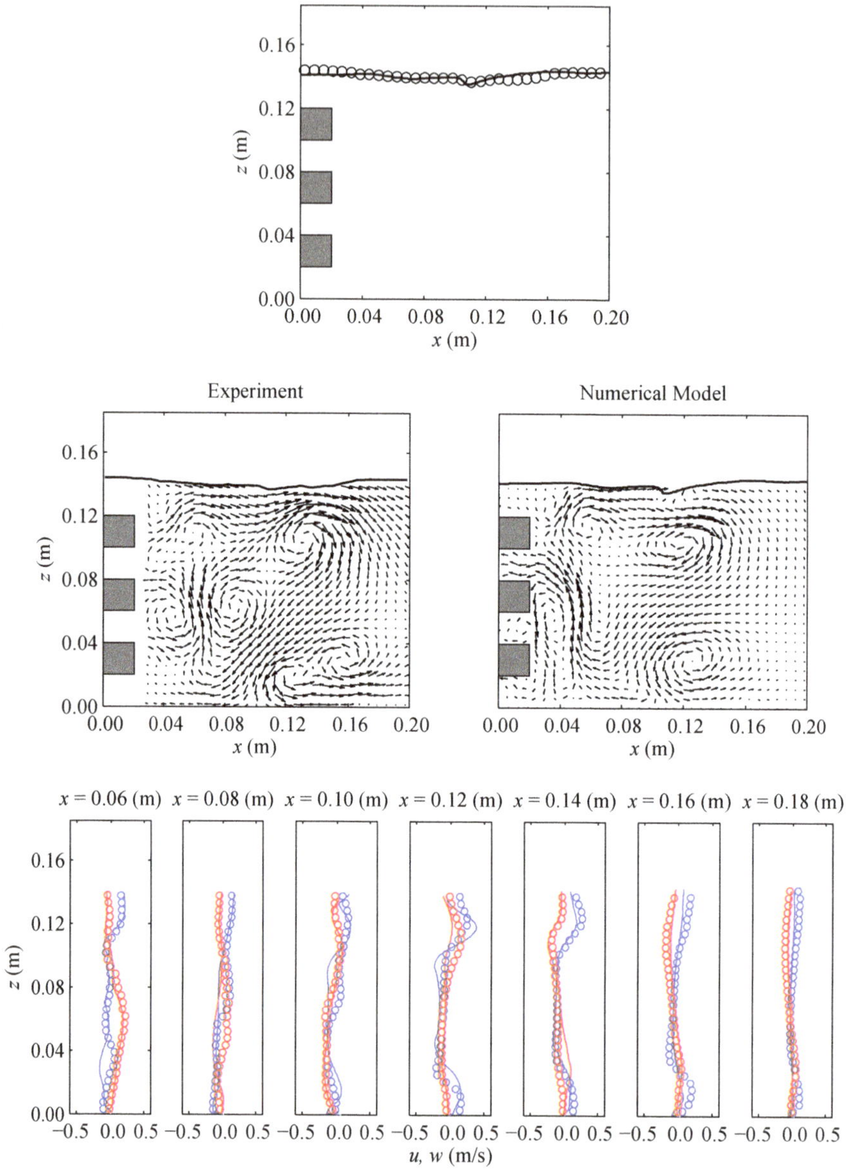

Figure 9. Comparison between measured (o) and modeled (–) results for the case of $H/h = 0.29$ at t = 2.16 s in terms of free surface elevation (**top panel**), velocity fields (**middle panel**), and corresponding velocity profiles (**lower panel**). For the lower panel, blue and red indicate horizontal and vertical velocities, respectively.

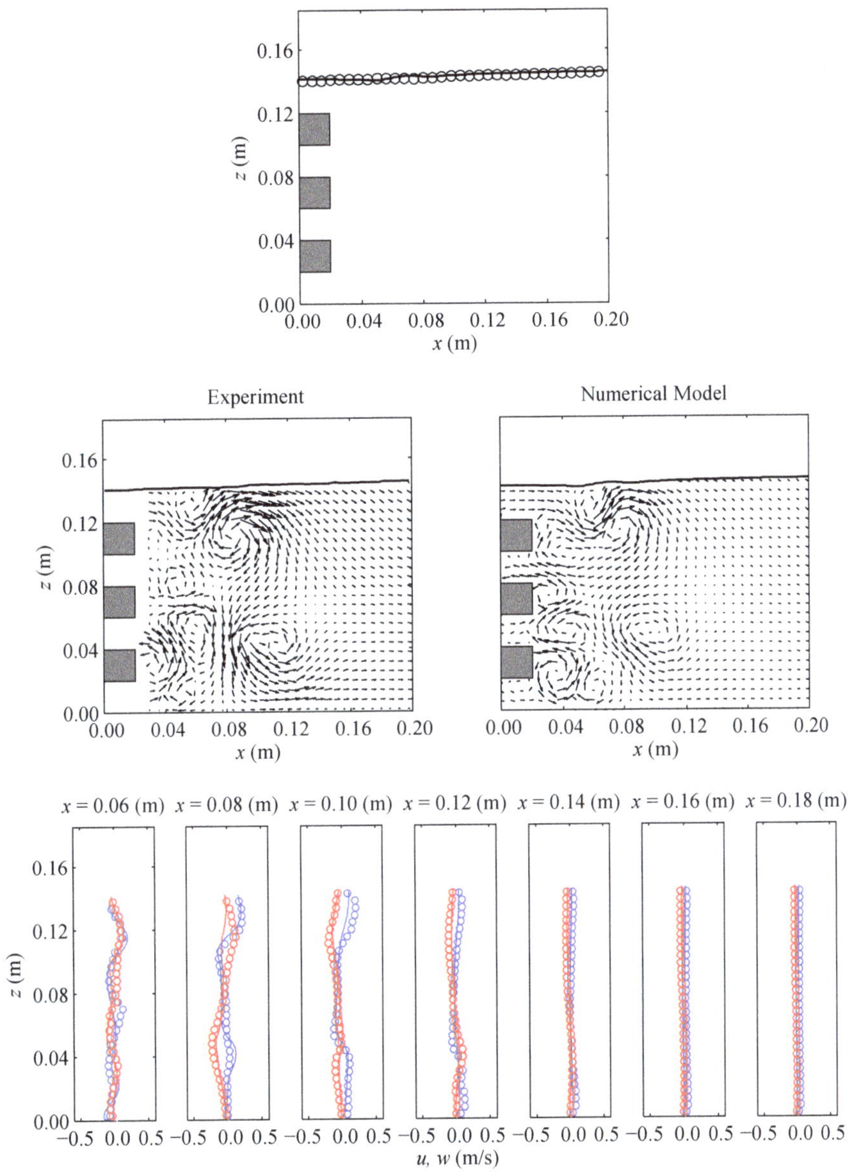

Figure 10. Comparison between measured (o) and modeled (–) results for the case of $H/h = 0.18$ at t = 2.32 s in terms of free surface elevation (**top panel**), velocity fields (**middle panel**), and corresponding velocity profiles (**lower panel**). For the lower panel, blue and red indicate horizontal and vertical velocities, respectively.

Figure 6 shows the time instant that the solitary wave crest is at the leading edge of the slotted barrier, belonging to the initial stage of flow separation. In Figure 7, the crest of the solitary wave has entirely passed over the slotted barrier and the velocity fields reveal the start of the vortex shedding process. The physical phase for the interaction of vortices is demonstrated in Figure 8. Distinct vortices A1, A2, A3, B1, C2, C3, and S1 are clearly visible and resolved by PIV measurements and RANS simulations. Figure 9 demonstrates the time instant of the strong upward velocity like a vertical jet shot from the bottom, which causes significant seeding particles suspended. These are initially accumulated near the seafloor. Model-data comparisons show that the free surface deformation during the interaction of a solitary wave and a submerged slotted barrier can be well-captured by the present RANS model, where the wave heights and phases at different time instants show satisfactory agreements. However, for the velocity fields, modeled results at some points fail to fit measured data. More specifically, the modeled results fit the measurements very well at the initial stage of flow separation, as shown in Figures 6 and 7. Significant discrepancies are observed during the vortex shedding process, which can be seen in Figures 8 and 9, and can be summarized into three main reasons. First, the positions of the induced vortices between the physical experiment and numerical model are not always the same, which causes significant variations after complicated vortices interact. Second, the vortices A1 and B1 at t = 2.16 s are not yet merged into a distinct vortex S2 in the measurement whereas the numerical results show those two vortices have been merged already at that time instant (see Figure 9). Third, it seems that the modeled flow fields are affected by the deceleration part of the solitary wave more than those of experiments, so that the entire flow fields are somewhat shifted to the opposite direction of the propagating wave (see Figures 8 and 9). Except those discrepancies, model-data comparisons are generally in satisfactory agreements. The main feature of the flow pattern can be simulated well by the model, including the initial stage of flow separation, an overall vortex shedding process, and an upward vertical velocity due to the interaction between vortices C3 and S1, which may further trigger sediment suspension locally.

Another model-data comparison is made for the case of H/h = 0.18. Since the overall flow pattern is mostly identical to the case of H/h = 0.29, only a few stages are slightly different from those of H/h = 0.29. As such, only one time instant for a model-data comparison is provided. Experimentally, one of the variations of varying wave conditions is that the time instant for merging vortices A1 and B1 into S1 is earlier than the case of H/h = 0.29. Figure 10 shows the time instant at t = 2.32 s, which is 0.6 s behind the wave crest arriving at the leading edge of the slotted barrier. The two vortices A1 and B1 have merged into a distinct vortex S1. In addition, the upward vertical jet is much closer to the barrier than the case of H/h = 0.29, which is around 0.04 m and 0.06 m for the cases of H/h = 0.18 and H/h = 0.29, respectively. In addition, modeled results again fit the measured data well for overall comparisons, but some detail flow fields cannot be simulated well, especially for the positions of the induced vortices and their sequential interactions. One of the reasons may be due to the use of a two-equation turbulence model, which may be oversimplified for representing the effects of turbulent flow fields. RANS equations with a two-equation turbulence closure model is useful for practical applications [37], as only mean flow fields are simulated. However, detailed flow fields due to complicated wave-structure interactions may not be accurately simulated. Various efforts have been made using different approaches prior to demonstrate the simulations presented herein, such as employing very fine mesh to resolve the flow characteristics near the solid boundaries instead of using the log-law approach. As one of the ongoing works, further attempts may be considered using different two-equation turbulence models, such as those used in Reference [37], and various concepts of turbulence representation, such as a LES (large-eddy-simulation) model [38].

3.4. Turbulent Kinetic Energy

According to References [39,40], meaningful turbulence characteristics can be estimated through ensemble averaging over 16 or more repetitions of the same experiments under identical initial and boundary conditions. As a result, only the case of H/h = 0.29 can be used to estimate the turbulent kinetic energy based on measured velocity information. Detailed information on how to obtain turbulence characteristics can be found in References [5,12,13,39,40]. Model-data comparison in terms of spatial distribution of turbulent kinetic energy (TKE) and its corresponding cross-sections is presented, where the profile locations of TKE are identical to those of velocity fields. Since TKE generation and evolution at the initial stage of flow separation are surrounded by each element of the slotted barrier also in order to reduce the number for figures, only the time instants showing significant TKE evolution, i.e., t = 1.96 s and 2.16 s, are demonstrated and the other physical phases are described concisely. In Figures 11 and 12, the TKE fields are superimposed with velocity maps to help identify the locations of induced vortices and TKE.

At the initial stage of flow separation, TKE is generated from the slots of two neighboring elements with a local maximum TKE around 0.005 m^2/s^2 for both measured and modeled results. At t = 1.76 s, the measured local maximum TKE induced by vortices A2 and B2 are around values of 0.006 m^2/s^2 and 0.005 m^2/s^2, respectively, whereas the modeled TKE induced by vortices A2 and B2 are around values of 0.005 m^2/s^2 and 0.002 m^2/s^2, respectively. As shown in Figure 11, the local maximum TKE is introduced by vortex A2 with a value of around 0.006 m^2/s^2 for measurement and around 0.005 m^2/s^2 for simulation. Another important phase of TKE is due to the upward vertical jet at t = 2.16 s, as shown in Figure 12, where the measured and modeled local maximum TKE are around 0.005 m^2/s^2 and 0.006 m^2/s^2, respectively. As mentioned in the previous section, the locations of induced vortices between measured and modeled results are not identical, which are also shown for the vortices induced by TKE, so that it is more appropriate to find a local maximum value of a specific vortex for a model-data comparison. Detailed comparisons are made for cross-section TKE. In general, the overall trend of TKE profiles fits the measurements with reasonable agreement.

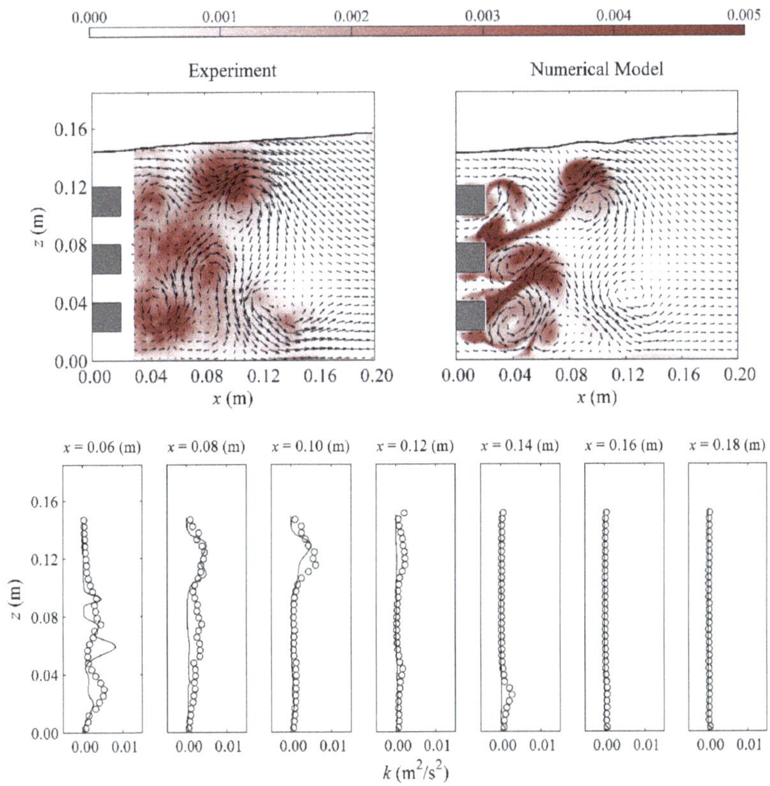

Figure 11. Comparison between measured (o) and modeled (–) results for the case of $H/h = 0.29$ at t = 1.96 s in terms of spatial distributions (**top panel**, unit: m^2/s^2) and corresponding vertical cross-sections (**lower panel**) of turbulent kinetic energy.

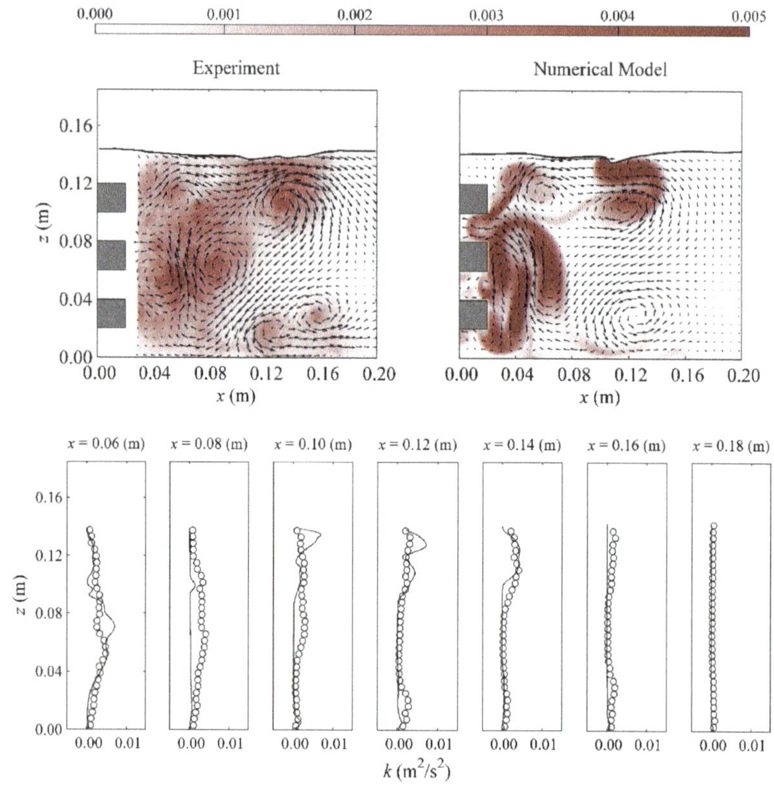

Figure 12. Comparison between measured (o) and modeled (–) results for the case of $H/h = 0.29$ at $t = 2.16$ s in terms of spatial distributions (**top panel**, unit: m^2/s^2) and corresponding vertical cross–sections (**lower panel**) of turbulent kinetic energy.

4. Conclusions

In this study, the interaction of solitary waves and a submerged slotted barrier was investigated experimentally and numerically. An experiment was conducted to measure free surface elevations using wave gauges and velocity fields using a time-resolved PIV. A 2D depth-resolving and phase-resolving wave model based on the 2D RANS equations and the non-linear k-ε turbulence closure model was employed to reproduce the experiments.

Due to the highly complex nature of flow fields from wave-structure interactions, flow visualization was also utilized to help identify the detailed flow characteristics of tiny vortices that may not be measured using present PIV. Numerically, very fine meshes were used to resolve the velocities near all solid boundaries instead of using a log-law approach. Model-data comparisons were performed in terms of free surface elevation, velocity, and turbulence characteristics. Overall, the comparisons were in satisfactory agreements. Expect few inaccurate predictions for the positions of induced vortices. Since most of the existing literature paid attention to evaluating the hydraulic performance of slotted barriers, to the best knowledge of the authors, no available study provides the detailed velocity and turbulence information of a slotted barrier under a water wave. This is the first dataset providing detailed measurements on solitary wave interactions with a submerged slotted barrier, which can be used as a benchmark case for further development of the numerical model and for model validation.

According to laboratory observations, the flow separation of each element of the slotted barrier and the vortex shedding process was triggered due to the interaction of those induced vortices to generate a complicated flow pattern. It was also found that, for this setup, an upward vertical velocity with a significant strength of velocity and turbulence was qualitatively observed and quantitatively measured, which showed a large number of suspended seeding particles. These were originally accumulated near the seafloor. This indicates that there may be potential scouring near the submerged slotted barrier and this may result in a local geometry change. Numerical simulation also confirmed those findings. As part of ongoing work, it will be interesting to conduct a mobile seabed experiment for a submerged slotted barrier under a solitary wave to investigate the possibility of scouring the seafloor. In addition, only one obstacle setup was considered in the experiment. The gap of each element and the designed water depth may also be critical factors affecting the formation of velocity fields as well as the vortex shedding process, which should be worthy of investigation.

Author Contributions: Y.-T.W. and S.-C.H. conceived the research topic. Y.-T.W. performed experiments, numerical simulations, and data analyses. Y.-T.W. and S.-C.H. wrote the paper. All authors have read and agreed to the published version of the manuscript.

Funding: The Ministry of Science and Technology, Taiwan (MOST 108-2218-E-006-053-MY3, 108-2221-E-006-087-MY3) and the Water Resources Agency of Ministry of Economic Affairs, Taiwan (MOEAWRA1090350) funded this research. Yun-Ta Wu appreciates the support of the Open Fund Research SKHL1813 of Sichuan University for a short-term visit in 2019.

Acknowledgments: The author sincerely appreciates the staffs of Tainan Hydraulics Laboratory of NCKU for their help in conducting the experiments.

Conflicts of Interest: The authors declare no conflict of interest.

References

1. Strusińska-Correia, A. Tsunami mitigation in Japan after the 2011 Tōhoku tsunami. *Int. J. Disaster Risk Reduct.* **2017**, *22*, 397–411. [CrossRef]
2. Ware, M.; Long, J.W.; Fuentes, M.M.P.B. Using wave runup modeling to inform coastal species management: An example application for sea turtle nest relocation. *Ocean Coast. Manag.* **2019**, *173*, 17–25. [CrossRef]
3. Lara, J.L.; Garcia, N.; Losada, I.J. RANS modelling applied to random wave interaction with submerged permeable structures. *Coast. Eng.* **2006**, *53*, 395–417. [CrossRef]
4. Kobayashi, N.; Meigs, L.; Ota, T.; Melby, J. Irregular breaking wave transmission over submerged porous breakwater. *J. Waterw. Port Coast. Eng.* **2007**, *133*, 104–116. [CrossRef]
5. Wu, Y.-T.; Hsiao, S.-C. Propagation of solitary waves over a submerged permeable breakwater. *Coast. Eng.* **2013**, *81*, 1–18. [CrossRef]
6. Pourteimouri, P.; Hejazi, K. Development of an integrated numerical model for simulating wave interaction with permeable submerged breakwaters using extended Navier–Stokes equations. *J. Mar. Sci. Eng.* **2020**, *8*, 87. [CrossRef]
7. Wu, Y.-T.; Hsiao, S.-C. Propagation of solitary waves over double submerged barriers. *Water* **2017**, *9*, 917. [CrossRef]
8. Yao, Y.; Tang, Z.; He, F.; Yuan, W. Numerical investigation of solitary wave interaction with double row of vertical slotted piles. *J. Eng. Mech.* **2018**, *144*, 04017147. [CrossRef]
9. Huang, Z.; Li, Y.; Liu, Y. Hydraulic performance and wave loadings of perforated/slotted coastal structures: A review. *Ocean Eng.* **2011**, *38*, 1031–1053. [CrossRef]
10. Meringolo, D.D.; Aristodemo, F.; Veltri, P. SPH numerical modeling of wave–perforated breakwater interaction. *Coast. Eng.* **2015**, *101*, 48–68. [CrossRef]
11. Wang, D.; Liu, P.L.F. An ISPH with k–ε closure for simulating turbulence under solitary waves. *Coast. Eng.* **2020**, *157*, 103657. [CrossRef]
12. Liu, P.L.-F.; Al-Banaa, K. Solitary wave runup and force on a vertical barrier. *J. Fluid Mech.* **2004**, *505*, 225–233. [CrossRef]
13. Wu, Y.-T.; Hsiao, S.-C.; Huang, Z.-C.; Hwang, K.-S. Propagation of solitary waves over a bottom-mounted barrier. *Coast. Eng.* **2012**, *62*, 31–47. [CrossRef]
14. Thomson, G.G. Wave Transmission through Multi-Layered Wave Screens. Master's Thesis, Queen's University, Kingston, ON, Canada, 2000.

15. Krishnakumar, C.; Sundar, V.; Sannasiraj, S.A. Hydrodynamic performance of single- and double-wave screens. *J. Waterw. Port Coast. Eng.* **2010**, *136*, 59–65. [CrossRef]
16. Lin, P.; Liu, P.L.-F. A numerical study of breaking waves in the surf zone. *J. Fluid Mech.* **1998**, *359*, 239–264. [CrossRef]
17. Lin, P.; Liu, P.L.-F. Turbulence transport, vorticity dynamics, and solute mixing under plunging breaking waves in surf zone. *J. Geophys. Res.* **1998**, *103*, 15677–15694. [CrossRef]
18. Lin, C.; Wong, W.-Y.; Kao, M.-J.; Tsai, C.-P.; Hwung, H.-H.; Wu, Y.-T.; Raikar, R.V. Evolution of velocity field and vortex structure during run-down of solitary wave over very steep beach. *Water* **2018**, *10*, 1713. [CrossRef]
19. Chen, Y.-Y.; Li, M.-S. Evolution of breaking waves on sloping beaches. *Coast. Eng.* **2015**, *95*, 51–65. [CrossRef]
20. Wu, Y.-T.; Hsiao, S.-C. Generation of stable and accurate solitary waves in a viscous numerical wave tank. *Ocean Eng.* **2018**, *167*, 102–113. [CrossRef]
21. Mori, N.; Chang, K.-A. Introduction to Mpiv. Available online: http://www.oceanwave.jp/softwares/mpiv/ (accessed on 1 February 2014).
22. Raffel, M.; Willert, C.E.; Kompenhans, J. *Particle Image Velocimetry*; Springer: Berlin/Heidelberg, Germany, 1998.
23. Choi, B.H.; Kim, D.C.; Pelinovsky, E.; Woo, S.B. Three-dimensional simulation of tsunami run-up around conical island. *Coast. Eng.* **2007**, *54*, 618–629. [CrossRef]
24. Kim, D.C.; Kim, K.O.; Pelinovsky, E.; Didenkulova, I.; Choi, B.H. Three-dimensional tsunami runup simulation for the port of Koborinai on the Sanriku coast of Japan. *J. Coast. Res.* **2013**, 266–271. [CrossRef]
25. Hsiao, Y.; Tsai, C.-L.; Chen, Y.-L.; Wu, H.-L.; Hsiao, S.-C. Simulation of wave-current interaction with a sinusoidal bottom using OpenFOAM. *Appl. Ocean Res.* **2020**, *94*, 101998. [CrossRef]
26. Pelinovsky, E.; Choi, B.H.; Talipova, T.; Woo, S.B.; Kim, D.C. Solitary wave transformation on the underwater step: Asymptotic theory and numerical experiments. *Appl. Math. Comput.* **2010**, *217*, 1704–1718. [CrossRef]
27. Moideen, R.; Ranjan Behera, M.; Kamath, A.; Bihs, H. Effect of girder spacing and depth on the solitary wave impact on coastal bridge deck for different airgaps. *J. Mar. Sci. Eng.* **2019**, *7*, 140. [CrossRef]
28. Chorin, A.J. Numerical solution of the Navier–Stokes equations. *Math. Comput.* **1968**, *22*, 745–762. [CrossRef]
29. Hirt, C.W.; Nichols, B.D. Volume of fluid (VOF) method for the dynamics of free boundaries. *J. Comput. Phys.* **1981**, *39*, 201–225. [CrossRef]
30. Boussinesq, M.J. Théorie de l'intumescence liquide, appelée onde solitaire ou de translation, se propageant dans un canal rectangulaire. *CR Acad. Sci. Paris* **1871**, *72*, 755–759.
31. Chang, K.-A.; Hsu, T.-J.; Liu, P.L.-F. Vortex generation and evolution in water waves propagating over a submerged rectangular obstacle: Part i. Solitary waves. *Coast. Eng.* **2001**, *44*, 13–36. [CrossRef]
32. Lin, C.; Hsieh, S.; Lin, W.; Raikar, R. Characteristics of recirculation zone structure behind an impulsively started circular cylinder. *J. Eng. Mech.* **2012**, *138*, 184–198. [CrossRef]
33. Tanaka, H.; Sumer, B.M.; Lodahl, C. Theoretical and experimental investigation on laminar boundary layers under cnoidal wave motion. *Coast. Eng. J.* **1998**, *40*, 81–98. [CrossRef]
34. Huang, C.-J.; Dong, C.-M. On the interaction of a solitary wave and a submerged dike. *Coast. Eng.* **2001**, *43*, 265–286. [CrossRef]
35. Lin, C.; Ho, T.-C.; Chang, S.-C.; Hsieh, S.-C.; Chang, K.-A. Vortex shedding induced by a solitary wave propagating over a submerged vertical plate. *Int. J. Heat Fluid Flow* **2005**, *26*, 894–904. [CrossRef]
36. Lin, C.; Chang, S.-C.; Ho, T.C.; Chang, K.-A. Laboratory observation of solitary wave propagating over a submerged rectangular dike. *J. Eng. Mech.* **2006**, *132*, 545–554. [CrossRef]
37. Higuera, P.; Lara, J.L.; Losada, I.J. Three-dimensional interaction of waves and porous coastal structures using OpenFOAM®. Part ii: Application. *Coast. Eng.* **2014**, *83*, 259–270. [CrossRef]
38. Wu, Y.-T.; Yeh, C.-L.; Hsiao, S.-C. Three-dimensional numerical simulation on the interaction of solitary waves and porous breakwaters. *Coast. Eng.* **2014**, *85*, 12–29. [CrossRef]
39. Chang, K.-A.; Liu, P.L.-F. Experimental investigation of turbulence generated by breaking waves in water of intermediate depth. *Phys. Fluids* **1999**, *11*, 3390–3400. [CrossRef]
40. Huang, Z.-C.; Hsiao, S.-C.; Hwung, H.-H.; Chang, K.-A. Turbulence and energy dissipations of surf-zone spilling breakers. *Coast. Eng.* **2009**, *56*, 733–746. [CrossRef]

Journal of
Marine Science and Engineering

MDPI

Article

Numerical Study of the Influence of Fishnet Mesh Size on a Floating Platform

Hung-Jie Tang [1], Chai-Cheng Huang [2] and Ray-Yeng Yang [3,*]

1 Tainan Hydraulics Laboratory, National Cheng Kung University, Tainan 709, Taiwan; hjtang@thl.ncku.edu.tw
2 Department of Marine Environment and Engineering, National Sun Yat-sen University, Kaohsiung 804, Taiwan; cchuang@mail.nsysu.edu.tw
3 Department of Hydraulic and Ocean Engineering, National Cheng Kung University, Tainan 701, Taiwan
* Correspondence: ryyang@mail.ncku.edu.tw; Tel.: +886-6-2757275 (ext. 63246)

Received: 30 March 2020; Accepted: 8 May 2020; Published: 12 May 2020

Abstract: This study aims to investigate the influence of fishnet mesh size on a floating platform. A self-developed, time-domain numerical model was used for the evaluation. This model is based on potential flow theory, uses the boundary element method (BEM) to solve nonlinear wave-body interactions, and applies the Morison equation to calculate the hydrodynamic forces exerted on fishnets. The mooring system is treated as a linear and symmetric spring. The results near the resonant frequency of the platform indicate that the smaller the fishnet mesh size, the lower the heave, pitch, and sea-side tension response amplitude operators (RAOs), but the higher the reflection coefficient. The results in the lower frequency region reveal that the smaller the fishnet mesh size, the lower the surge and heave RAOs, but the higher the pitch and tension RAOs. Meanwhile, the time-domain results at the resonant frequency of heave motion are shown to indicate the influences of a platform with various fishnets mesh sizes on the rigid body motion, mooring line tension, and transmitted wave heights. In addition, a comparison of nonlinear effects indicates that, after reducing the fishnet mesh size, the second-order RAOs of heave, pitch, and sea-side tension decrease, but the changes are minor against the first-order results.

Keywords: floating platform; fishnet mesh size; frequency-domain; time-domain; nonlinear waves; BEM

1. Introduction

In recent years, owing to environmental impacts and spatial conflicts with other industries, the development of marine cage aquaculture moved towards deep sea operations. For the better utilization of ocean space, as well as reduced construction costs, the multi-purpose floating platform is becoming a popular research topic around the world. For example, a European Union (EU) project called "The Blue Growth Farm" was proposed to develop a multi-purpose floating platform that is intended to combine blue energy (renewable energy) and fish farming and use renewable energy to supply the power required for intelligent farming equipment. Nevertheless, the fishnet used has not yet been addressed in current research ([1,2]).

In recent decades, many researchers adopted the Morison-type numerical model to study the hydrodynamic characteristics of marine fish cages [3–10]. This net cage structure has been treated as a so-called small-body in order to ignore the wave–body interaction. Conversely, many studies considered the floating structure as a so-called large-body in order to analyze the nonlinear wave–body interaction by means of potential flow theory [11–15]. However, studies of marine aquaculture structure that include small-body and large-body structures remain few and far [16,17]. Therefore, a number of issues that must be resolved still exist. For example, the influences of different fishnet mesh sizes on marine cage systems have been investigated [18,19]. Moreover, fishnet biofouling in the real sea is inevitable and can result in the fishnet exhibiting a narrowed mesh size and significant mass increment [20,21]. In the real world, replacing the fishnet with different mesh sizes according to the size of fish during fish farming is also necessary. Therefore, studying the effect of fishnet mesh size on floating platform dynamics is important.

In our previous study [16], a two-dimensional, nonlinear numerical wave tank incorporating potential flow theory was developed to investigate the dynamic interaction between waves and a floating platform with a fishnet. This model has been validated by physical model tests and shows good agreement. The research [16] studied related conditions such as net depth, net width, and the nonlinearity of dynamic response. In this study, we continue to explore the impact of fishnet mesh size on the hydrodynamic characteristics of an aquaculture platform, with a view of providing a reference for the development of a net-type aquaculture floating platform.

2. Description of the Numerical Model

An aquaculture-purposed floating platform (see Figure 1a) consisting of a pair of floating rectangular pontoons and restrained by a linear symmetric mooring system is shown in Figure 1b, where a is the width of each pontoon, b is the spacing between the two pontoons, d is the draught, (x_G, z_G) is the position of the center of gravity, l_G is the pitch moment arm, θ_0 is the mooring line angle, and l_0 is the original length of the mooring line. The floating structure was deployed in a numerical wave tank with a constant water depth, h. A numerical damping zone was used at each end of the wave tank to dissipate the reflected and transmitted waves, where x_{d1} and x_{d2} are the entrance positions.

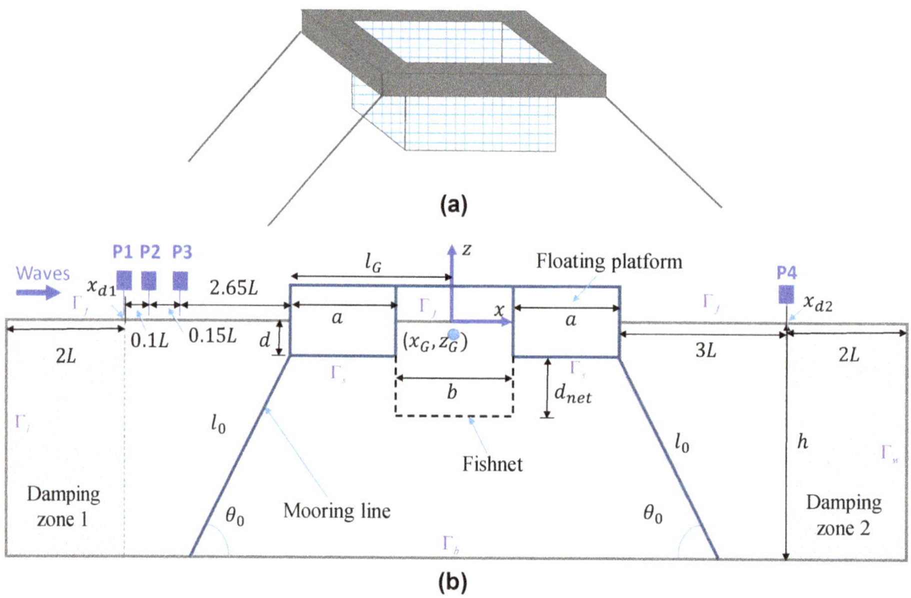

Figure 1. (**a**) The concept design and (**b**) definition sketch of the floating platform with a fishnet in a numerical wave tank.

2.1. Governing Equation

The two-dimensional flow field is assumed to be incompressible, inviscid, and irrotational. Thus, a velocity potential exists and satisfies the Laplace equation:

$$\nabla^2\phi = \frac{\partial^2\phi}{\partial x^2} + \frac{\partial^2\phi}{\partial z^2} = 0. \tag{1}$$

Incorporating Equation (1) into the Green second identity, the velocity potential in the fluid domain can be determined by solving the following boundary integral equation (BIE):

$$\alpha\phi_i = \int_{\Gamma_j}\left(\frac{\partial G_{ij}}{\partial n}\phi_j - G_{ij}\frac{\partial \phi_j}{\partial n}\right)\mathrm{d}\Gamma_j \tag{2}$$

where $G_{ij} = \ln r_{ij}/2\pi$ is the fundamental solution to the Laplace equation and represents a flow field generated by a concentrated unit source acting at the ith source point, r_{ij} is the distance from source point (x_i, z_i) to field point (x_j, z_j), and α is the internal solid angle between the two boundary elements. In this model, the linear element scheme and six-point Gaussian quadrature integration method are applied to solve the BIE.

2.2. Inflow Boundary Condition

On the basis of the continuity of velocity, a theoretical particle velocity profile can be used to specify the boundary value along the inflow boundary. For nonlinear regular waves, the second-order Stokes wave is used to prevent a mismatch between the input velocity profiles and real water particle velocity, as described in [22,23] and expressed below:

$$\frac{\partial\phi}{\partial n} = -\begin{bmatrix} \frac{gAk}{\sigma}\frac{\cosh k(z+h)}{\cosh kh}\cos(kx-\sigma t) \\ +\frac{3}{4}A^2k\sigma\frac{\cosh 2k(z+h)}{\sinh^4 kh}\cos 2(kx-\sigma t) \end{bmatrix} f_m \quad \text{on } \Gamma_I\ , \tag{3}$$

where A, k, and σ are the amplitude, wave number, and angular frequency, respectively; g is the gravitational acceleration; and t is the time. The modulation function f_m is used to prevent impulse-like behavior of a wave maker and is written as

$$f_m(t) = \begin{cases} \frac{1}{2}\left[1-\cos\left(\frac{\pi t}{T_m}\right)\right] & for\ t < T_m, \\ 1 & for\ t \geq T_m, \end{cases} \tag{4}$$

where T_m is the modulation duration that depends on wave steepness. For a steeper wave, the modulation duration is usually twice as long as a regular wave period.

2.3. Free Surface Boundary Condition

One of the most popular and successful approaches to the fully nonlinear free surface simulation is the mixed Eulerian and Lagrangian (MEL) method, which was first presented by Longuet–Higgins and Cokelet [24]. In this method, the kinematic and dynamic free surface boundary conditions are transformed into the Lagrangian framework. To obtain numerical solutions for wave propagation in a wave tank, the scheme used numerical damping zones at both ends of the wave tank to absorb the transmitted wave energy at the end of the tank, as well as to dissipate the reflected waves in front of the input boundary. The numerical damping zones [25,26] are incorporated into the free surface boundary conditions as

$$\begin{cases} \frac{\mathrm{d}x}{\mathrm{d}t} = \frac{\partial\phi}{\partial x} \\ \frac{\mathrm{d}z}{\mathrm{d}t} = \frac{\partial\phi}{\partial z} - \nu(x)(z-z_e) \\ \frac{\mathrm{d}\phi}{\mathrm{d}t} = -gz + \frac{1}{2}\left|\nabla\phi\right|^2 - \nu(x)(\phi-\phi_e) \end{cases} \quad \text{on } \Gamma_f, \tag{5}$$

where $\nu(x)$ is the damping coefficient of the numerical damping zone, given by

$$\nu(x) = \begin{cases} \alpha_d\sigma[(x_{d1}-x)/L]^2 & x \leq x_{d1}, \\ \alpha_d\sigma[(x-x_{d2})/L]^2 & x \geq x_{d2}, \\ 0 \end{cases} \tag{6}$$

where α_d is the dimensionless parameter for the strength of the damping zone; after several tests, we found that α_d set to 1 is adequate for obtaining accurate results. L is the wavelength of the input wave, while x_{d1} and x_{d2} are the entrance positions of each damping zone shown in Figure 1b. Meanwhile, z_e and ϕ_e in Equation (5) are the entrance wave elevation and potential function in the front damping zone, only existing in $x \leq x_{d1}$. Tanizawa [26] applied this damping zone technique to dissipate the wave energy reflected from the structure, but without disturbing the outgoing incident waves. For practical purposes, the nonlinear analytical solution of the second-order Stokes wave theory was adopted in damping zone 1 to improve the computational process. In this model, the entrance potential and wave elevation are written as follows:

$$\begin{cases} \phi_e = \frac{Ag}{\sigma}\frac{\cosh k(z+h)}{\cosh kh}\sin(kx-\sigma t) + \frac{3}{8}A^2\sigma\frac{\cosh 2k(z+h)}{\sinh^4 kh}\sin 2(kx-\sigma t), \\ z_e = A\cos(kx-\sigma t) + \frac{kA^2\cosh kh}{4\sinh^3 kh}(2+\cosh 2kh)\cos 2(kx-\sigma t). \end{cases} \tag{7}$$

Additionally, the nodal velocities in Equation (5) are obtained by using the cubic spline scheme in the curvilinear coordinate system, as described in Section 2.6. The corner problem between the free surface and body surface is treated according to [27], as described in Section 2.7.

2.4. Body Surface Boundary Condition

In this model, the body surface (Γ_s) is impermeable. Therefore, the fluid velocity is equal to the normal velocity on the body surface:

$$\frac{\partial\phi}{\partial n} = n_1\dot{x}_G + n_2\dot{z}_G + n_3\dot{\theta}_G \quad \text{on } \Gamma_s, \tag{8}$$

where $(n_1, n_2, n_3) = (n_x, n_z, r_z n_x - r_x n_z)$ is the unit normal vector on the body boundary, and $(r_x, r_z) = (x - x_G, z - z_G)$ is the position vector from the body surface to the gravity center. Subscript G designates the gravity center of the body; $(\dot{x}_G, \dot{z}_G)$ are the translational velocities in the x and z axes (surge and heave motions), respectively; and $\dot{\theta}_G$ is the angular velocity about the y axis (pitch motion).

2.5. Rigid Boundary Condition

At the end-wall (Γ_w) and bottom (Γ_b) of the wave tank, the boundary conditions are considered impermeable. The normal velocities are then set to zero

$$\frac{\partial\phi}{\partial n} = 0 \text{ on } \Gamma_b \text{ and } \Gamma_w. \tag{9}$$

2.6. Curvilinear Coordinate System

In this paper, a curvilinear coordinate system and the cubic spline scheme are adopted to solve the spatial derivatives of velocity potential in Equation (5) on the free surface boundary. The relationship between the velocity components in Cartesian and curvilinear coordinates is written as

$$\begin{cases} \frac{\partial\phi_f}{\partial x} = \frac{\partial\phi_f}{\partial s}\cos\beta_f - \frac{\partial\phi_f}{\partial n}\sin\beta_f, \\ \frac{\partial\phi_f}{\partial z} = \frac{\partial\phi_f}{\partial s}\sin\beta_f + \frac{\partial\phi_f}{\partial n}\cos\beta_f, \end{cases} \tag{10}$$

where ϕ_f represents the potential function on the free surface and β_f is the angle between s, a section of the free surface, and the x axis. The normal velocities of the free surface $\partial\phi_f/\partial n$ are obtained after solving the BIE, while the angle β_f is determined from the following equation:

$$\tan\beta_f = \frac{\sin\beta_f}{\cos\beta_f} = \frac{\partial z/\partial s}{\partial x/\partial s}, \tag{11}$$

where, $\partial\phi_f/\partial s$, $\partial x/\partial s$, and $\partial z/\partial s$ are calculated by using cubic spline interpolation in curvilinear coordinates along the free surface.

Once the values of the time derivative of the potential function on the right side of Equation (5) are known, the substantial derivative equations on the left side can be used to predict the new nodal position and its corresponding potential on the free surface boundary by employing the fourth-order Runge–Kutta (RK4) method as a time marching scheme. This process was repeated until the simulation reached a steady-state condition.

Note that the node-regridding and smoothing technique is also applied in the present model using the cubic spline interpolation on the curvilinear coordinate system in order to prevent free surface nodes from moving too close to one another and to prevent the occurrence of the saw-tooth condition, which may lead to numerical instability.

2.7. Corner Problem between the Free Surface and Body Surface

At the intersection of the body surface and free surface, the discontinuity of the flux occurs as a result of the discontinuity of the normal direction. Although the cubic spline scheme is accurate for determining the tangential slope at the end-point with the natural condition (curvature equal to zero) and the Lagrangian polynomial method, the requirement of continuity of flux at the corner is still difficult to achieve. To deal with this discontinuity, the double collocation node technique is often used. Grilli and Svendsen [27] proposed a treatment for the corner problem at the intersection based on the continuity flux as follows:

$$\frac{\partial\phi_f}{\partial s} = \frac{\partial\phi_f}{\partial n}\frac{\cos(\beta_b - \beta_f)}{\sin(\beta_b - \beta_f)} - \frac{\partial\phi_b}{\partial n}\frac{1}{\sin(\beta_b - \beta_f)}, \tag{12}$$

where the subscripts b and f denote the body and water free surface, respectively. $\partial\phi_b/\partial n$ and $\partial\phi_f/\partial n$ are the normal velocities on the free surface and body surface, respectively; and $\partial\phi_f/\partial s$ is the modified tangential velocity on the free surface and will be used in Equation (10) when dealing with the corner problem. In this model, the input boundary angle at the front of the tank is $\beta_b = \pi/2$, while that of the wall boundary at the end of the wave tank is $\beta_b = 3\pi/2$.

2.8. Wave Forces on the Body

The hydrodynamic forces on the body can be calculated by integrating the pressure around the wetted body surface as

$$\begin{cases} \boldsymbol{F} = \int_{\Gamma_s} -\rho\left(\phi_t + gz + \frac{1}{2}|\nabla\phi|^2\right)\boldsymbol{n}ds, \\ \boldsymbol{M} = \int_{\Gamma_s} -\rho\left(\phi_t + gz + \frac{1}{2}|\nabla\phi|^2\right)\boldsymbol{r}\times\boldsymbol{n}ds, \end{cases} \tag{13}$$

where $\boldsymbol{F}$ and $\boldsymbol{M}$ are the hydrodynamic force and moment, respectively; ρ is the water density; $\boldsymbol{n}$ is the normal unit vector on the body surface and points into the body; and $\boldsymbol{r}$ is the position vector from the gravity center to body's surface.

2.9. Acceleration Potential Method

In order to evaluate the hydrodynamic forces on the floating body using Equation (13), both gradients of velocity potential ($\nabla\phi$) and unsteady ϕ_t terms on the wetted body surface must be determined beforehand. $\nabla\phi$ is evaluated by the regular boundary element method (BEM), while ϕ_t is determined by an acceleration potential method proposed by [28], taking the advantage of the feature that ϕ_t also satisfies the Laplace equation:

$$\nabla^2\phi_t = \frac{\partial^2\phi_t}{\partial x^2} + \frac{\partial^2\phi_t}{\partial z^2} = 0. \tag{14}$$

In accordance with [27], the body surface–surface boundary condition in the acceleration field is described as

$$\frac{\partial\phi_t}{\partial n} = n_1\ddot{x}_G + n_2\ddot{z}_G + n_3\ddot{\theta}_G + q \text{ on } \Gamma_s, \tag{15}$$

where $(\ddot{x}_G, \ddot{z}_G)$ are the translational accelerations in the x- and z-axes, and $\ddot{\theta}_G$ is the angular acceleration around the y−axis. In turn, q is defined as

$$\begin{aligned} q &= n_1\dot{\theta}_G\left(r_x\dot{\theta}_G - 2\dot{z}_G + 2\tfrac{\partial\phi}{\partial z}\right) + n_2\dot{\theta}_G\left(r_z\dot{\theta}_G + 2\dot{x}_G - 2\tfrac{\partial\phi}{\partial x}\right) \\ &+k_n\left[\left(\tfrac{\partial\phi}{\partial x} - \dot{x}_G - \dot{\theta}_G r_z\right)^2 + \left(\tfrac{\partial\phi}{\partial z} - \dot{z}_G + \dot{\theta}_G r_x\right)^2\right] \\ &-k_n\left[\left(\tfrac{\partial\phi}{\partial x}\right)^2 + \left(\tfrac{\partial\phi}{\partial z}\right)^2\right] - \tfrac{\partial\phi}{\partial s}\tfrac{\partial^2\phi}{\partial n\partial s} + \tfrac{\partial\phi}{\partial n}\tfrac{\partial^2\phi}{\partial s^2}, \end{aligned} \tag{16}$$

where $k_n = 1/\rho^*$ is the normal curvature along the s direction of the body surface and ρ^* is the radius of the curvature. Note that $\partial\phi/\partial s$, $\partial^2\phi/(\partial s\partial n)$, and $\partial^2\phi/\partial s^2$ are calculated using cubic spline interpolation in the curvilinear coordinates along the body surface. The above variables at the corners of the structure are considered the natural condition (curvature equal to zero), while those between the structure and free surface are modified by the same method, as described in Section 2.7.

For solving ϕ_t in the acceleration field, four methods are available, which include (1) the iterative method [29,30], (2) the modal decomposition method [31], (3) the implicit boundary condition method [28], and (4) the indirect method [32]. Detailed descriptions of these can be found in [22,33,34].

2.10. Modal Decomposition Method

The modal decomposition method was first introduced in [31]. This approach solves the BIE for the acceleration field. The acceleration potential function is decomposed into four modes corresponding to three unit accelerations for surge-heave-pitch motions (radiation problem) and acceleration due to the incident wave field (diffraction problem). Using these four modes in Equation (17) and the boundary conditions listed in Equations (18)–(21), the unknown mode's amplitude can be determined by solving its respective BIE. The ϕ_t is given by

$$\phi_t = \sum_{m=1}^{3} a_m\varphi_m + \varphi_4, \tag{17}$$

where a_m is the mth mode component of generalized body acceleration (1 = surge, 2 = heave, 3 = pitch, 4 = diffraction mode).

The boundary conditions in the acceleration field for each mode are given as

$$\frac{\partial\varphi_m}{\partial n} = \begin{cases} n_m & m = 1,2,3 \\ q & m = 4 \end{cases} \quad \text{on } \Gamma_s, \tag{18}$$

$$\varphi_m = \begin{cases} 0 & m = 1,2,3 \\ -gz - \frac{1}{2}|\nabla\phi|^2 & m = 4 \end{cases} \quad \text{on } \Gamma_f, \tag{19}$$

$$\frac{\partial \varphi_m}{\partial n} = 0\ (m = 1, 2, 3) \quad \text{on } \Gamma_I, \tag{20}$$

$$\frac{\partial \varphi_m}{\partial n} = 0 (\, m = 1 \sim 4) \quad \text{on } \Gamma_b \text{ and } \Gamma_w. \tag{21}$$

The inflow boundary condition for mode 4 is obtained from the second-order Stokes wave theory:

$$\frac{\partial \varphi_4}{\partial n} = -\left[\begin{array}{l} gAk\frac{\cosh k(z+h)}{\cosh kh} \sin(kx - \sigma t) \\ +\frac{3}{2}A^2 k\sigma^2 \frac{\cosh 2k(z+h)}{\sinh^4 kh} \sin 2(kx - \sigma t) \end{array} \right] \quad \text{on } \Gamma_I. \tag{22}$$

After solving the BIE for each mode, the values of φ_m and $\partial\varphi_m/\partial n$ for all of the boundaries are obtained, with the remaining unknown in Equation (17) being a_m.

Substituting Equation (17) into Equation (13), combined with Newton's second law including the wave hydrodynamic and other forces (with subscript e; for example, gravitational force, damping force, restoring force, and force on net), the equation of motion becomes

$$\begin{cases} ma_1 = \int_{\Gamma_s} -\rho\left(a_1\varphi_1 + a_2\varphi_2 + a_3\varphi_3 + \varphi_4 + gz + \frac{1}{2}|\nabla\phi|^2\right) n_1 \mathrm{d}s + F_{ex}, \\ ma_2 = \int_{\Gamma_s} -\rho\left(a_1\varphi_1 + a_2\varphi_2 + a_3\varphi_3 + \varphi_4 + gz + \frac{1}{2}|\nabla\phi|^2\right) n_2 \mathrm{d}s + F_{ez}, \\ I_G a_3 = \int_{\Gamma_s} -\rho\left(a_1\varphi_1 + a_2\varphi_2 + a_3\varphi_3 + \varphi_4 + gz + \frac{1}{2}|\nabla\phi|^2\right) n_3 \mathrm{d}s + M_{ey}. \end{cases} \tag{23}$$

After solving Equation (23), the generalized acceleration a_m can be obtained. The details of the extra forces are described below.

2.11. Mooring Force

The mooring system is considered linear and symmetrical (see Figure 1b) with the spring constant K, while the wave hydrodynamic forces on the mooring line are negligible in comparison with the force exerted on the floating dual pontoon. The pre-tension force of this mooring system is written as

$$F_{T_0} = \frac{2\rho g a d\lambda - mg}{2\sin\theta_0}, \tag{24}$$

where λ is the total length of the floating dual pontoon structure in the direction of y.

2.12. Wave Forces on Fishnet

In this model, a fishnet is set up between the pontoons and secured by a steel frame. The net is assumed not to deform. By applying the lumped mass method discussed in [5], the fishnet panel is divided into several elements and nodes, while a modified Morison equation [35] to calculate the drag and inertia forces on the net elements is given as

$$\boldsymbol{F}_{\text{net}} = \frac{1}{2}\rho C_D A_{\text{net}} \boldsymbol{V}_R |\boldsymbol{V}_R| + \rho\delta_{\text{net}} C_M \frac{\mathrm{d}\boldsymbol{V}}{\mathrm{d}t} - \rho\delta_{\text{net}} K_M \frac{\mathrm{d}\dot{\boldsymbol{R}}}{\mathrm{d}t}, \tag{25}$$

where ρ is water density, C_D is the drag coefficient, $C_M = 1 + K_M$ is the inertia coefficient, and K_M is the added mass coefficient. In this model, $C_M = 2.0$ is assumed, which is generally between 1.0 and 2.0. A_{net} is the projected area of the net element, δ_{net} is the volume of the net element, and $\boldsymbol{V}_R = \boldsymbol{V} - \dot{\boldsymbol{R}}$ the relative velocity between the flow field and net element. In turn, $\boldsymbol{V}$ is the flow velocity at the center of a net element, $\dot{\boldsymbol{R}}$ is the central velocity of the net element, $d\boldsymbol{V}/dt$ is the fluid particle acceleration at the center of the net element, and $d\dot{\boldsymbol{R}}/dt$ is the central acceleration of the net element. Details about fluid particle velocity and acceleration are described in [16].

According to Loland's empirical formula [36], the drag force, which is parallel to the fluid motion, and the lift force, which is perpendicular to the fluid motion, are as follows:

$$\begin{cases} F_D = \frac{1}{2}\rho C_D(\alpha) A_{\text{net}} |\boldsymbol{V_R}|^2, \\ F_L = \frac{1}{2}\rho C_L(\alpha) A_{\text{net}} |\boldsymbol{V_R}|^2, \end{cases} \tag{26}$$

where $C_D(\alpha)$ and $C_L(\alpha)$ are coefficients related to the angle of α between the fluid particle velocity vector and the normal vector of the net element:

$$\begin{cases} C_D(\alpha) = 0.04 + (-0.04 + 0.33S_n + 6.54{S_n}^2 - 4.88{S_n}^3)\cos(\alpha), \\ C_L(\alpha) = (-0.05S_n + 2.3{S_n}^2 - 1.76{S_n}^3)\sin(2\alpha), \end{cases} \tag{27}$$

where S_n is the solidity ratio, which is defined as the ratio between the area covered by the threads and the total area of the net panel.

2.13. Consideration of Damping Effects

Experimental testing in a physical wave tank revealed that dynamic responses near the resonant frequency of body motions significantly dampen; similar phenomena were also identified by [37] and may be attributed to the fluid viscous effect. The flow field is assumed to be inviscid in the present model and, at present, determining the damping coefficients of a floating structure is nearly impossible. To simplify this problem, an uncoupled damping coefficient matrix is incorporated into the equation of motion of Equation (22) to represent the damping forces. The equation is then rewritten as

$$\begin{bmatrix} m & & \\ & m & \\ & & I_G \end{bmatrix} \begin{bmatrix} \ddot{x}_G \\ \ddot{z}_G \\ \ddot{\theta}_G \end{bmatrix} + \begin{bmatrix} C_{xx} & & \\ & C_{zz} & \\ & & C_{\theta\theta} \end{bmatrix} \begin{bmatrix} \dot{x}_G \\ \dot{z}_G \\ \dot{\theta}_G \end{bmatrix} + [K] \begin{bmatrix} x_G \\ z_G \\ \theta_G \end{bmatrix} = \begin{bmatrix} F_x \\ F_z \\ M_y \end{bmatrix}, \tag{28}$$

where $[K]$ is the stiffness matrix, and the components are determined numerically at each time step according to the mooring angle. F_x, F_z, and M_y are the resultant hydrodynamic forces acting on the platform, respectively, while C_{xx}, C_{zz}, and $C_{\theta\theta}$ are the damping coefficients of surge, heave, and pitch motion, respectively. Furthermore, these damping coefficients are assumed to be equal and denoted by C, and are obtained from the damping ratio [38]:

$$\zeta = \frac{C}{2\sqrt{Km}}. \tag{29}$$

In this study, the damping ratio is set to 0.1 in order to fit our experimental data, while K is the spring constant and m the total mass of the floating structure.

3. Numerical Model Test

In our previous research [16], the numerical model under discussion here was verified by physical model tests in a wave flume. Table 1 lists the physical model data of the aquaculture platform and its corresponding materials. According to previous results [16], apart from the resonant frequency region, the influence of the fishnet on the platform is small. The conclusion was that the mass and drag force of the fishnet is too small to affect the motion of the platform. Thus, in this study, we decrease the fishent mesh size to increase the total mass and drag force, but keep the twine diameter constant. Table 2 shows the mesh size, the solidity ratio [36], the total mass of the fishnet (obtained according to the solidity ratio), and the mass ratio between the fishnet and platform.

Table 1. Characteristics of the floating platform with a fishnet.

Specifications	Sizes
Wave amplitude (A)	0.02 m
Wave period (T)	0.73–2.54 s
Water depth (h)	0.80 m
Total mass of platform (m)	62.58 kg
Width of platform (a)	0.25 m
Spacing between pontoons (b)	0.50 m
Draft (d)	0.153 m
Pitch moment arm (l_G)	0.50 m
Moment of inertia (I_G)	8.93 kg·m^2
Spring constant (K)	674.93 N/m
Mooring angle (θ_0)	52°
Gravity center (x_G, z_G)	(0.0 m, −0.0861 m)
Net depth (d_{net})	0.455 m
Twine diameter (D_{line})	0.175 cm
Specific gravity of net (Nylon)	1.14

Table 2. Numerical model test conditions.

Half Mesh Size (cm) λ	Solidity Ratio S_n	Total Mass (kg) m_{net}	Mass Ratio (%) m_{net}/m
2.0	0.179	1.645	2.6
1.0	0.365	3.361	5.4
0.5	0.761	7.003	11.2

In addition, the appropriate grid size and time step for the numerical model are obtained through a preliminary convergence test. The grid size is $L/30$ (L is the wave length) on the free surface, the space between each pontoon is 10 elements, $h/20$ on both sides of the NWT, $L/10$ on the bottom, and 40 elements are present on each pontoon surface. The appropriate marching time step of the RK4 is $T/32$, while the total simulated time is $50T$. The input wave periods are in the range of 0.73 s and 2.54 s, while the input wave heights are all 4 cm. The simulation was executed on a personal computer (Intel i5 CPU, 16 GB RAM), with a calculation time for each simulation of about 30 min.

4. Results and Discussion

4.1. Frequency-Domain Results

In general, the response amplitude operator (RAO) is used to describe the first-order dynamic response of body motions related to incident wave amplitudes through an FFT (Fast Fourier Transform) analysis, as is shown in Table 3. The second-order dynamic responses can also be calculated using the same analysis, as the dynamic responses of wave-body interactions are harmonic. Therefore, in this study, the second-order RAOs are adopted to describe the nonlinear dynamic responses. The incident and reflected waves are separated by Mansard and Funke's method [39] using three wave gauges installed in front of the structure (see Figure 1b).

Table 3. Definition of the response amplitude operators (RAOs) and frequency.

Normalized Parameters	Definition
Normalized surge RAO	x_G/A
Normalized heave RAO	z_G/A
Normalized pitch RAO	$l_G\theta_G/A$
Normalized tension RAO	F_T/KA
Normalized angular frequency	$\sigma^2 h/g$

Figures 2–4 show a comparison of surge, heave, and pitch RAO for a platform with different fishnet mesh sizes (λ). The results include previous measurements and simulations from [16] to ensure that the present simulations fall within a reasonable area. Unsurprisingly, the results indicate that, when the fishnet mesh sizes are different, the RAOs of platform motion will change significantly. The surge RAO exhibits good reduction in the low-frequency region ($\sigma^2 h/g < 2$), as λ decreases from 2 cm to 0.5 cm. The low-frequency response is usually related to the restoring mooring force, except that the differences in surge of the RAO among the simulated cases are small. As for the heave RAO, in both the low-frequency region ($\sigma^2 h/g < 2$) and the resonant frequency region (near $\sigma^2 h/g = 3.9$), the heave RAO results are greatly reduced with decreasing λ. Moreover, a slight phase difference of heave RAO can be observed in the resonant frequency region. Finally, for the pitch RAO, the response increases greatly near $\sigma^2 h/g = 2$, but decreases in the frequency region between $\sigma^2 h/g = 3$ and 6. Overall, reducing the mesh size of the net is helpful for mitigating the dynamic response.

Figure 5 presents a comparison of sea-side tension RAO across different fishnet mesh sizes (λ). Because the mooring system is deployed to limit the movement of the platform, the response of the tension RAO should be closely related to the surge-heave-pitch motion. Around $\sigma^2 h/g = 4$, the response of the tension RAO may be dominated by the heave RAO. In this region, the tension RAO decreases with λ, but no phase difference occurs that can be observed in the heave RAO. In the region of $\sigma^2 h/g < 2$, the response of the tension RAO relates to the combination of surge-heave-pitch motion, but the pitch RAO seems to dominate the tension response. In this region, the tension RAO generally increases as λ decreases.

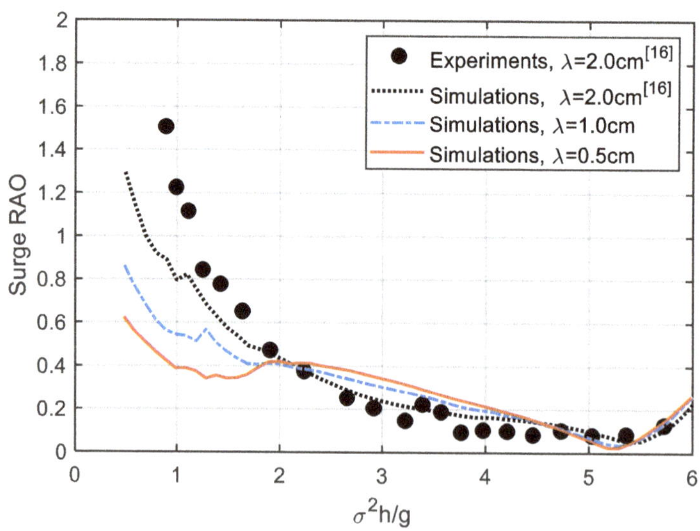

Figure 2. Comparison of the surge response amplitude operator (RAO) for the platform with various fishnet mesh sizes.

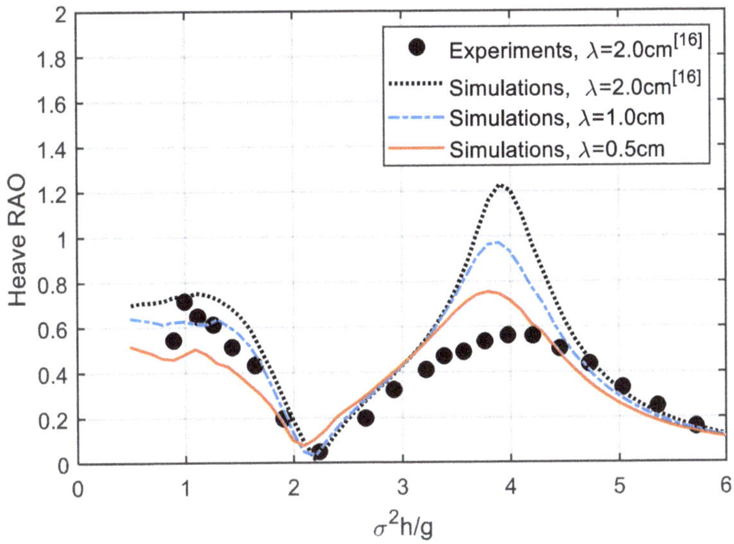

Figure 3. Comparison of the heave RAO for the platform with various fishnet mesh sizes.

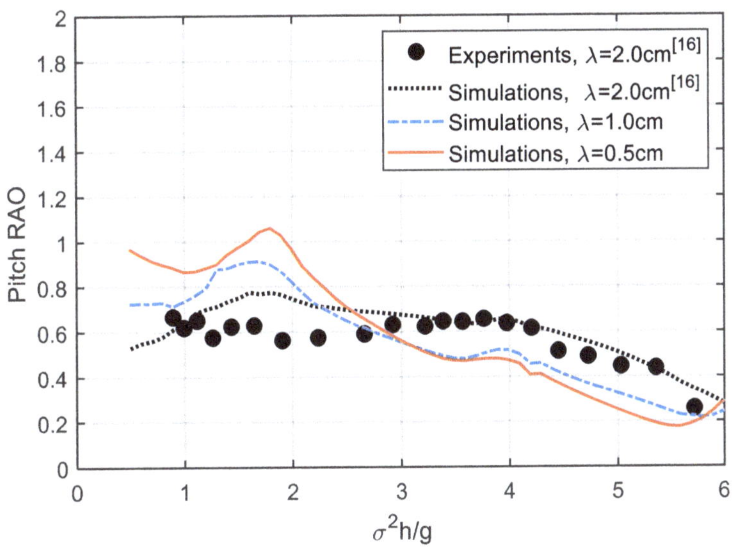

Figure 4. Comparison of the pitch RAO for the platform with various fishnet mesh sizes.

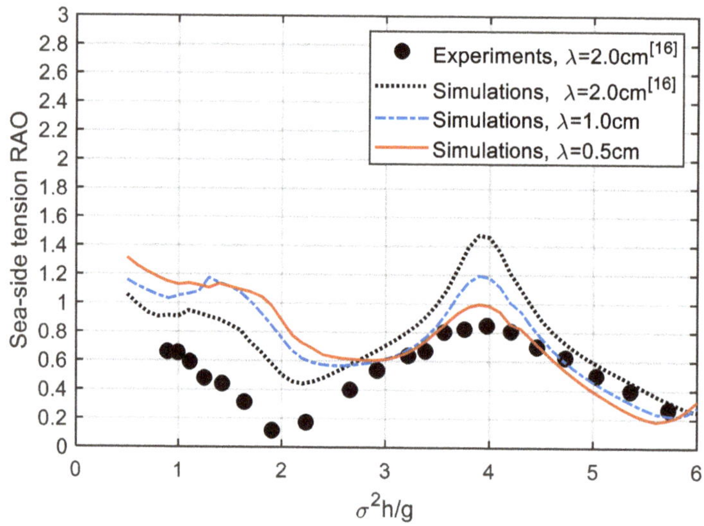

Figure 5. Comparison of the sea-side tension RAO for the platform with various fishnet mesh sizes.

In some studies [40,41], the floating structure is used as the floating breakwater to protect the facilities behind it. Similarly, this aquaculture-purposed platform may demonstrate good performance as a breakwater. Furthermore, the drag force on the net and its mass may affect wave–platform interactions, in spite of the fact that the wave–net interaction is ignored in this model. Figure 6 shows a comparison of the reflection coefficient under different fishnet mesh sizes (λ). In addition, the result includes previous data from [16] in order to compare with the present simulations. A remarkable increment in the reflection coefficient is found in the range of $\sigma^2 h/g$ from 3.7 to 6.0 when λ decreases from 2 cm to 0.5 cm, especially in the resonant frequency region of heave motion ($\sigma^2 h/g = 3.9$). However, a not to be ignored reduction in the reflection coefficient appears in the range of $\sigma^2 h/g$, from 2.0 to 3.7. Overall, the platform with a fishnet with a smaller mesh size exhibits better performance in terms of the reflection coefficient.

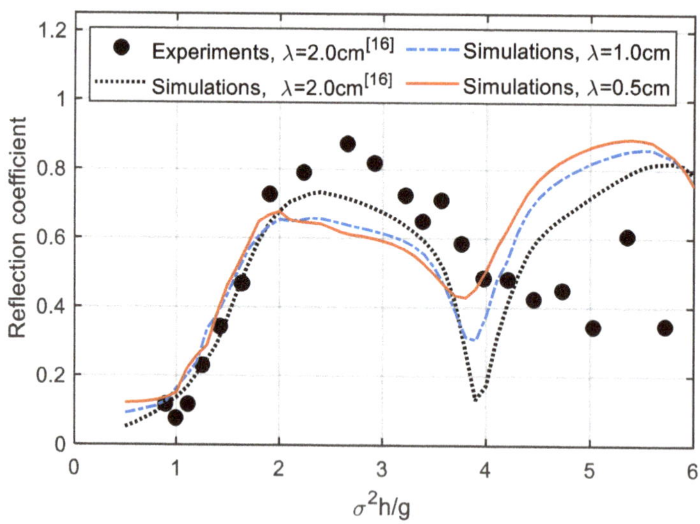

Figure 6. Comparison of the reflection coefficient for the platform with various fishnet mesh sizes.

4.2. Time-Domain Results

In this section, the time-domain results at the resonant frequency of heave motion ($\sigma^2 h/g$ = 3.9) are chosen for discussion. This is because the main difference in the frequency-domain results occurs here.

Figures 7–9 show variations in the surge, heave, and pitch motions of the platform with different fishnet mesh sizes (λ) during the last 10 waves. All of the results reach steady-state conditions, which means that the simulation results are convergent. In the comparison of surge motion, the results show that both the response amplitude and mean value increase as λ decreases. In addition, a short time delay appears as λ decreases. In the comparison with the heave motion, the results indicate that the peak value of the vibration decreases greatly with λ, while the valley value does not. This causes the response amplitude of the heave motion to decrease as λ decreases. In the comparison of the pitch motion, the results show that both the peak value and valley value of the vibration decrease as λ decreases, but the peak value demonstrates a larger reduction. Thus, the response amplitude of the pitch motion decreases with λ. In addition, as λ decreases, an obvious time delay observed in the pitch motion.

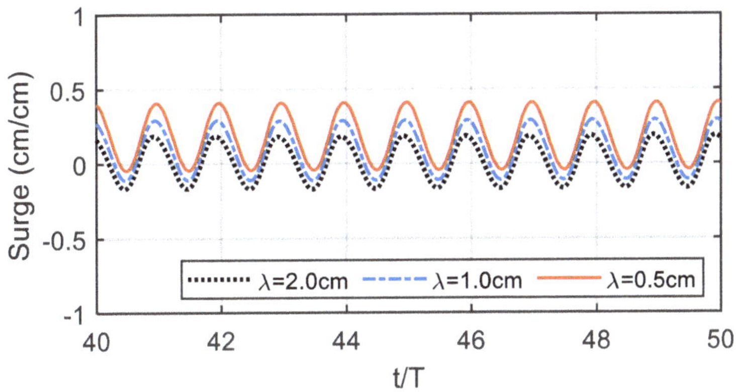

Figure 7. Variations in the surge motion for the platform with various fishnet mesh sizes (H = 4 cm, $T = 0.91$ s, and $\sigma^2 h/g = 3.9$).

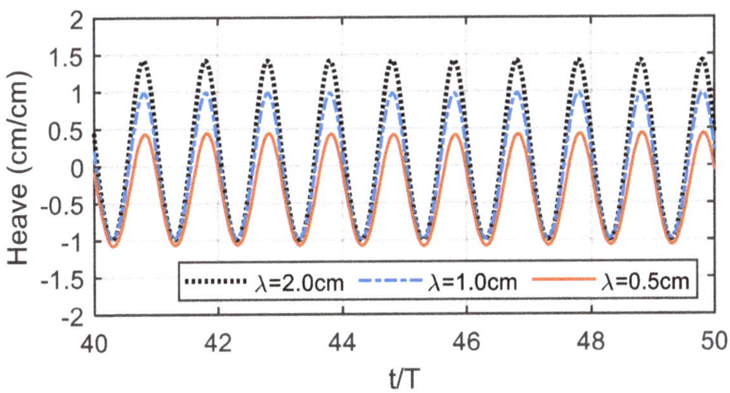

Figure 8. Variations in the heave motion for the platform with various fishnet mesh sizes (H = 4 cm, $T = 0.91$ s, and $\sigma^2 h/g = 3.9$).

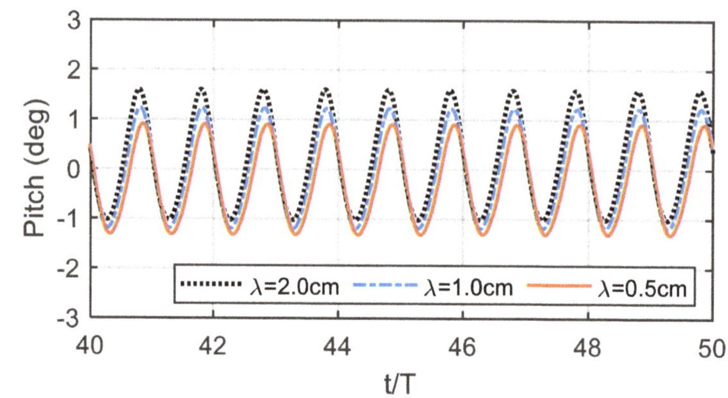

Figure 9. Variations in the pitch motion for the platform with various fishnet mesh sizes (H = 4 cm, T = 0.91 s, and $\sigma^2 h/g = 3.9$.

Figure 10 displays a comparison of sea-side tension for different fishnet mesh sizes (λ). The results show that as λ decreases, the peak value of the tension increases, while the change in the valley value is not obvious. In addition, as λ decreases, a slight time delay occurs. Comparing the results to the surge-heave-pitch motions, the phase of peak and valley values of tension are consistent with the phase of heave and pitch motion, but inconsistent with the phase of surge motion. This may be because of the surge motion being associated with restoring mooring force (wave drift forces) rather than wave forces.

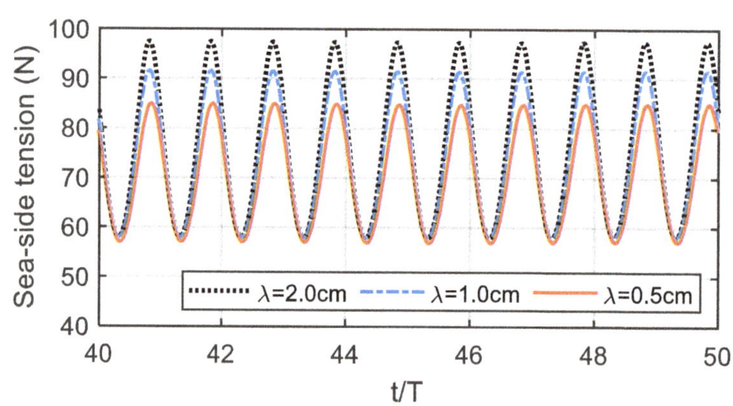

Figure 10. Variations in the sea-side tension for the platform with various fishnet mesh sizes (H = 4 cm, T = 0.91 s, and $\sigma^2 h/g = 3.9$).

Figure 11 illustrates a comparison of wave elevation at gauge P4 (see Figure 1b) for different fishnet mesh sizes (λ). The results show that as λ decreases, the wave amplitude is greatly reduced, which means many incident waves do not pass through the platform as λ decreases. Instead, they reflect why the reflection coefficient increases greatly at $\sigma^2 h/g = 3.9$ in Figure 6.

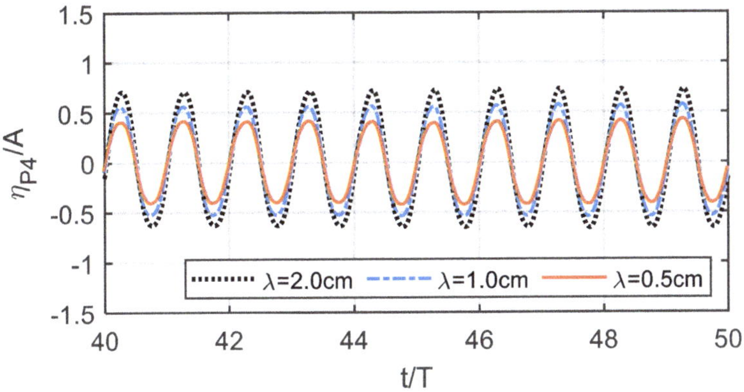

Figure 11. Variations in the wave elevation at gauge P4 for the platform with various fishnet mesh sizes (H = 4 cm, T = 0.91 s, and $\sigma^2 h/g$ = 3.9).

4.3. Nonlinear Dynamic Properties

In this study, the input wave height is 4 cm, and the wave steepness (wave height/wavelength) is in the range of 0.006 to 0.048. In fact, in this situation, the nonlinearity of waves is not significant. However, in our previous study [16], the second-order RAOs are still easy to observe at the resonant frequencies of the mooring system and platform. Thus, in this section, the second-order RAO of the platform motion and mooring system under different mesh size will be discussed.

Figures 12–14 show the second-order RAO of the surge, heave, and pitch motions of the platform with different fishnet mesh sizes (λ). In general, the second-order RAO is much smaller than the first-order RAO. However, it can still be observed, especially near the resonant frequencies of the platform and mooring system. Firstly, in the comparison of second-order surge RAO, as λ decreases, the RAO increases near the resonance frequency of the mooring system (between $\sigma^2 h/g$ = 1.0 and 2.0), but no significant change near the resonance frequency of the surge motion ($\sigma^2 h/g$ = 4.1) occurs. Next, compared with the second-order heave RAO, as λ decreases, the RAO is greatly reduced near the resonance frequency of the heave motion ($\sigma^2 h/g$ = 3.9). However, near the resonance frequency of the mooring system ($\sigma^2 h/g$ = 1), the changes between RAO and λ are irregular. As λ decreases from 2.0 cm to 1.0 cm, the RAO greatly decreases and the peak value shifts to a higher frequency. In contrast, as λ decreases from 1.0 cm to 0.5 cm, the RAO increases, and the peak value shifts to a lower frequency. Finally, in comparison with the second-order pitch RAO, as λ decreases, the peaks around the resonant frequency of the pitch motion ($\sigma^2 h/g$ = 4.1) hardly differ, but the frequency band is wider. Around the resonance frequency of the mooring system (between $\sigma^2 h/g$ = 1.0 and 2.0), it seems that the smaller the mesh size, the lower the RAO.

Figure 15 displays the second-order sea-side tension RAO of the platform with different fishnet mesh sizes (λ). In fact, the second-order tension RAO is very small by comparison with the first-order tension RAO. The results show that, near the frequency of $\sigma^2 h/g$ = 4.0, the RAO decreases with λ. Apart from that, the relationship between RAO and λ is irregular.

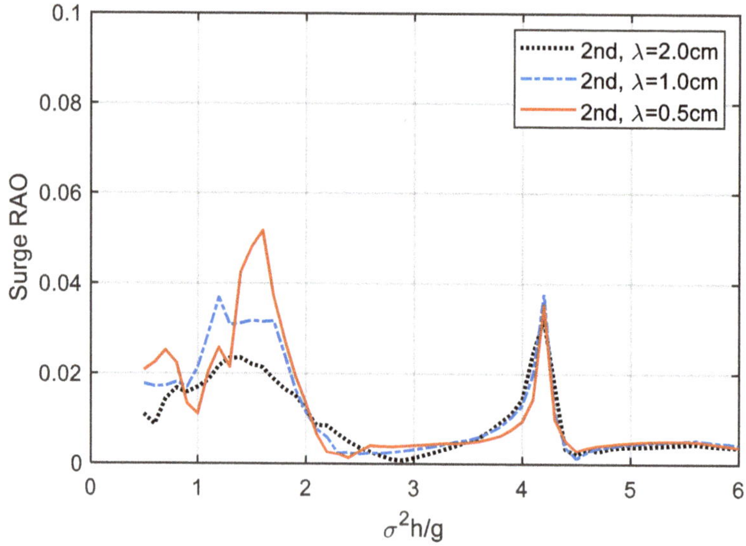

Figure 12. Comparison of the second-order surge RAO of the floating platform with different fishnet mesh sizes.

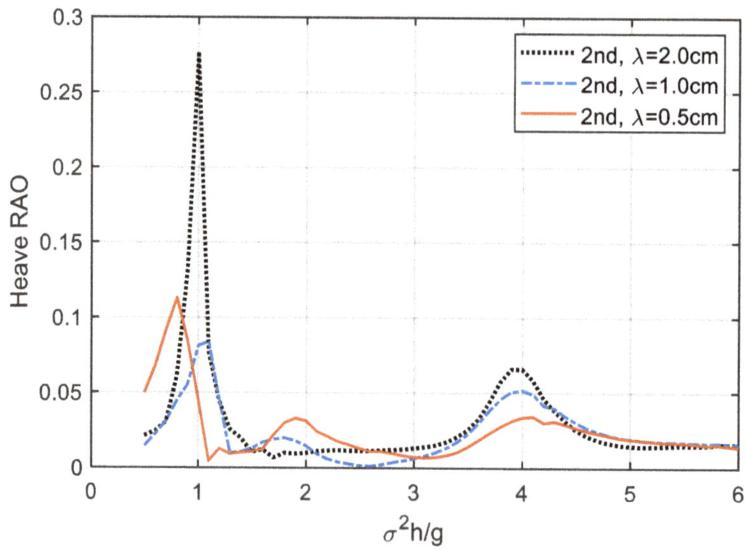

Figure 13. Comparison of the second-order heave RAO of the floating platform with different fishnet mesh sizes.

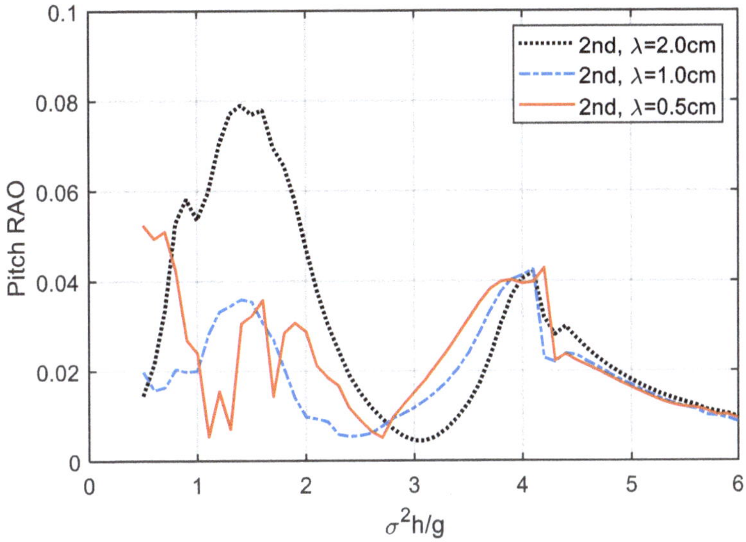

Figure 14. Comparison of the second-order pitch RAO of the floating platform with different fishnet mesh sizes.

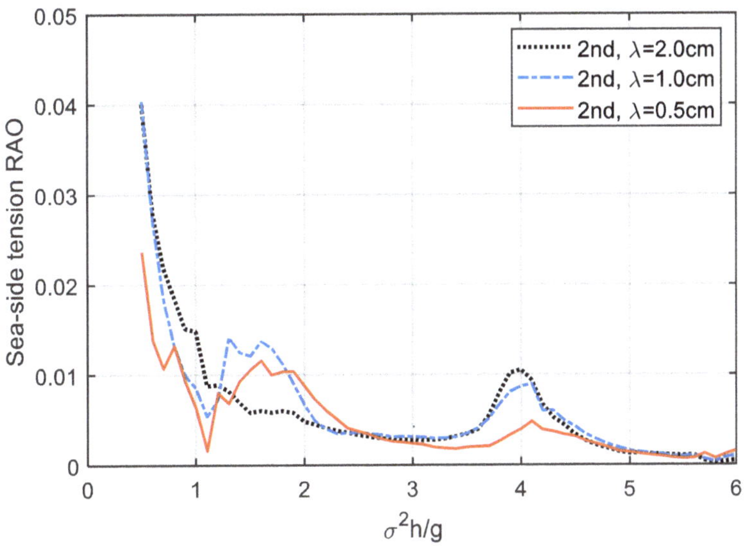

Figure 15. Comparison of the second-order sea-side tension RAO of the floating platform with different fishnet mesh sizes.

5. Conclusions

In this study, a fully nonlinear numerical wave tank was developed to investigate the dynamic response of a floating platform with three different fishnet mesh sizes. The mooring system was considered to be linearly elongated and symmetrically installed. This model is solved by the BEM, while the free surface nodes were tracked by the MEL approach with a cubic spline scheme and the RK4 method. Damping zones were arranged at both ends of the tank to absorb reflected wave energy and dissipate the transmitted wave energy. The instantaneous floating body motion was calculated by means of an acceleration potential method and a modal decomposition method.

The frequency-domain results show that as the fishnet mesh size decreases, all of the surge, heave, pitch, and sea-side tension RAO reduce, apart from the pitch and tension RAOs in the lower frequency region. In addition, the most significant reduction is observed at the resonant frequency of heave motion. At this frequency, the time-domain results indicate that major reductions in platform motions, tension forces, and transmitted wave heights will appear when decreasing the fishnet mesh size and may even occur with a slight time delay. Moreover, the second-order RAOs for the platform motion and mooring tension are observed in the simulation. However, these are very small compared with the first-order RAO. Essentially, most of them are decreased with the fishnet mesh size.

In the future, attaching fishnets to any kind of floating platform is not necessarily for aquaculture purposes only. It can also be a damper or breakwater because it can not only stabilize the movement of the structure, but also reduce the impact of waves.

Author Contributions: Conceptualization, H.-J.T. and R.-Y.Y.; methodology, H.-J.T. and C.-C.H.; software, H.-J.T.; validation, H.-J.T., C.-C.H. and R.-Y.Y.; formal analysis, H.-J.T.; investigation, H.-J.T.; resources, H.-J.T. and R.-Y.Y.; data curation, H.-J.T.; writing—original draft preparation, H.-J.T.; writing—review and editing, H.-J.T., C.-C.H. and R.-Y.Y.; visualization, H.-J.T.; supervision, R.-Y.Y. and C.-C.H.; project administration, H.-J.T. and R.-Y.Y.; funding acquisition, H.-J.T. and R.-Y.Y. All authors have read and agreed to the published version of the manuscript.

Funding: This research was funded by the Ministry of Science and Technology of Taiwan, grant number MOST 109-3116-F-006-013-CC1 and MOST 109-2222-E-006-003-MY2. The APC was funded by the Ministry of Science and Technology of Taiwan (MOST 109-3116-F-006-013-CC1).

Acknowledgments: The authors would like to thank the reviewers for their constructive comments and suggestions and the great support from the Ministry of Science and Technology of Taiwan.

Conflicts of Interest: The authors declare no conflict of interest.

References

1. Li, L.; Collu, M.; Ruzzo, C.; Failla, G.; Abhinav, K.A.; Arena, F. Analysis of the Coupled Dynamics of an Offshore Floating Multi-Purpose Platform: Part A—Rigid Body Analysis. In Proceedings of the ASME 2019 38th International Conference on Ocean, Offshore and Arctic Engineering (OMAE2019), Glasgow, Scotland, UK, 9–14 June 2019.
2. Ruzzo, C.; Failla, G.; Arena, F.; Li, L.; Collu, M. Analysis of the Coupled Dynamics of an Offshore Floating Multi-Purpose Platform: Part B—Hydro-Elastic Analysis with Flexible Support Platform. In Proceedings of the ASME 2019 38th International Conference on Ocean, Offshore and Arctic Engineering (OMAE2019), Glasgow, Scotland, UK, 9–14 June 2019.
3. Tsukrov, I.; Eroshkin, O.; Fredriksson, D.; Swift, M.; Celikkol, B. Finite element modeling of net panels using a consistent net element. *Ocean Eng.* **2003**, *30*, 251–270. [CrossRef]
4. Fredriksson, D.W.; Swift, M.; Irish, J.D.; Tsukrov, I.; Celikkol, B. Fish cage and mooring system dynamics using physical and numerical models with field measurements. *Aquac. Eng.* **2003**, *27*, 117–146. [CrossRef]
5. Huang, C.-C.; Tang, H.-J.; Liu, J.-Y. Dynamical analysis of net cage structures for marine aquaculture: Numerical simulation and model testing. *Aquac. Eng.* **2006**, *35*, 258–270. [CrossRef]
6. Huang, C.-C.; Tang, H.-J.; Liu, J.-Y. Modeling volume deformation in gravity-type cages with distributed bottom weights or a rigid tube-sinker. *Aquac. Eng.* **2007**, *37*, 144–157. [CrossRef]
7. Zhao, Y.-P.; Li, Y.; Dong, G.; Gui, F.-K.; Teng, B. A numerical study on dynamic properties of the gravity cage in combined wave-current flow. *Ocean Eng.* **2007**, *34*, 2350–2363. [CrossRef]
8. Kristiansen, T.; Faltinsen, O.M. Experimental and numerical study of an aquaculture net cage with floater in waves and current. *J. Fluids Struct.* **2015**, *54*, 1–26. [CrossRef]
9. Moe-Føre, H.; Lader, P.F.; Lien, E.; Hopperstad, O.S. Structural response of high solidity net cage models in uniform flow. *J. Fluids Struct.* **2016**, *65*, 180–195. [CrossRef]
10. Cifuentes, C.; Kim, M. Hydrodynamic response of a cage system under waves and currents using a Morison-force model. *Ocean Eng.* **2017**, *141*, 283–294. [CrossRef]
11. Bai, W.; Taylor, R.E. Higher-order boundary element simulation of fully nonlinear wave radiation by oscillating vertical cylinders. *Appl. Ocean Res.* **2006**, *28*, 247–265. [CrossRef]

12. Bai, W.; Taylor, R.E. Numerical simulation of fully nonlinear regular and focused wave diffraction around a vertical cylinder using domain decomposition. *Appl. Ocean Res.* **2007**, *29*, 55–71. [CrossRef]
13. Bai, W.; Taylor, R.E. Fully nonlinear simulation of wave interaction with fixed and floating flared structures. *Ocean Eng.* **2009**, *36*, 223–236. [CrossRef]
14. Boo, S. Linear and nonlinear irregular waves and forces in a numerical wave tank. *Ocean Eng.* **2002**, *29*, 475–493. [CrossRef]
15. Gaeta, M.G.; Segurini, G.; Moreno, A.; Archetti, R. Implementation and Validation of a Potential Model for a Moored Floating Cylinder under Waves. *J. Mar. Sci. Eng.* **2020**, *8*, 131. [CrossRef]
16. Tang, H.-J.; Huang, C.-C.; Chen, W.-M. Dynamics of dual pontoon floating structure for cage aquaculture in a two-dimensional numerical wave tank. *J. Fluids Struct.* **2011**, *27*, 918–936. [CrossRef]
17. Guo, Y.; Mohapatra, S.; Soares, C.G. Review of developments in porous membranes and net-type structures for breakwaters and fish cages. *Ocean Eng.* **2020**, *200*, 107027. [CrossRef]
18. Bi, C.-W.; Zhao, Y.-P.; Dong, G.; Cui, Y.; Gui, F.-K. Experimental and numerical investigation on the damping effect of net cages in waves. *J. Fluids Struct.* **2015**, *55*, 122–138. [CrossRef]
19. Zhou, C.; Xu, L.; Hu, F.; Qu, X. Hydrodynamic characteristics of knotless nylon netting normal to free stream and effect of inclination. *Ocean Eng.* **2015**, *110*, 89–97. [CrossRef]
20. Gansel, L.C.; Plew, D.R.; Endresen, P.C.; Olsen, A.I.; Misimi, E.; Guenther, J. Jensen, Østen Drag of Clean and Fouled Net Panels – Measurements and Parameterization of Fouling. *PLoS ONE* **2015**, *10*, e0131051. [CrossRef]
21. Bi, C.-W.; Zhao, Y.-P.; Dong, G.-H.; Wu, Z.-M.; Zhang, Y.; Xu, T.-J. Drag on and flow through the hydroid-fouled nets in currents. *Ocean Eng.* **2018**, *161*, 195–204. [CrossRef]
22. Koo, W.; Kim, M.-H. Freely floating-body simulation by a 2D fully nonlinear numerical wave tank. *Ocean Eng.* **2004**, *31*, 2011–2046. [CrossRef]
23. Tang, H.-J.; Huang, C.-C. Bragg reflection in a fully nonlinear numerical wave tank based on boundary integral equation method. *Ocean Eng.* **2008**, *35*, 1800–1810. [CrossRef]
24. Longuet-Higgins, M.S.; Cokelet, E. The Deformation of Steep Surface Waves on Water: I. a Numerical Method of Computation. *Proc. Royal. Soc. Lond.* **1976**, *A350*, 1–26.
25. Cointe, R. Numerical simulation of a wave channel. *Eng. Anal. Bound. Elem.* **1990**, *7*, 167–177. [CrossRef]
26. Tanizawa, K. Long Time Fully Nonlinear Simulation of Floating Body Motions with Artificial Damping Zone. *J. Soc. Nav. Arch. Jpn.* **1996**, *1996*, 311–319. [CrossRef]
27. Grilli, S.; Svendsen, I. Corner problems and global accuracy in the boundary element solution of nonlinear wave flows. *Eng. Anal. Bound. Elem.* **1990**, *7*, 178–195. [CrossRef]
28. Tanizawa, K. A Nonlinear Simulation Method of 3-D Body Motions in Waves (1st Report). *J. Soc. Nav. Arch. Jpn.* **1995**, *1995*, 179–191. [CrossRef]
29. Cao, Y.; Beck, R.F.; Schultz, W.W. Nonlinear Computation of Wave Loads and Motions of Floating Bodies in Incident Waves. In Proceedings of the 9th Workshop on Water Waves and Floating Bodies, Kuju, Oita, Japan, 17–20 April 1994; pp. 33–37.
30. Yan, S.; Ma, Q. Numerical simulation of fully nonlinear interaction between steep waves and 2D floating bodies using the QALE-FEM method. *J. Comput. Phys.* **2007**, *221*, 666–692. [CrossRef]
31. Vinje, T.; Brevig, P. Nonlinear Ship Motions. In Proceedings of the 3rd International Conference on Numerical Ship Hydrodynamics, Paris, France, 16–19 June 1981; Volume 10.
32. Wu, G.X.; Eatock Taylor, R. Transient Motion of a Floating Body in Steep Water Waves. In Proceedings of the Eleventh International Workshop on Water Waves and Floating Bodies, Hamburg, Germany, 17–20 March 1996.
33. Tanizawa, K. The State of the Art on Numerical Wave Tank. In Proceedings of the 4th Osaka Colloquium on Seakeeping Performance of Ships, Osaka, Japan, 17–21 October 2000; pp. 95–114.
34. Ma, Q.; Yan, S. QALE-FEM for numerical modelling of non-linear interaction between 3D moored floating bodies and steep waves. *Int. J. Numer. Methods Eng.* **2009**, *78*, 713–756. [CrossRef]
35. Brebbia, C.; Walker, S. Dynamic analysis of offshore structures. *Appl. Ocean Res.* **1981**, *3*, 205. [CrossRef]
36. Loland, G. *Current Forces On and Flow through Fish Farms, Division of Marine Hydrodynamics*; Norwegian Institute of Technology: Trondheim, Norway, 1991; pp. 85–95.
37. Yamamoto, T.; Yoshida, A.; Ijima, T. Dynamics of elastically moored floating objects. *Appl. Ocean Res.* **1980**, *2*, 85–92. [CrossRef]

38. Chopra, A.K. *Dynamics of Structures: Theory and Application to Earthquake Engineering*; Prentice-Hall Inc.: Upper Saddle River, NJ, USA, 2001.
39. Mansard, E.P.D.; Funke, E.R. The Measurement of Incident and Reflected Spectra Using a Least Squares Method. In Proceedings of the 17th Coastal Engineering Conference, ASCE, Sydney, Australia, 23–28 March 1980; pp. 154–172.
40. Williams, A.; Abul-Azm, A. Dual pontoon floating breakwater. *Ocean Eng.* **1997**, 24, 465–478. [CrossRef]
41. Williams, A.; Lee, H.; Huang, Z. Floating pontoon breakwaters. *Ocean Eng.* **2000**, 27, 221–240. [CrossRef]

Journal of
Marine Science and Engineering

MDPI

Article

Mooring Drag Effects in Interaction Problems of Waves and Moored Underwater Floating Structures

Cheng-Tsung Chen [1,2], Jaw-Fang Lee [3,*] and Chun-Han Lo [4]

1 Department of Civil Engineering, Yancheng Institute of Technology, Yancheng 224051, China; ctchen42@yahoo.com.tw
2 Center for Innovative Research on Aging Society, National Chung Cheng University, Chia-yi 621, Taiwan
3 Department of Hydraulic and Ocean Engineering, National Cheng Kung University, Tainan 701, Taiwan
4 Water Resources Planning Institute, Water Resources Agency, Ministry of Economic Affairs, Taichung City 413, Taiwan; a6bennylo@gmail.com
* Correspondence: jflee@mail.ncku.edu.tw

Received: 28 December 2019; Accepted: 21 February 2020; Published: 25 February 2020

Abstract: In contrast to either considering structures with full degrees of freedom but with wave force on mooring lines neglected or with wave scattering and radiation neglected, in this paper, a new analytic solution is presented for wave interaction with moored structures of full degrees of freedom and with wave forces acting on mooring lines considered. The linear potential wave theory is applied to solve the wave problem. The wave fields are expressed as superposition of scattering and radiation waves. Wave forces acting on the mooring lines are calculated using the Morison equation with relative motions. A coupling formulation among water waves, underwater floating structure, and mooring lines are presented. The principle of energy conservation, as well as numerical results, are used to verify the present solution. With complete considerations of interactions among waves and moored structures, the characteristics of motions of the structure, the wave fields, and the wave forces acting on the mooring lines are investigated.

Keywords: analytic solution; water waves; underwater floating structure; mooring forces; interaction

1. Introduction

With increasing development of ocean wave energy extraction, various types of underwater ocean structures were used in these aspects [1]. As incident waves acting on underwater floating structures, the wave forces in fact act on both main floating objects as well as the mooring deployments. The reactive motions of the structure systems then feedback into the surrounding wave fields, thus forming wave and structure interaction systems. The floating structures offer scattering and radiation phenomena on water waves, whereas wave forces on mooring lines induce motions of the mooring lines and cause motions of the floating structures and surrounding wave fields. In literature, most studies on interactions of floating structures and incident waves consider either problems neglecting wave forces on mooring lines [2] or problems with wave forces on mooring lines but without scattering and radiation structural effects [3].

Numerical simulations are commonly used to calculate problems of ocean structures subjected to incident waves. A three-dimensional finite element method was developed by Huang et al. [4] to calculate wave diffraction, wave radiation, and body responses of multiple bodies of arbitrary shape. Sannasiraj et al. [5] applied both experimental and finite element methods to study behaviors of pontoon-type floating structures in waves. The slack mooring lines were idealized as spring coefficients calculated from the catenary equation of cables. Chen et al. [6] used a boundary integral and Green's function to investigate floating breakwaters consisting of rectangular pontoon and horizontal plates. The mooring lines were calculated using the static catenary equation. Mohapatra and Sahoo [7] used a Green's integral to study the interaction problem of oblique surface gravity waves with a floating flexible plate. In Cao and Zhao [8], a Computational fluid dynamics (CFD) numerical method was used to study nonlinear dynamic behaviors of a two-dimensional box-shaped floating structure in focused waves. Mohapatra and Soares [9] used linearized Boussinesq equations to study the wave forces acting on a floating structure over a flat bottom. Kao et al. [10] used a boundary element method to solve the problem of floating structures subjected to incident waves. Rivera-Arreba [11] studied the dynamic response of the floating wind turbine subjected to wind and waves. On the other hand, Guo et al. [12] used a linear wave theory to calculate the wave forces acting on the structure. In numerical simulations for waves acting on floating structures with moorings, mostly the mooring mechanisms were included and wave interferences were considered. However, wave forces acting on mooring lines were not considered. The reason was that the calculation of wave forces on mooring lines includes wave kinematics and motions of mooring lines, which complicated mathematical formulation in the problem.

As for the analytic approach solving for interaction problems of wave and mooring floating structures, in the analysis of waves interacting with moored floating structures, most researches neglected wave forces acting on the mooring lines [2]. Lee [13] first considered a tension-leg floating structure subjected to wave actions, in which the floating structure was assumed to have surge motion only. The wave force acting on the tension leg was calculated by a linearized Morison equation, and an analytic solution was proposed for the entire interacted problem. Lee et al. [14] applied the interaction methodology of large and small structures, extending to tethered mooring tension leg floating structures. Lee and Wang [15] extended the same technique to problems of tension leg platforms with net cages. In the articles mentioned above, the floating structures allow only surge motion; therefore, analytic solutions could be obtained without any difficulty.

If the structures are allowed to include heave and surge motions, one would then encounter the typical heave radiation problem. Lee [16] proposed an easy-to-follow derivation to obtain an analytic solution. Other than that, a particular solution approach has to be applied that was not convenient to use in obtaining the solution. It could be said that, using Lee's method, the radiation problem of the two-dimensional structure can be obtained completely. Chen et al. [17] then investigated the problem of wave interaction with a floating structure with moorings, and wave forces acting on the mooring lines were included. In the two-dimensional problem, the floating structure had complete three degrees of freedom, namely, surge, heave, and pitch.

In this study, an underwater floating structure with moorings subjected to incident waves is considered. Zheng et al. [18] presented an analytic solution for oblique waves passing an underwater floating rectangular structure. A particular solution was used to satisfy the nonhomogeneous boundary value problem. However, the analytic solution could not be simplified to the case of normal incident wave, as in the two-dimensional problem. The intention of this paper is to present a new analytic solution to the problem. The significance of this paper is that the wave forces acting on the mooring lines are considered in the coupling problems of waves and floating structures and the problems solved analytically. The floating structure has motions with three degrees of freedom. The effects of the mooring lines subjected to wave forces on the hydrodynamics of the wave and structure interaction system are investigated.

2. Problem Description and Solution

The problem considered is an underwater floating structure moored to the sea bottom and subjected to the action of incident waves, as shown in Figure 1. A Cartesian coordinate system is adopted with the positive x pointed to the right and positive z pointed upward. The constant water depth is h, the width of the structure is 2ℓ, the structural submergence is d_1, and the distance from the structural bottom to the sea bottom is d_2. The incident wave η^I is propagating from $-x$ to the $+x$ direction. With the action of incident waves, the structure system does respond accordingly and also interferes with the surrounding wave field. In this two-dimensional problem, the floating structure has, in general, three degrees of freedom, namely, surge, heave, and pitch motions. With the prerequisite periodic motion, the displacement functions of the structure can be expressed as

$$\xi_j = s_j \cdot e^{-i\omega t},\ j = 1,2,3 \tag{1}$$

where the subscripts 1, 2, and 3 represent surge, heave, and pitch, respectively. s_j represents amplitudes of the structural motions. Since an analytic solution is pursued, and with a rectangular shape of the structure, the method of separation of variables is used to solve the problem, and the domain is divided into four regions, as indicated in Figure 1. Region 1 is in front of the structure, regions 2 and 4 are above and beneath the structure, and region 3 is behind the structure.

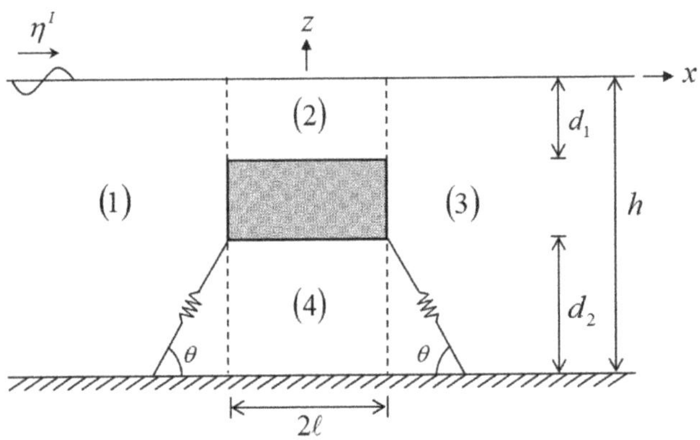

Figure 1. Definition sketch of waves incidents on an underwater floating structure with moorings.

The interference wave field surrounding the floating structure needs to be solved, in addition to the known incident wave, so that wave forces acting on the floating structure and the moorings can be calculated, and so, to calculate structural motions.

A linear potential wave theory is used to describe the wave problem. The definition of the velocity $\vec{V}$ related to the wave potential function Φ is written as

$$\vec{V} = -\nabla\Phi \tag{2}$$

where ∇ is the gradient operator. Since steady and periodic problems are considered, the periodic time function can be factored out and wave potential be expressed as

$$\Phi(x,z,t) = \phi(x,z) \cdot e^{-i\omega t} \tag{3}$$

where $\omega = 2\pi/T$, T is the wave period, and $i = \sqrt{-1}$. The incident wave potential is given as

$$\Phi^I(x,z,t) = \frac{igA^I}{\omega} \cdot \frac{\cosh K(z+h)}{\cosh Kh} \cdot e^{i\,(Kx-\omega t)} \tag{4}$$

where g is the gravitational constant, A^I is the wave amplitude, and K is the wave number that is calculated by the dispersion equation

$$\omega^2 = gK\tanh Kh \tag{5}$$

The entire problem is decomposed into a scattering problem and radiation problems in the three degrees of freedom [2]; i.e., the interference wave potential is expressed as

$$\Phi(x,z,t) = \Phi^D + \sum_{j=1}^{3} s_j \cdot \Phi^j \tag{6}$$

where Φ^D is the scattering wave produced by the incident wave acting on the structure with the structure held fixed. On the other hand, Φ^j corresponds to the radiation wave for the structure having motion in the j-direction and with unit amplitude. For the problem here, surge, heave, and pitch motions.

Now, the task becomes to obtain analytical solutions for the scattering wave and the radiation waves and, particularly, the solution expression for each individual region. Since analytic solutions for the problem of a surface-floating structure have been presented by Chen et al. [17], the solutions for divided regions can be followed, except the region 2 for heave and pitch radiation problems. Further, solutions for the pitch radiation problem are an application of heave and surge problems; therefore, only derivation details of the region 2 in the heave problem will be shown here.

The boundary value problem for region 2 in the heave radiation problem can be written as:

The governing equation:

$$\nabla^2 \phi_2^2 = 0, \ \ -d_1 < z < 0, \ \ -\ell < x < \ell \tag{7}$$

The upper free surface condition:

$$\frac{\partial \phi_2^2}{\partial z} = \frac{\omega^2}{g} \phi_2^2, \ \ z = 0 \tag{8}$$

The lower boundary condition:

$$\frac{\partial \phi_2^2}{\partial z} = i\omega s_2, \ \ z = -d_1 \tag{9}$$

The left boundary condition:

$$\phi_2^2 = \phi_1^2, \ \ x = -\ell \tag{10}$$

The right boundary condition:

$$\phi_2^2 = \phi_3^2, \ \ x = \ell \tag{11}$$

Note that it is the nonhomogeneous form shown in Equation (1) that poses the difficulty in obtaining the solution. In this study, a method proposed by Lee [6] is used to derive the solution. In Equations (7)–(11), the wave potential is divided into two parts:

$$\phi_2^2 = \widetilde{\phi}_2^2 + \hat{\phi}_2^2 \tag{12}$$

where $\widetilde{\phi}_2^2$ and $\hat{\phi}_2^2$ satisfy vertically and horizontally homogeneous conditions, respectively, as shown in Figure 2.

Following the standard method of separation of variables, one can obtain the solutions

$$\begin{aligned} \phi_2^2 &= \widetilde{\phi}_2^2 + \hat{\phi}_2^2 \\ &= \sum_{n=0}^{\infty} \left[A_{2n}^2 e^{-k_{2n}(x+\ell)} + B_{2n}^2 e^{k_{2n}(x-\ell)} \right] \cos[k_{2n}(z+d_1)] \\ &+ \sum_{n=1}^{\infty} D_{2n}^2 \left(\mu_{n1} e^{\gamma_n(z-h)} + \mu_{n2} e^{-\gamma_n(z+h)} \right) \sin \gamma_n(x+\ell) \end{aligned} \tag{13}$$

in which the coefficients k_{2n}, γ_n, μ_{n1}, μ_{n2}, and D^2_{2n} are given in Appendix A. The way of obtaining the solution for region 2 in the heave radiation problem can also be applied to obtain the solution for the same region 2 in the pitch radiation problem.

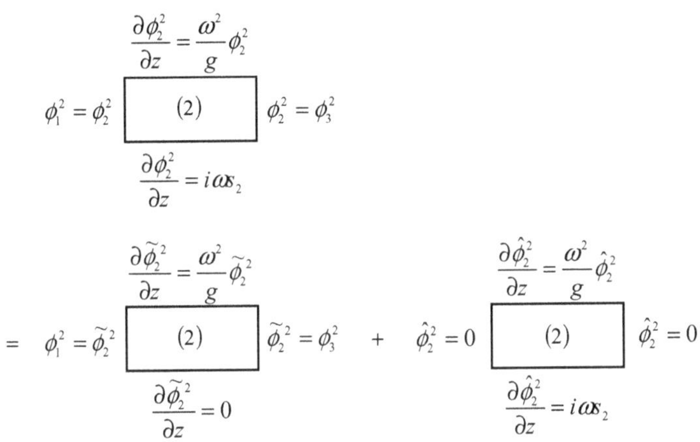

Figure 2. Decomposition of nonhomogeneous boundary conditions.

As for the rest of the scattering and the radiation problems, one can easily obtain the solutions. For completeness, they are also listed here. For the wave scattering problem, solutions for the four regions can be written as:

$$\phi_1^D = \sum_{n=0}^{\infty} B_{1n}^D \cos[k_n(z+h)] e^{k_n(x+\ell)} \tag{14}$$

$$\phi_2^D = \sum_{n=0}^{\infty} \left[A_{2n}^D e^{-k_{2n}(x+\ell)} + B_{2n}^D e^{k_{2n}(x-\ell)} \right] \cos[k_{2n}(z+d_1)] \tag{15}$$

$$\phi_3^D = \sum_{n=0}^{\infty} A_{3n}^D \cos[k_n(z+h)] e^{-k_n(x-\ell)} \tag{16}$$

$$\phi_4^D = \left(A_{40}^D x + B_{40}^D \right) + \sum_{n=1}^{\infty} \left[A_{4n}^D e^{-k_{4n}(x+\ell)} + B_{4n}^D e^{k_{4n}(x-\ell)} \right] \cos[k_{4n}(z+h)] \tag{17}$$

For the radiation wave problems, the three radiations surge, heave, and pitch are expressed, respectively, as:

Surge radiation:

$$\phi_1^1 = \sum_{n=0}^{\infty} B_{1n}^1 \cos[k_n(z+h)] e^{k_n(x+\ell)} \tag{18}$$

$$\phi_2^1 = \sum_{n=0}^{\infty} \left[A_{2n}^1 e^{-k_{2n}(x+\ell)} + B_{2n}^1 e^{k_{2n}(x-\ell)} \right] \cos[k_{2n}(z+d_1)] \tag{19}$$

$$\phi_3^1 = \sum_{n=0}^{\infty} A_{3n}^1 \cos[k_n(z+h)] e^{-k_n(x-\ell)} \tag{20}$$

$$\phi_4^1 = \left(A_{40}^1 x + B_{40}^1 \right) + \sum_{n=1}^{\infty} \left[A_{4n}^1 e^{-k_{4n}(x+\ell)} + B_{4n}^1 e^{k_{4n}(x-\ell)} \right] \cos[k_{4n}(z+h)] \tag{21}$$

Heave radiation:

$$\phi_1^2 = \sum_{n=0}^{\infty} B_{1n}^2 \cos[k_n(z+h)]e^{k_n(x+\ell)} \tag{22}$$

$$\begin{aligned}\phi_2^2 = & \sum_{n=0}^{\infty}\left[A_{2n}^2 e^{-k_{2n}(x+\ell)} + B_{2n}^2 e^{k_{2n}(x-\ell)}\right]\cos[k_{2n}(z+d_1)] \\ & + \sum_{n=1}^{\infty} D_{2n}^2\left(\mu_{n1}e^{\gamma_n(z-h)} + \mu_{n2}e^{-\gamma_n(z+h)}\right)\sin\gamma_n(x+\ell)\end{aligned} \tag{23}$$

$$\phi_3^2 = \sum_{n=0}^{\infty} A_{3n}^2 \cos[k_n(z+h)]e^{-k_n(x-\ell)} \tag{24}$$

$$\begin{aligned}\phi_4^2 = & \left(A_{40}^2 x + B_{40}^2\right) + \sum_{n=1}^{\infty}\left[A_{4n}^2 e^{-k_{4n}(x+\ell)} + B_{4n}^2 e^{k_{4n}(x-\ell)}\right]\cos[k_{4n}(z+h)] \\ & + \sum_{n=1}^{\infty} F_{4n}^2 \cosh\gamma_n(z+h)\sin\gamma_n(x+\ell)\end{aligned} \tag{25}$$

Pitch radiation:

$$\phi_1^3 = \sum_{n=0}^{\infty} B_{1n}^3 \cos[k_n(z+h)]e^{k_n(x+\ell)} \tag{26}$$

$$\begin{aligned}\phi_2^3 = & \sum_{n=0}^{\infty}\left[A_{2n}^3 e^{-k_{2n}(x+\ell)} + B_{2n}^2 e^{k_{2n}(x-\ell)}\right]\cos[k_{2n}(z+d_1)] \\ & + \sum_{n=1}^{\infty} D_{2n}^3\left(\mu_{n1}e^{\gamma_n(z-h)} + \mu_{n2}e^{-\gamma_n(z+h)}\right)\sin\gamma_n(x+\ell)\end{aligned} \tag{27}$$

$$\phi_3^3 = \sum_{n=0}^{\infty} A_{3n}^3 \cos[k_n(z+h)]e^{-k_n(x-\ell)} \tag{28}$$

$$\begin{aligned}\phi_4^3 = & \left(A_{40}^3 x + B_{40}^3\right) + \sum_{n=1}^{\infty}\left[A_{4n}^3 e^{-k_{4n}(x+\ell)} + B_{4n}^3 e^{k_{4n}(x-\ell)}\right]\cos[k_{4n}(z+h)] \\ & + \sum_{n=1}^{\infty} F_{4n}^3 \cosh\gamma_n(z+h)\sin\gamma_n(x+\ell)\end{aligned} \tag{29}$$

where k_n, k_{4n}, F_{4n}^2, D_{2n}^3, and F_{4n}^3 are listed in Appendix A. The undetermined coefficient shown in Equations (18)–(29) are then obtained by solving simultaneous equations obtained from matching the velocity and pressure conditions at the interfacial boundary of two neighboring regions and integration of associated water depth multiplied by the orthogonal functions.

Once the decomposed wave scattering problem and the wave radiation problem of unit amplitude are obtained, the unknown variables shown in the interference wave potential, Equation (6), are amplitudes of the structural motion, which can then be solved by the equations of motion of the structure.

The equations of motion of the underwater floating structure can be written as [19]:

$$[M]\begin{Bmatrix} \ddot{\xi}_1 \\ \ddot{\xi}_2 \\ \ddot{\xi}_3 \end{Bmatrix} = \begin{Bmatrix} F_1 \\ F_2 \\ F_3 \end{Bmatrix} - \begin{Bmatrix} T_1 \\ T_2 \\ T_3 \end{Bmatrix} + \begin{Bmatrix} F_1^M \\ F_2^M \\ F_3^M \end{Bmatrix} \tag{30}$$

where $[M]$ is the mass matrix, $\{F\}$ is the wave forces acting on the floating structure, $\{T\}$ represents the restoring force of the mooring springs, and $\{F^M\}$ is the wave forces acting on the mooring lines. The mass matrix can be expressed as:

$$[M] = \begin{bmatrix} m & 0 & 0 \\ 0 & m & 0 \\ 0 & 0 & I_0 \end{bmatrix} \tag{31}$$

in which m is mass of the structure and I_0 is moment of inertia.

Wave forces acting on the floating structure can be calculated using wave potentials surrounding the structure, and be expressed as:

$$\left\{\begin{array}{c} F_1 \\ F_2 \\ F_3 \end{array}\right\} = \left[f^R\right]\left\{\begin{array}{c} s_1 \\ s_2 \\ s_3 \end{array}\right\} + \left\{\begin{array}{c} f_1^D \\ f_3^D \\ f_3^D \end{array}\right\} \tag{32}$$

where $\left[f^R\right]$ and $\left\{f^D\right\}$ are calculated from radiated potentials of unit amplitudes and diffracted potentials, respectively. Detailed expressions are given in Appendix B.

The restoring forces produced by the mooring springs can be calculated according to orientations of the springs $\overline{AB}$ and $\overline{CD}$, and be expressed as:

$$T^{\overline{AB}} = \left[\begin{array}{ccc} K_{\underline{11}}^{\overline{AB}} & K_{\underline{12}}^{\overline{AB}} & K_{\underline{13}}^{\overline{AB}} \\ K_{\underline{21}}^{\overline{AB}} & K_{\underline{22}}^{\overline{AB}} & K_{\underline{23}}^{\overline{AB}} \\ K_{31}^{\overline{AB}} & K_{32}^{\overline{AB}} & K_{33}^{\overline{AB}} \end{array}\right]\left[\begin{array}{c} \xi_1 \\ \xi_2 \\ \xi_3 \end{array}\right] \tag{33}$$

$$T^{\overline{CD}} = \left[\begin{array}{ccc} K_{\underline{11}}^{\overline{CD}} & K_{\underline{12}}^{\overline{CD}} & K_{\underline{13}}^{\overline{CD}} \\ K_{\underline{21}}^{\overline{CD}} & K_{\underline{22}}^{\overline{CD}} & K_{\underline{23}}^{\overline{CD}} \\ K_{31}^{\overline{CD}} & K_{32}^{\overline{CD}} & K_{33}^{\overline{CD}} \end{array}\right]\left[\begin{array}{c} \xi_1 \\ \xi_2 \\ \xi_3 \end{array}\right] \tag{34}$$

in which the expressions of the stiffness matrices are given in Appendix C.

The wave forces acting on the mooring lines are calculated using a linearized Morison [3]. Using the present analytic solutions for the wave fields and the associated geometrical deployments of the mooring lines, the wave forces can be calculated. Detailed derivations are given in Appendix D. The induced forces acting on the floating structure can then be written as

$$\left\{\begin{array}{c} F_1^M \\ F_2^M \\ F_3^M \end{array}\right\} = \left[f^{MR}\right]\left\{\begin{array}{c} s_1 \\ s_2 \\ s_3 \end{array}\right\} + \left\{\begin{array}{c} f_1^{MD} \\ f_3^{MD} \\ f_3^{MD} \end{array}\right\} \tag{35}$$

in which $\left[f^{MR}\right]$ is the radiation wave generated coefficient matrix.

Having the required expressions of all forces acting on the floating structure, including scattering and radiation waves, mooring restoring forces, and effects of wave forces on mooring lines, the equations of motion of the structure, Equation (30), can be solved and expressed as:

$$\left\{\begin{array}{c} s_1 \\ s_2 \\ s_3 \end{array}\right\} = \left[\widetilde{K}\right]^{-1}\left(-i\omega\rho\left\{\begin{array}{c} f_1^D \\ f_3^D \\ f_3^D \end{array}\right\} + \left\{\begin{array}{c} f_1^{MD} \\ f_2^{MD} \\ f_3^{MD} \end{array}\right\}\right) \tag{36}$$

where the general stiffness matrix is

$$\left[\widetilde{K}\right] = \left(-\omega^2[M] + i\omega\rho\left[f^R\right] + [K] - \left[f^{MR}\right]\right) \tag{37}$$

Once amplitudes of the structural motion can be calculated, then the wave potentials of the entire problem domain can then be determined via Equation (6). The reflected wave in front of the structure η^R and the transmitted wave behind the structure η^T can then be calculated using the Bernoulli's equation. So far, the entire coupling problem is solved. A consideration of wave forces calculated from incident wave, scattering wave, radiation wave, and wave forces on the mooring lines, then the motions of the structure with moorings, are solved.

3. Results and Discussion

In this paper, the problem of moored underwater floating structures with motions of full degrees of freedom subjected to incident waves is investigated, and an analytic solution is presented. The present analytic solution is first validated by conservation of wave energy with no energy dissipation. In the present theory, the only energy loss in the problem is the drag forces acting on the mooring lines; therefore, if the drag coefficient is specified zero, $C_D = 0$, then the energy conservation of the system should satisfy. Figure 3 shows reflection and transmission coefficients and total wave energy, $K_r^2 + K_t^2$, versus relative water depth, Kh, and as is expected, the wave energy conserved to unity. The conditions used are water depth, $h = 10$ m, and incident wave amplitude, $A^I = 0.5$ m; other parameters used are $d_1/h = 0.2$, $\ell/h = 0.5$, $a/h = 0.3$, and the virtual coefficient $C_M = 2.0$. The corresponding result for the case considering the drag coefficient, $C_D = 2.0$, is shown in Figure 4. It is reasonable to identify that, with the drag effect, the total energy indicates dissipation. With energy dissipation, the total energy curve decreases about 10% at resonant frequency, the reflection coefficient decreases from 1.0 to 0.94, and the transmission coefficient increases from zero to 0.13. The present analytic solution is further applied to calculate a wave scattering problem of an underwater plate, and the results compared with that calculated using a numerical finite element method (Cheong et al. [20]). The conditions used are water depth, $h = 10$ m, and incident wave amplitude, $A^I = 0.5$ m; other parameters used are $d_1/h = 0.3$, $\ell/h = 0.5$, and $a/h = 0.025$. The comparisons of the reflection and transmission coefficients versus dimensionless water depth are shown in Figure 5, in which good agreements are indicated.

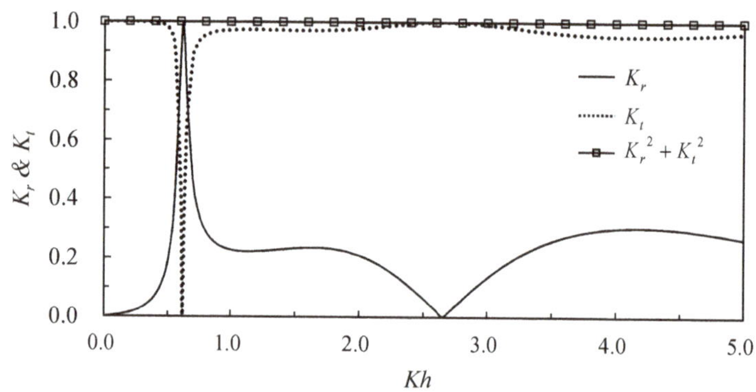

Figure 3. Reflection coefficient; transmission coefficient; and total energy versus relative water depth, Kh ($C_D = 0.0$, $d_1/h = 0.2$, $\ell/h = 0.5$, $a/h = 0.3$, and $C_M = 2.0$).

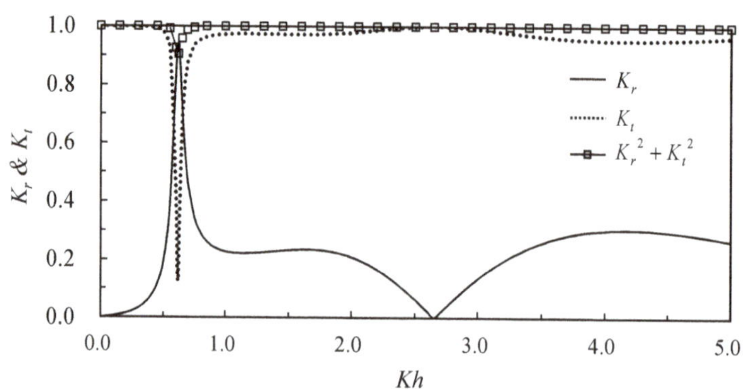

Figure 4. Reflection coefficient; transmission coefficient; and total energy versus relative water depth, Kh ($C_D = 2.0$, $d_1/h = 0.2$, $\ell/h = 0.5$, $a/h = 0.3$, and $C_M = 2.0$).

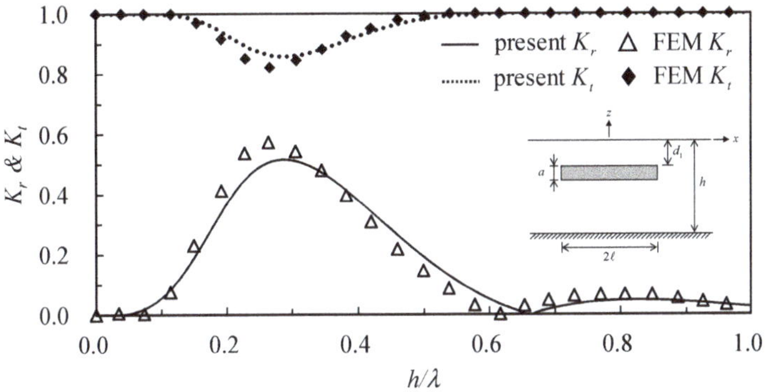

Figure 5. Reflection coefficient versus dimensionless water depth, h/λ, for a fixed underwater plate ($h = 10$ m, $d_1/h = 0.3$, $\ell/h = 0.5$, and $a/h = 0.025$).

In the present theory, the wave forces acting on the mooring lines are considered. To comply with motions of the mooring lines, in the wave force calculation, a relative flow velocity is used in the Morison equation. Since the motions of the mooring lines are not known a priori, an iteration algorithm is used in the calculation. Figure 6 shows the iteration times versus relative water depth, Kh. The conditions used are water depth, $h = 10$ m, and incident wave amplitude, $A^I = 0.5$ m; other parameters used are $d_1/h = 0.25$, $\ell/h = 0.2$, and $a/h = 0.3$, and the drag coefficient and the virtual mass coefficient, $C_D = 2.0$ and $C_M = 2.0$, respectively. The iteration number can be as high as 14 times at the resonant frequency and only one time at other frequencies. Additionally, with a higher drag coefficient, a higher iteration number is required.

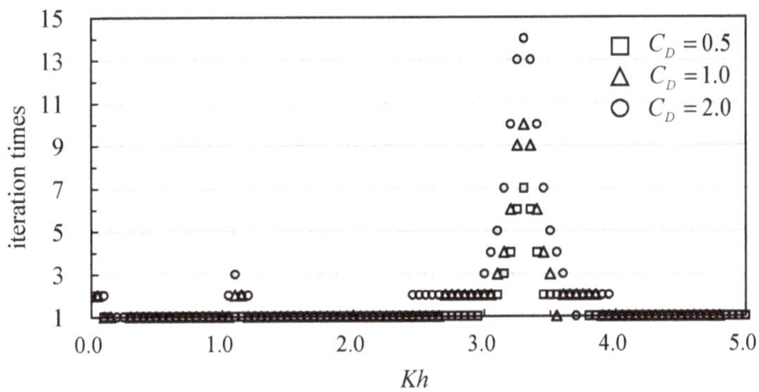

Figure 6. Iteration number versus relative water depth for different drag coefficients ($d_1/h = 0.25$, $\ell/h = 0.2$, $a/h = 0.3$, $C_D = 2.0$, and $C_M = 2.0$).

Using the same conditions as those in Figure 6, effects of the drag coefficient on wave reflection and wave transmission are shown in Figures 7 and 8, respectively. With the increase of the drag coefficient from zero to 2.0, the reflection coefficient at the first resonant peak at a lower frequency drops from 1.0 to 0.94, while the second peak at a higher frequency drops from 1.0 to 0.48. It is reasonable that the drag force acting on mooring lines can dampen only high-frequency short waves; rather, it is not sufficient in reducing wave energy of long waves. Therefore, high-frequency waves at resonant peaks are damped and decrease the reflection coefficient. The results also indicate that the increase of the drag coefficient from zero to 2.0 can dampen out 50% of the reflected waves. The tendency reverses for the transmission coefficient. The effects of the drag coefficient on structural motions are shown in Figures 9–11 for surge, heave, and pitch motions, respectively. Similar trends can be observed.

The increase of the drag coefficient reduces amplitudes of surge and pitch motions at high-frequency peaks, whereas it is not obvious at low-frequency resonant peaks. In this study, since it is difficult to obtain experimental results for comparison, numerical results using a boundary element method [21] for the case of $C_D = 2.0$ are also plotted for comparison. The comparisons also validate the present analytic solution for the problem.

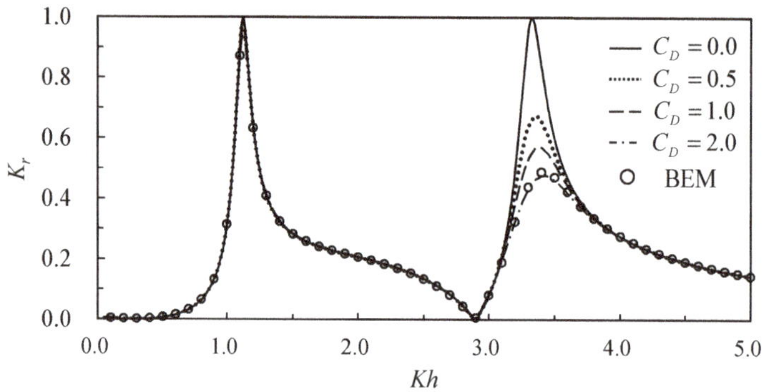

Figure 7. Reflection coefficient versus relative water depth for various drag coefficients ($d_1/h = 0.25$, $\ell/h = 0.2$, $a/h = 0.3$, and $C_M = 2.0$).

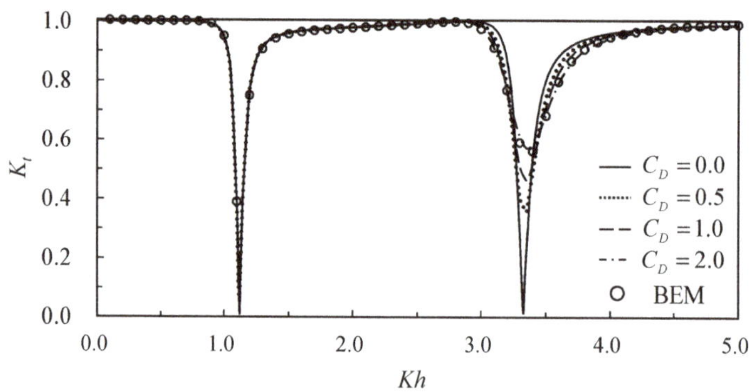

Figure 8. Transmission coefficient versus relative water depth for various drag coefficients ($d_1/h = 0.25$, $\ell/h = 0.2$, $a/h = 0.3$, and $C_M = 2.0$).

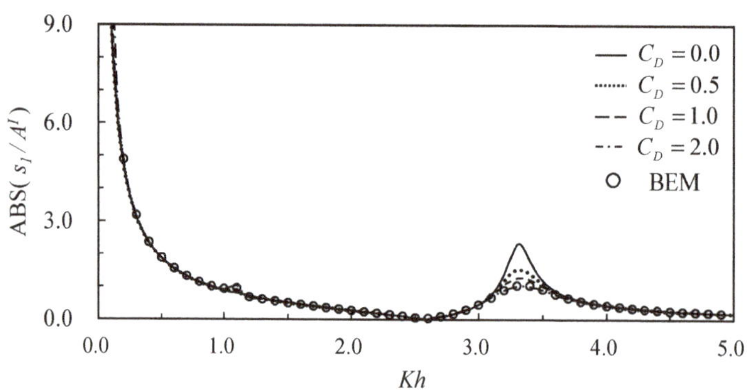

Figure 9. Surge motion amplitude versus relative water depth for various drag coefficients ($d_1/h = 0.25$, $\ell/h = 0.2$, $a/h = 0.3$, and $C_M = 2.0$).

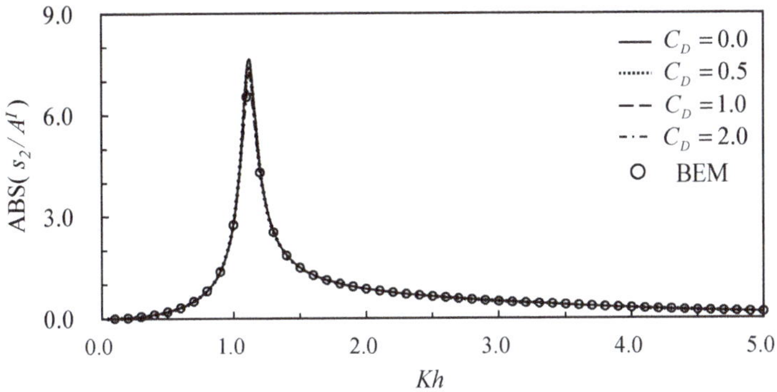

Figure 10. Heave motion amplitude versus relative water depth for various drag coefficients ($d_1/h = 0.25$, $\ell/h = 0.2$, $a/h = 0.3$, and $C_M = 2.0$).

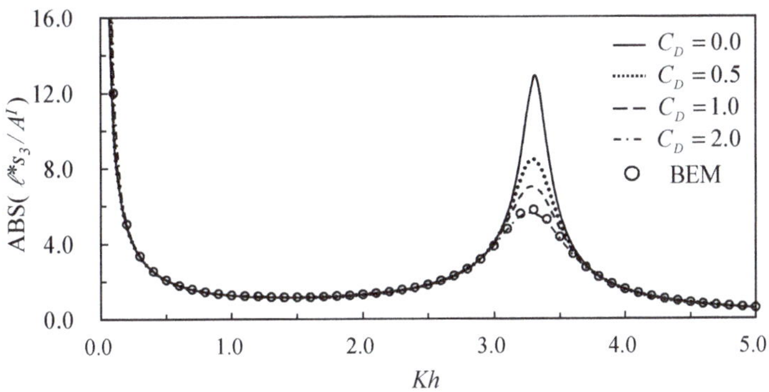

Figure 11. Pitch motion amplitude versus relative water depth for various drag coefficients ($d_1/h = 0.25$, $\ell/h = 0.2$, $a/h = 0.3$, and $C_M = 2.0$).

Figure 12 shows dimensionless horizontal and vertical forces versus relative water depth, that the horizontal force is divided by $\rho g A^I a$ and the vertical force is divided by $2\rho g A^I \ell$. $F^M_{\overline{AB},1}$ represents the horizontal force, while $F^M_{\overline{AB},2}$ represents the vertical force on the spring $\overline{AB}$. $F^M_{\overline{CD},1}$ represents the horizontal force, while $F^M_{\overline{CD},2}$ represents the vertical force on the spring $\overline{CD}$. For the given conditions $d_1/h = 0.25$, $\ell/h = 0.2$, $a/h = 0.3$, $C_D = 2.0$, and $C_M = 2.0$, wave forces acting on the upwind mooring lines, $\overline{AB}$, are obviously smaller than the mooring line, $\overline{CD}$, on the lee side. Additionally, the horizontal component of the forces are bigger than vertical ones. The maximum wave forces can reach up to 12% of the incident wave forces.

The present analytic solution is used to study the submerged depth of the structure on reflection and transmission coefficients and motions of the structure. The conditions used are water depth, $h = 10$ m, incident wave amplitude, $A^I = 0.5$ m, and width and height of the structure, $\ell/h = 0.4$ and $a/h = 0.3$. The submerged depths considered are near the water surface; one-fourth the water depth; and one-half the water depth ($d_1/h = 0.10$, $d_1/h = 0.25$, and $d_1/h = 0.50$). The dimensionless water depth related to the wave length covers the range from shallow water, $Kh < \pi/10$, up to the deep water, $Kh < \pi$. Variations of the reflection coefficient and the transmission coefficient versus Kh for various structural submergences $d_1/h = 0.10$, 0.25, and 0.50 are shown in Figures 13 and 14, respectively. It can be expected that, since the underwater structure is blocking the incident wave while located under the water surface, the nearer the structure is close to the water surface, the structure can block more surface waves, except the longer waves can have less effect from the structure. Figure 13 also shows the structure can have a total reflection for short waves and near the free surface. Furthermore, at the resonant frequency, there exists a total reflection. Figure 14 indicates a reverse tendency for the transmission coefficient versus dimensionless water depth.

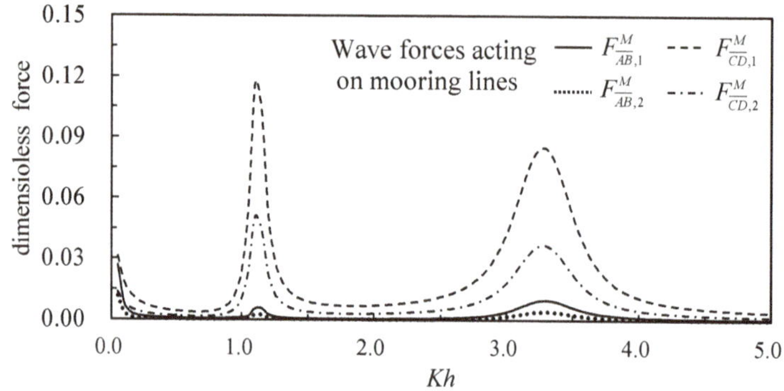

Figure 12. Wave forces on the mooring lines versus relative water depth ($d_1/h = 0.25$, $\ell/h = 0.2$, $a/h = 0.3$, $C_D = 2.0$, and $C_M = 2.0$).

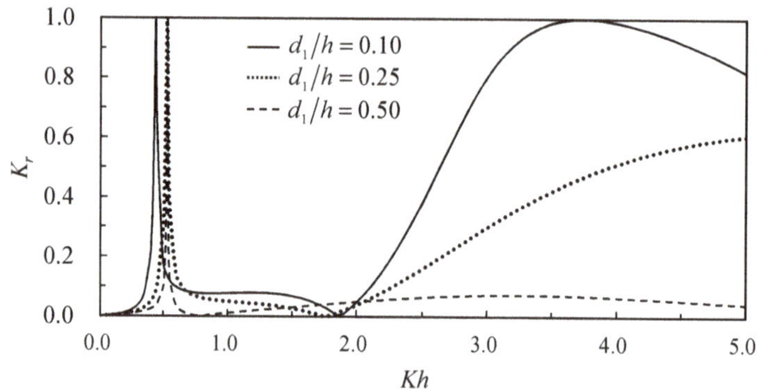

Figure 13. Reflection coefficient versus relative water depth for various submerged depths of the structure ($d_1/h = 0.10$, 0.25, and 0.50).

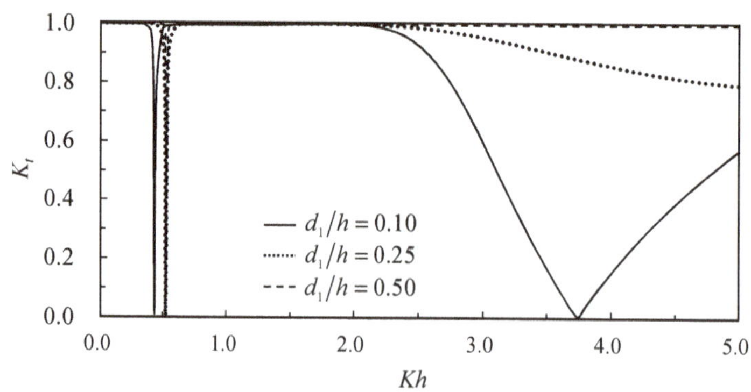

Figure 14. Transmission coefficient versus relative water depth for various submerged depths of the structure ($d_1/h = 0.10$, 0.25, and 0.50).

Effects of various submerged depths of the structure on the motions of the structure, namely, surge, heave, and pitch, are shown in Figures 15–17, respectively. The surge and pitch amplitudes decrease with the increasing relative water depth (the shorter waves), as the shorter waves induce less structural motions. In general, the structure located near the free surface can get bigger motions. The same tendency applies to the heave motion, but there exists a resonant frequency due to the hydrostatic restoring force of the water buoyancy. The resonant frequency shifted for different structural submergence due to different hydrodynamic forces acting on the structure.

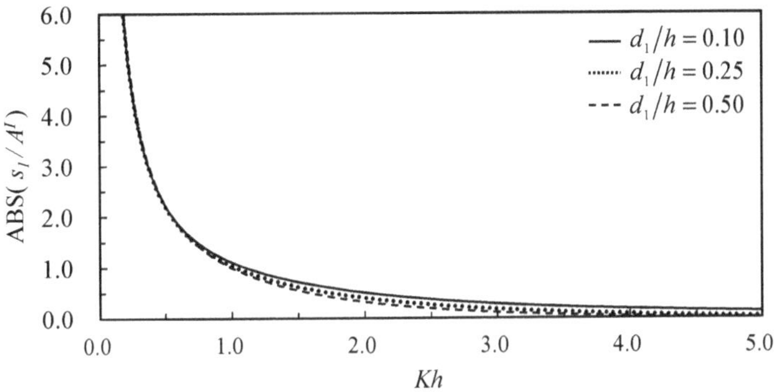

Figure 15. Dimensionless surge amplitude versus relative water depth for various submerged depths of the structure (d_1/h = 0.10, 0.25, and 0.50).

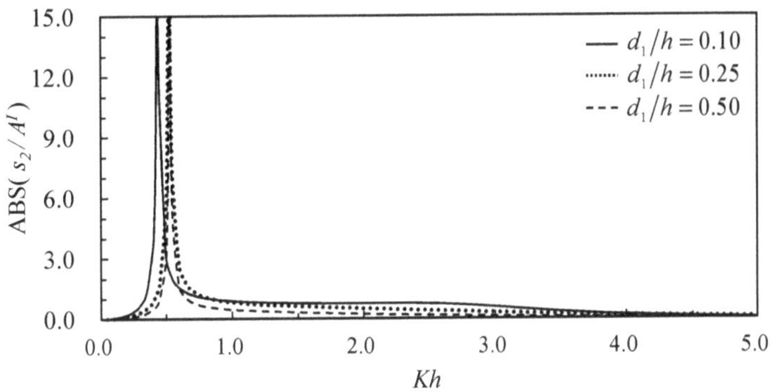

Figure 16. Heave amplitude versus relative water depth for various submerged depths of the structure (d_1/h = 0.10, 0.25, and 0.50).

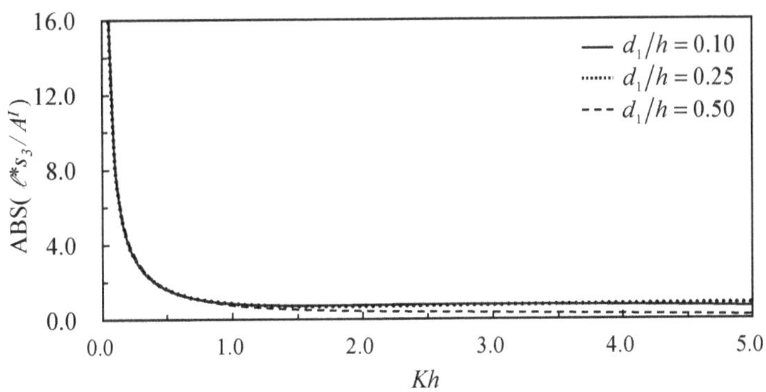

Figure 17. Pitch amplitude versus relative water depth for various submerged depths of the structure (d_1/h = 0.10, 0.25, and 0.50).

In this study, we emphasize that the moored underwater structures with motions of full degrees of freedom subjected to actions of incident waves and present an analytic solution. However, the solution is restricted to geometrical deployment and a linear assumption. Nevertheless, the methodology can be expanded to nonlinear problems via a higher-order solution; the linear solution can be applied as a preliminary identification of the characteristics of the practical problems.

4. Conclusions

With considerations of wave forces acting on mooring lines, a new analytic solution is presented for the problem of an underwater moored floating structure with motions of full degrees of freedom subjected to incident waves. A coupling formulation among water waves, underwater floating structure, and mooring lines is presented. Iterations for the drag coefficients, energy conservation without drag loss, reflection and transmission coefficients, and comparisons of wave scatterings with a finite elements result, as well as motion amplitudes in comparison with a numerical boundary element model, provide valid validation of the present solution. With additional considerations of wave forces acting on the mooring lines, the drag dampening significantly decreases wave reflections and the motions of the structure at the high-frequency resonance. The magnitudes of the wave forces acting on the mooring lines can reach up to 12% of the incident wave forces. The study of the submerged depth of the structure indicates that the structure deployed nearer the free surface can induce bigger motions. The analytic solution is very much dependent on the geometry of the structure; however, the interaction formulation in this paper can be applied to practical problems for more complete considerations.

Author Contributions: Conceptualization, J.-F.L., C.-H.L.; methodology, J.-F.L., C.-T.C.; validation, C.-H.L.; writing—original draft preparation, J.-F.L., C.-H.L.; writing—review and editing, C.-T.C.; funding acquisition, C.-T.C., J.-F.L. All authors have read and agreed to the published version of the manuscript.

Funding: This research was funded by Ministry of Science and Technology, Taiwan, grant number MOST 104-2221-E-006-188.

Acknowledgments: Financial support partially by the Yancheng Institute of Technology under Grant Number XJ201750 and partially by the Ministry of Science and Technology, Taiwan, under Grant Number MOST 104-2221-E-006-188 are gratefully acknowledged.

Conflicts of Interest: The authors declare no conflict of interest.

Appendix A. Definitions of Coefficients

Definitions of coefficients $k_n, k_{2n}, k_{4n}, \gamma_n, \mu_{n1}, \mu_{n2}, D^2_{2n}, F^2_{4n}, D^3_{2n}$, and F^3_{4n}.

$$\omega^2 = -gk_n \tan k_n h,\ n = 1, 2, \cdots \infty,\ k_0 = -iK \tag{A1}$$

$$\omega^2 = -gk_{2n} \tan k_{2n} d_1, n = 1, 2, \cdots \infty \tag{A2}$$

$$k_{4n} = n\pi/d_2, n = 1, 2, \cdots \infty \tag{A3}$$

$$\gamma_n = n\pi/2\ell, n = 1, 2, \cdots \infty \tag{A4}$$

$$\mu_{n1} = \gamma_n + \omega^2/g, n = 1, 2, \cdots \infty \tag{A5}$$

$$\mu_{n2} = \gamma_n - \omega^2/g, n = 1, 2, \cdots \infty \tag{A6}$$

$$D^2_{2n} = \frac{i\omega s_2(1 - \cos 2\gamma_n \ell)}{\ell \gamma_n{}^2\left(\mu_{n1} e^{\gamma_n(-d_1-h)} - \mu_{n2} e^{-\gamma_n(-d_1+h)}\right)}, n = 1, 2, \cdots \infty \tag{A7}$$

$$F^2_{4n} = \frac{i\omega s_2(1 - \cos 2\gamma_n \ell)}{\ell \gamma_n{}^2 \sinh \gamma_n d_2}, n = 1, 2, \cdots \infty \tag{A8}$$

$$D^2_{3n} = \frac{-i\omega s_3(1 + \cos 2\gamma_n \ell)}{\gamma_n{}^2\left(\mu_{n1} e^{\gamma_n(-d_1-h)} - \mu_{n2} e^{-\gamma_n(-d_1+h)}\right)}, n = 1, 2, \cdots \infty \tag{A9}$$

$$F_{4n}^{3} = \frac{-i\omega s_3(1+\cos 2\gamma_n \ell)}{\gamma_n{}^2 \sinh \gamma_n d_2}, n = 1,2,\cdots\infty \tag{A10}$$

Appendix B. Wave Forces Acting on the Floating Structure

Wave forces acting on the floating structure in the direction of each degree of freedom can be calculated as:

$$\begin{aligned} f_1 = -i\omega\rho\cdot e^{-i\omega t}\Big\{&\int_{-h+d_2}^{-d_1}\Big[\big(\phi^I+\phi_1^D\big)\Big|_{x=-\ell}-\phi_3^D\big|_{x=\ell}\Big]dz \\ &+s_1\cdot\int_{-h+d_2}^{-d_1}\Big[\phi_1^1\big|_{x=-\ell}-\phi_3^1\big|_{x=\ell}\Big]dz+s_3\cdot\int_{-h+d_2}^{-d_1}\Big[\phi_1^3\big|_{x=-\ell}-\phi_3^3\big|_{x=\ell}\Big]dz\Big\} \end{aligned} \tag{A11}$$

$$\begin{aligned} f_2 = -i\omega\rho\cdot e^{-i\omega t}\Big[&\int_{-\ell}^{\ell}\Big(\phi_4^D\big|_{z=-h+d_2}-\phi_2^D\big|_{x=-d_1}\Big)dx \\ &+s_2\cdot\int_{-\ell}^{\ell}\Big(\phi_4^2\big|_{z=-h+d_2}-\phi_2^2\big|_{z=-d_1}\Big)dx\Big] \end{aligned} \tag{A12}$$

$$\begin{aligned} f_3 = \ & -i\omega\rho\cdot e^{-i\omega t}\Big\{\int_{-h+d_2}^{-d_1}(z-z_0)\Big[\big(\phi^I+\phi_1^D\big)\Big|_{x=-\ell}-\phi_3^D\big|_{x=\ell}\Big]dz \\ & -\int_{-\ell}^{\ell}x\Big[\phi_4^D\big|_{z=-h+d_2}-\phi_2^D\big|_{z=-d_1}\Big]dx+s_1\cdot\int_{-h+d_2}^{-d_1}(z-z_0)\Big[\phi_1^1\big|_{x=-\ell}-\phi_3^1\big|_{x=\ell}\Big]dz \\ & -s_1\cdot\int_{-\ell}^{\ell}x\Big[\phi_4^1\big|_{z=-h+d_2}-\phi_2^1\big|_{z=-d_1}\Big]dx+s_3\cdot\int_{-h+d_2}^{-d_1}(z-z_0)\Big[\phi_1^3\big|_{x=-\ell}-\phi_3^3\big|_{x=\ell}\Big]dz \\ & -s_3\cdot\int_{-\ell}^{\ell}x\Big[\phi_4^3\big|_{z=-h+d_2}-\phi_2^3\big|_{z=-d_1}\Big]dx\Big\} \end{aligned} \tag{A13}$$

Integrations shown in Equations (A11)–(A13) can be calculated, and the equations be rewritten, as:

$$\left\{\begin{array}{c} f_1 \\ f_2 \\ f_3 \end{array}\right\} = \left[f^R\right]\left\{\begin{array}{c} s_1 \\ s_2 \\ s_3 \end{array}\right\} + \left\{\begin{array}{c} f_1^D \\ f_3^D \\ f_3^D \end{array}\right\} \tag{A14}$$

where $\left[f^R\right]$ and $\left\{f^D\right\}$ can be expressed as:

$$\left[f^R\right] = \left[\begin{array}{ccc} f_{11}^R & 0 & f_{13}^R \\ 0 & f_{22}^R & 0 \\ f_{31}^R & 0 & f_{33}^R \end{array}\right] \tag{A15}$$

with the components expressed as:

$$f_{11}^R = \sum_{n=0}^{\infty}\left(B_{1n}^1 - A_{3n}^1\right)\frac{\sin k_n(h-d_1) - \sin k_n d_2}{k_n} \tag{A16}$$

$$f_{13}^R = \sum_{n=0}^{\infty}\left(B_{1n}^3 - A_{3n}^3\right)\frac{\sin k_n(h-d_1) - \sin k_n d_2}{k_n} \tag{A17}$$

$$\begin{aligned} f_{22}^R &= \left(2\ell B_{40}^2\right) + \sum_{n=1}^{\infty}\left(A_{4n}^2 + B_{4n}^2\right)\cos k_{4n}d_2\frac{\left(1-e^{-2k_{4n}\ell}\right)}{k_{4n}} \\ &+ \sum_{n=1}^{\infty}F_{4n}^2\cosh \kappa_{4n}d_2\frac{(1-\cos 2\kappa_{4n}\ell)}{\kappa_{4n}} - \sum_{n=0}^{\infty}\left(A_{2n}^2 + B_{2n}^2\right)\frac{\left(1-e^{-2k_{2n}\ell}\right)}{k_{2n}} \\ &- \sum_{n=1}^{\infty}D_{2n}^2\left(\mu_{n1}e^{-\kappa_{2n}(d_1+h)} + \mu_{n2}e^{-\kappa_{2n}(-d_1+h)}\right)\frac{(1-\cos 2\kappa_{2n}\ell)}{\kappa_{2n}} \end{aligned} \tag{A18}$$

$$\begin{aligned} f_{31}^R &= \sum_{n=0}^{\infty}\left(B_{1n}^1 - A_{3n}^1\right)C_n^{16} - \tfrac{2}{3}\ell^3 A_{40}^1 \\ &- \sum_{n=1}^{\infty}\left(A_{4n}^1 - B_{4n}^1\right)\cos k_{4n}d_2\cdot C_{4n}^{17} + \sum_{n=0}^{\infty}\left(A_{2n}^1 - B_{2n}^1\right)C_{2n}^{17} \end{aligned} \tag{A19}$$

$$
\begin{aligned}
f_{33}^{R} = & \sum_{n=0}^{\infty}\left(B_{1n}^{3}-A_{3n}^{3}\right)C_{n}^{16}-\tfrac{2}{3}\ell^{3}A_{40}^{3} \\
& -\sum_{n=1}^{\infty}\left(A_{4n}^{3}-B_{4n}^{3}\right)\cos k_{4n}d_{2}\cdot C_{4n}^{17}-\sum_{n=1}^{\infty}F_{4n}^{3}\cosh \kappa_{n}d_{2}\cdot C_{n}^{18} \\
& +\sum_{n=0}^{\infty}\left(A_{2n}^{3}-B_{2n}^{3}\right)C_{2n}^{17}-\sum_{n=1}^{\infty}D_{2n}^{3}\left(\mu_{n1}e^{-\kappa_{2n}(d_{1}+h)}+\mu_{n2}e^{-\kappa_{2n}(-d_{1}+h)}\right)C_{n}^{18}
\end{aligned}
\tag{A20}
$$

$$
\begin{aligned}
f_{1}^{D} = & \frac{igA^{I}e^{-iK\ell}}{\omega\cosh Kh}\frac{\sinh K(h-d_{1})-\sinh Kd_{2}}{K} \\
& +\sum_{n=0}^{\infty}\left(B_{1n}^{D}-A_{3n}^{D}\right)\cdot\frac{\sin k_{n}(h-d_{1})-\sin k_{n}d_{2}}{k_{n}}
\end{aligned}
\tag{A21}
$$

$$
\begin{aligned}
f_{2}^{D} = & 2\ell B_{40}^{D}+\sum_{n=1}^{\infty}\left(A_{4n}^{D}+B_{4n}^{D}\right)\cos k_{4n}d_{2}\frac{\left(1-e^{-2k_{4n}\ell}\right)}{k_{4n}} \\
& -\sum_{n=0}^{\infty}\left(A_{2n}^{D}+B_{2n}^{D}\right)\frac{\left(1-e^{-2k_{2n}\ell}\right)}{k_{2n}}
\end{aligned}
\tag{A22}
$$

$$
\begin{aligned}
f_{3}^{D} = & \frac{igA^{I}e^{-iK\ell}}{\omega\cosh Kh}C_{0}^{16}+\sum_{n=0}^{\infty}\left(B_{1n}^{D}-A_{3n}^{D}\right)C_{n}^{16}-\tfrac{2}{3}\ell^{3}A_{30}^{D} \\
& +\sum_{n=1}^{\infty}\left(A_{4n}^{D}-B_{4n}^{D}\right)\cos k_{4n}d_{2}\cdot C_{4n}^{17}+\left(A_{2n}^{3}-B_{2n}^{3}\right)C_{2n}^{17}
\end{aligned}
\tag{A23}
$$

And the constants C_{n}^{16}, C_{2n}^{17}, C_{4n}^{17}, and C_{n}^{18} are:

$$
\begin{aligned}
C_{n}^{16} & = \int_{-d_{1}}^{0}(z-z_{0})\cos[k_{n}(z+h)]dz \\
& = \frac{(h-d_{2}-z_{0})\sin k_{n}d_{2}-(d_{1}+z_{0})\sin k_{n}(h-d_{1})}{k_{n}} \\
& +\frac{\cos k_{n}(h-d_{1})-\cos k_{n}d_{2}}{k_{n}^{2}}
\end{aligned}
\tag{A24}
$$

$$
\begin{aligned}
C_{2n,4n}^{17} & = \int_{-\ell}^{\ell}xe^{-k_{2n,4n}(x+\ell)}dx = -\int_{-\ell}^{\ell}xe^{k_{2n,4n}(x-\ell)}dx \\
& = \frac{1-e^{-2k_{2n,4n}\ell}}{k_{2n,4n}{}^{2}}-\frac{\ell\left(1+e^{-2k_{2n,4n}\ell}\right)}{k_{2n,4n}}
\end{aligned}
\tag{A25}
$$

$$
\begin{aligned}
C_{n}^{18} & = \int_{-\ell}^{\ell}x\sin\gamma_{n}(x+\ell)dx \\
& = \frac{\sin 2\gamma_{n}\ell}{\gamma_{n}{}^{2}}-\frac{\ell(1+\cos 2\gamma_{n}\ell)}{\gamma_{n}{}^{2}}
\end{aligned}
\tag{A26}
$$

Appendix C. Stiffness Matrix of the Mooring Springs

The components of the stiffness matrices for the springs $\overline{AB}$ and $\overline{CD}$ are calculated according to the geometrical orientations of the springs and can be expressed as:

$$K_{11}^{\overline{AB}} = K_{11}^{\overline{CD}} = K_{s}\cos^{2}\theta \tag{A27}$$

$$K_{12}^{\overline{AB}} = K_{21}^{\overline{AB}} = -K_{12}^{\overline{CD}} = -K_{21}^{\overline{CD}} = K_{s}\cos\theta\sin\theta \tag{A28}$$

$$K_{13}^{\overline{AB}} = K_{31}^{\overline{AB}} = K_{13}^{\overline{CD}} = K_{31}^{\overline{CD}} = K_{s}\left[0.5(h-d_{1}-d_{2})\cos^{2}\theta-\ell\cos\theta\sin\theta\right] \tag{A29}$$

$$K_{22}^{\overline{AB}} = K_{22}^{\overline{CD}} = K_{s}\sin^{2}\theta \tag{A30}$$

$$K_{33}^{\overline{AB}} = K_{33}^{\overline{CD}} = K_{s}[0.5(h-d_{1}-d_{2})\cos\theta-\ell\sin\theta]^{2} \tag{A31}$$

Appendix D. Wave Forces Acting on Mooring Lines

Wave forces acting on the mooring springs are calculated using the linearized Morison equation (Lee, 1994):

$$dF^{M} = \frac{\rho C_{D\ell}D_{S}}{2}\left(U-\dot{\varsigma}\right)dS+\frac{\rho\pi C_{M}D_{S}^{2}}{4}\left(\dot{U}-\ddot{\varsigma}\right)dS \tag{A32}$$

where ρ is fluid density, D_S is the diameter of the spring, U and $\dot{U}$ are the flow velocity and acceleration in the direction normal to the mooring line, C_M is the added mass coefficient, $\dot{\varsigma}$ and $\ddot{\varsigma}$ are the velocity and acceleration of the mooring line, and the linear drag coefficient is expressed as:

$$C_{D\ell} = \frac{4C_D}{3\pi\omega}\frac{\int_{-h}^{-h+d_2}\left|U-\dot{\varsigma}\right|^3 dz}{\int_{-h}^{-h+d_2}\left|U-\dot{\varsigma}\right|^2 dz} \tag{A33}$$

Note that motions of the mooring lines are not known in priori until the problem solved: therefore, a complete solution will contain an iteration procedure until a 0.5% convergent criteria is reached. Furthermore, the springs are not subjected to forces in a transverse direction; therefore, the wave forces calculated are transferred to the attached points A and C. Thus,

$$F_A^M = \frac{\csc\theta}{d_2}\int_{-h}^{-h+d_2}\left[\frac{\rho C_{D\ell}D_S}{2}\left(U_1-\dot{\varsigma}_{\overline{AB}}\right)+\frac{\rho\pi C_M D_S^2}{4}\left(\dot{U}_1-\ddot{\varsigma}_{\overline{AB}}\right)\right]dz \tag{A34}$$

$$F_C^M = \frac{\csc\theta}{d_2}\int_{-h}^{-h+d_2}\left[\frac{\rho C_{D\ell}D_S}{2}\left(U_3-\dot{\varsigma}_{\overline{CD}}\right)+\frac{\rho\pi C_M D_S^2}{4}\left(\dot{U}_3-\ddot{\varsigma}_{\overline{CD}}\right)\right]dz \tag{A35}$$

where the subscripts 1 and 3 stand for regions 1 and 3, while subscripts $\overline{AB}$ and $\overline{CD}$ stand for the spring AB and CD. The corresponding expressions are:

$$\begin{aligned} U_1 = & -\left(\Phi_x^I+\Phi_{1x}^D+s_1\cdot\Phi_{1x}^1+s_2\cdot\Phi_{1x}^2+s_3\cdot\Phi_{1x}^3\right)\sin\theta \\ & -\left(\Phi_z^I+\Phi_{1z}^D+s_1\cdot\Phi_{1z}^1+s_2\cdot\Phi_{1z}^2+s_3\cdot\Phi_{1z}^3\right)\cos\theta \end{aligned} \tag{A36}$$

$$\begin{aligned} U_3 = & -\left(\Phi_{3x}^D+s_1\cdot\Phi_{3x}^1+s_2\cdot\Phi_{3x}^2+s_3\cdot\Phi_{3x}^3\right)\sin\theta \\ & -\left(\Phi_{3z}^D+s_1\cdot\Phi_{1z}^1+s_2\cdot\Phi_{3z}^2+s_3\cdot\Phi_{3z}^3\right)\cos\theta \end{aligned} \tag{A37}$$

With substitutions of Equations (A36) and (A37) into Equation (A33) and Equation (A35), one can obtain

$$F_A^M = F_A^{Mw} + F_A^{Ms} \tag{A38}$$

$$F_C^M = F_C^{Mw} + F_C^{Ms} \tag{A39}$$

in which

$$\begin{aligned} F_A^{Mw} = & \frac{\csc\theta}{d_2}\left(\frac{\rho C_{D\ell}D_S}{2}-\frac{i\omega\rho\pi D_S^2 C_M}{4}\right)\left\{\left(\frac{KgA^I e^{-iK\ell}\sin\theta}{\omega\cosh Kh}\right)\left(\frac{Kd_2\sinh Kd_2-\cosh Kd_2+1}{K^2}\right)\right. \\ & +\left(\frac{iKgA^I e^{-iK\ell}\cos\theta}{\omega\cosh Kh}\right)\left(\frac{Kd_2\cosh Kd_2-\sinh Kd_2}{K^2}\right) \\ & +\sin\theta\left[\sum_{n=0}^{\infty}\left(B_{1n}^D+s_1B_{1n}^1+s_2B_{1n}^2+s_3B_{1n}^3\right)\left(\frac{-k_nd_2\sin k_nd_2-\cos k_nd_2-1}{k_n}\right)\right] \\ & \left.+\cos\theta\left[\sum_{n=0}^{\infty}\left(B_{1n}^D+s_1B_{1n}^1+s_2B_{1n}^2+s_3B_{1n}^3\right)\left(\frac{-\sin k_nd_2+k_nd_2\cos k_nd_2}{k_n}\right)\right]\right\} \end{aligned} \tag{A40}$$

$$\begin{aligned} F_A^{Ms} = & \frac{\csc\theta}{d_2}\left(\frac{i\omega\rho C_{D\ell}D_S^2}{2}+\frac{\pi\rho C_M\omega^2 D_S^2}{4}\right) \\ & \left\{s_1\left(\frac{\sin\theta}{d_2}\right)\frac{d_2{}^3}{3}-s_2\frac{d_2{}^2}{2}\cos\theta-s_3\left[\frac{(h-d_1-d_2)\sin\theta}{2d_2}\frac{d_2{}^3}{3}+\ell\frac{d_2{}^2}{2}\cos\theta\right]\right\} \end{aligned} \tag{A41}$$

$$\begin{aligned} F_C^{Mw} = & \frac{\csc\theta}{d_2}\left(\frac{\rho C_{D\ell}D_S}{2}-\frac{i\omega\rho\pi D_S^2 C_M}{4}\right) \\ & \cdot\left\{\sin\theta\left[\sum_{n=0}^{\infty}\left(A_{3n}^D+s_1A_{3n}^1+s_2A_{3n}^2+s_3A_{3n}^3\right)\left(\frac{k_nd_2\sin k_nd_2+\cos k_nd_2-1}{k_n}\right)\right]\right. \\ & \left.+\cos\theta\left[\sum_{n=0}^{\infty}\left(A_{3n}^D+s_1A_{3n}^1+s_2A_{3n}^2+s_3A_{3n}^3\right)\left(\frac{\sin k_nd_2-k_nd_2\cos k_nd_2}{k_n}\right)\right]\right\} \end{aligned} \tag{A42}$$

$$F_C^{Ms} = \frac{\csc\theta}{d_2}\left(\frac{i\omega\rho C_{D\ell} D_S}{2} + \frac{\pi\rho C_M \omega^2 D_S^2}{4}\right)$$
$$\left\{ s_1\left(\frac{\sin\theta}{d_2}\right)\frac{{d_2}^3}{3} + s_2\frac{{d_2}^2}{2}\cos\theta - s_3\left[\frac{(h-d_1-d_2)\sin\theta}{2d_2}\frac{{d_2}^3}{3} + \ell\frac{{d_2}^2}{2}\cos\theta\right]\right\} \quad \text{(A43)}$$

Note that the spring $\overline{CD}$ is located at the lee side of the structure; therefore, there is no incident wave in the expression.

References

1. Rusu, E.; Onea, F. A review of the technologies for wave energy extraction. *Clean Energy* **2018**, *2*, 10–19. [CrossRef]
2. Sarpkaya, T.; Isaacson, M. *Mechanics of Wave Forces on Offshore Structures*; Van Norstrand Reinhold Co.: New York, NY, USA, 1981.
3. Lee, J.-F.; Tu, L.-F. Finite element modeling of a single-point multi-segment mooring in water waves. *Ocean Eng.* **2018**, *160*, 461–470. [CrossRef]
4. Huang, M.-C.; Hudspeth, R.T.; Leonard, J.W. FEM Solution of 3-D Wave Interference Problems. *J. Waterw. Port Coast. Ocean Eng.* **1985**, *3*, 661–677. [CrossRef]
5. Sannasiraj, S.A.; Sundar, V.; Sundaravadivelu, R. Mooring forces and motion responses of pontoon-type floating breakwaters. *Ocean Eng.* **1998**, *25*, 27–48. [CrossRef]
6. Chen, Z.J.; Wang, Y.X.; Dong, H.Y.; Zheng, B.X. Time-domain hydrodynamic analysis of pontoon-plate floating breakwater. *Water Sci. Eng.* **2012**, *5*, 291–303.
7. Mohapatra, S.C.; Sahoo, T. Oblique wave diffraction by a flexible floating structure in the presence of a submerged flexible structure. *Geophys. Astrophys. Fluid Dyn.* **2014**, *108*, 615–638. [CrossRef]
8. Cao, F.-F.; Zhao, X.-Z. Nonlinear dynamic behaviors of a floating structure in focused waves. *China Ocean Eng.* **2015**, *29*, 807–820. [CrossRef]
9. Mohapatra, S.C.; Guesdes Soares, C. *Wave Forces on a Floating Structure over Flat Bottom Based on Boussinesq Formulation, Renewable Energies Offshore*; Guedes, S., Ed.; Taylor & Francis Group: London, UK, 2015.
10. Kao, J.H.; Lee, J.F.; Cheng, Y.C. A direct coupling numerical method for solving three-dimensional interaction problems of wave and floating structures. *Eng. Anal. Bound. Elem.* **2015**, *55*, 10–27. [CrossRef]
11. Rivera-Arreba, I. Computation of Nonlinear Wave Loads on Floating Structures. Master's Thesis, Delft University of Technology, Delf, The Netherlands, 2017.
12. Guo, Y.; Mohapatra, S.C.; Soares, C. *Guesdes Wave Interaction with a Rectangular Long Floating Structure over Flat Bottom*; Soares, C.G., Santos, T.A., Eds.; Progress in Maritime Energy and Technology, Taylor & Francis Group: London, UK, 2018.
13. Lee, C.P. Dragged surge motion of a tension leg structure. *Ocean Eng.* **1994**, *21*, 311–328. [CrossRef]
14. Lee, H.H.; Wang, P.W.; Lee, C.P. Dragged surge motion of tension leg platforms and strained elastic tethers. *Ocean Eng.* **1999**, *26*, 575–594. [CrossRef]
15. Lee, H.H.; Wang, P.W. Dynamic behavior of tension-leg platform with net-cage system subjected to wave forces. *Ocean Eng.* **2001**, *28*, 179–200. [CrossRef]
16. Lee, J.F. On the heave radiation of a rectangular structure. *Ocean Eng.* **1995**, *22*, 19–34. [CrossRef]
17. Chen, P.I.; Chen, C.T.; Lee, J.F. On wave interaction with floating structures with dragged moorings. *J. Mar. Sci. Technol.* **2016**, *24*, 530–538.
18. Zheng, Y.H.; Liu, P.F.; Shen, Y.M.; Wu, B.J.; Sheng, S.W. On the radiation and diffraction of linear water waves by and infinitely long rectangular structure submerged on oblique seas. *Ocean Eng.* **2007**, *34*, 436–450. [CrossRef]
19. Mei, C.C. *The Applied Dynamics of Ocean Surface Wave*; John Wiley & Sons, Inc.: New York, NY, USA, 1983.
20. Cheong, H.F.; Shankar, N.J.; Nallayarasu, S. Analysis of submerged platform breakwater by eigenfunction expansion method. *Ocean Eng.* **1996**, *23*, 649–666. [CrossRef]
21. Lo, C.H. Analytic Solutions of Wave Action on Underwater Floating Structures. Master's Thesis, Department of Hydraulic and Ocean Engineering, National Cheng Kung University, Tainan, Taiwan, 2016.

Journal of
Marine Science and Engineering

Article

Estimation of the Motion Performance of a Light Buoy Adopting Ecofriendly and Lightweight Materials in Waves

Se-Min Jeong [1], Bo-Hun Son [2] and Chang-Yull Lee [3,*]

1 Department of Naval Architecture and Ocean Engineering, Chosun University, Gwangju 61452, Korea; smjeong@chosun.ac.kr
2 Department of Naval Architecture and Ocean Engineering, Graduate School of Chosun University, Gwangju 61452, Korea; bohun1213@gmail.com
3 Department of Aerospace Engineering, Chosun University, Gwangju 61452, Korea
* Correspondence: cylee@chosun.ac.kr; Tel.: +82-62-230-7955

Received: 14 January 2020; Accepted: 19 February 2020; Published: 20 February 2020

Abstract: In this study, the methods and results of numerical simulations to estimate the motion performance of a newly developed lightweight light buoy in waves and to check the effect of conceptually developed appendages on that performance were introduced. The results from a potential-based motion analysis with viscous damping coefficients obtained from free decay tests using computational fluid dynamics (CFD) and those obtained from wave motion simulation using CFD were compared. From these results, it was confirmed that viscous damping should be considered when the frequency of an incoming wave is close to the natural frequency of the buoy. It was estimated that the pitch and heave motions of the light buoy became smaller when the developed appendages were adopted. Although the quantitative superiority of the appendages was different, the qualitative superiority was similar between both results.

Keywords: light buoy; motion performance in waves; potential-based simulations; viscous damping coefficients; free decay tests; computational fluid dynamics

1. Introduction

Light buoys are equipped with a lighting function and navigation sign (Figure 1). The buoy guides vessels sailing nearby in the daytime with its shape and color, and at night with its light. It also plays a role in notifying vessels about the presence of obstacles such as reefs and shallows.

Figure 1. Light buoys.

Because conventional large buoys are mainly made of steel, they are heavy and vulnerable to corrosion and erosion by seawater. This makes the installation and maintenance of the buoys difficult.

Moreover, vessel collision accidents with buoys and damage to vessels due to the light buoys' material (e.g., steel) are reported every year in Korea. Recently, light buoys adopting eco-friendly and lightweight materials have come into the spotlight to solve the previously mentioned problems. In Korea, a new lightweight light buoy with a 7-nautical-mile lantern, adopting an expanded polypropylene (EPP) and aluminum buoyant body and tower structure was developed by Jeong et al. [1]. Figure 2 shows the comparison of the conventional light buoy with the newly developed one. The total weight and manufacturing cost of the new buoys are approximately 40% and 27% lower than the conventional ones, respectively.

Conventional Light buoy		Model	Newly-developed One	
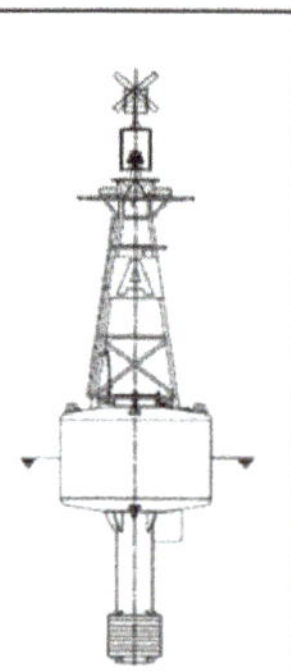	Separated Power	**Lantern Type**	Self-contained Power	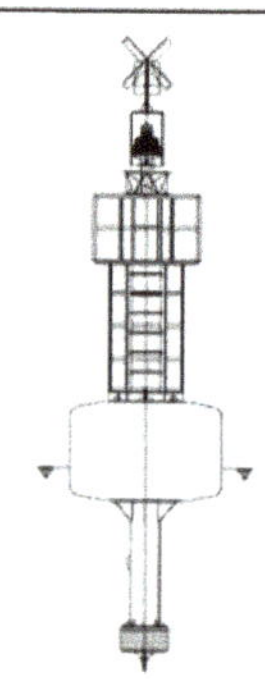
	Steel (SS-400)	**Tower Structure**	Aluminum (Al-6061 Alloy)	
	Steel (SS-400)	**Buoyant Body**	Expanded Poly-propylene	
	5,000 kg	**Weight**	3,000 kg	
	82 mil. KRW	**Price**	60 mil. KRW	

Figure 2. Comparison between a conventional light buoy and the lightweight one developed by Jeong et al. [1].

When the light buoy operates on the ocean, the visibility and angle of light from its lantern changes, which may cause it to function improperly. From this point of view, the pitch and roll motions of a light buoy are important. Moreover, large heave motions may cause structural damage to the mooring system. The motion of a floating body is greatly affected by external environmental loads, especially waves. To ensure motion stability and structural reliability, the natural frequency of the floating body needs to be very different from that of the dominant waves at the installation site. Because the mass distribution and center of gravity of the lightweight buoy are different from those of conventional one, the motion performance of the new type of buoy in waves should be assessed.

Therefore, after checking the static stability, Son et al. [2] carried out a motion analysis of a newly developed lightweight light buoy under various environmental conditions using potential-based commercial software ANSYS AQWA(Ansys, In., Canonsburg, PA, U.S.) that considers wind and current loads estimated by numerical simulations using the commercial computational fluid dynamics (CFD) software Siemens STAR-CCM+ (Siemens Industry Software Ltd., Plano, TX, U.S.) to increase the accuracy of the motion analysis. As a result, it was predicted that the pitch and roll motion were large and did not meet the design targets in specific conditions. As mentioned in several studies [3–7], one of the reasons might be that the viscous damping effect is ignored in the potential-based simulations commonly performed for the motion analysis.

A widely used way to consider the viscous effect in potential-based motion analyses is to evaluate and apply a viscous damping coefficient using the free decay test or force harmonic oscillation test [8,9]. These tests can also be conducted through CFD simulations. Wassermann et al. [10] estimated the roll damping of ships using CFD simulations of free decay and harmonic excited roll motion tests. They compared the advantages and disadvantages of both techniques. Wilson et al. [11] performed free roll decay tests on a surface combatant ship using CFD and compared the estimated damping coefficients with experimental results. Irkal et al. [12] carried out experiments and CFD simulations of free roll decay tests with different dimensions of a bilge keel. Song et al. [13] and Kianejad et al. [14] estimated the roll damping coefficient of a 2D section of a floating body and container ships through harmonic excited roll motion tests using CFD simulations.

Some studies based on CFD simulations were conducted to determine the resonance condition of a floating body as an energy harvesting device. Zhang et al. [15] investigated the influences of selected parameters such as incident wave condition, submerged depth, and power take off damping on the hydrodynamic performances of 2-D sections of rectangular heaving buoys. Luan et al. [16] estimated the hydrodynamic performance of a wave energy converter under various wave conditions and confirmed the relationship between optimal linear damping and incident wave conditions. Mohapatra et al. [17] formulated the mathematical modelling of wave diffraction by a floating fixed truncated vertical cylinder based on Boussinesq-type equations in the application range of weakly dispersive Boussinesq model, and showed the fidelity of the model by comparing the results from the developed analytical model with those from experiments and their CFD simulations using OpenFOAM.

The adoption of a proper appendage, such as a (bilge) keel or a heave (damping) plate, is one of the options for improving the motion performance of a floating body in waves. Although it is difficult to find motion reduction devices specifically designed for a light buoy, a heave damping plate for a vertical circular cylinder or a spar platform would be effective for a light buoy because of its geometrical similarity. Research on the effect of heave damping plates can be found in the following papers. Koh and Cho [18] carried out analytical and experimental studies to investigate the heave motion response of a circular cylinder according to the characteristics of dual damping plates as heave motion reduction appendages. Tao and Cai [19] investigated the vortex shedding pattern and hydrodynamics forces arising from the flow separation and vortex shedding around a damping plate of a circular cylinder. Through a series of experiments, some approximation equations were developed to calculate the added mass of the floating cylinder with a separate heave plate, and the motion response of a vertical circular cylinder with a heave plate to a series of regular waves was examined by Zhu and Lim [20]. Sudhakar and Nallayarasu [21,22] investigated the influence of single and double damping plates on the hydrodynamic response of a spar in regular and irregular waves by experimental studies. Koh et al. [23] performed free decay tests by experiments to obtain the viscous damping coefficients of a circular cylinder with a heave damping plate changing the porosity of the damping plate. From their experiments for regular and irregular waves, the pronounced motion reduction was observed by applying a porous plate.

In this study, motion analyses of a newly developed lightweight light buoy in waves were performed to predict the motion performance and to check the effects of the conceptually developed appendages intended for improving the motion performance. First, free decay tests including benchmark cases using CFD were carried out to estimate the viscous damping coefficients that cannot be obtained by potential-based simulations. Then, the results from potential-based simulations considering the viscous damping coefficients estimated by CFD were compared with the results of motion simulations in regular waves using CFD simulations (Figure 3).

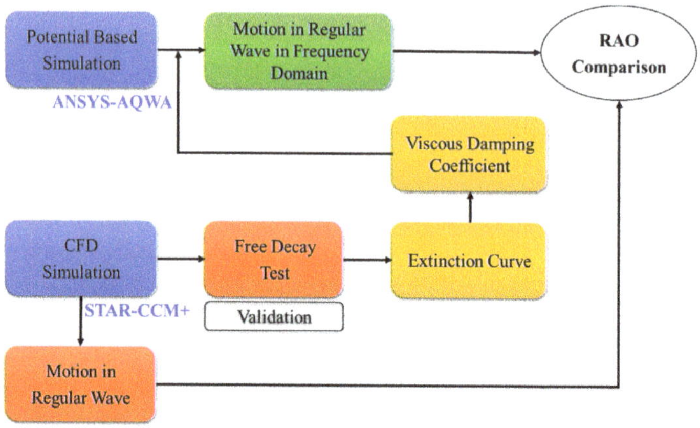

Figure 3. Scope and process of present study.

2. Problem Formulation

2.1. Governing Equation

For incompressible turbulent flows, the governing equations are the continuity and Reynolds-averaged Navier-Stokes equations, as shown in Equations (1) and (2), respectively.

$$\frac{\partial u_i}{\partial x_i} = 0 \tag{1}$$

$$\frac{\partial u_i}{\partial t} + \frac{\partial (u_i u_j)}{\partial x_j} = -\frac{1}{\rho}\frac{\partial p}{\partial x_i} + \frac{\partial}{\partial x_j}\left\{(\nu + \nu_t)\frac{\partial u_i}{\partial x_j}\right\} + f_i, \tag{2}$$

where u_i and x_i are the velocity component and coordinate in the i-direction; ρ is the density; p is the pressure; ν is the kinematic viscosity; ν_t is the eddy viscosity; and f_i is the external force per unit mass.

2.2. Estimation Procedure of Viscous Damping Coefficients from a Free Decay Test

The 1-degree of freedom (DOF) motion equations of pitch and heave of a floating body are as follows.

$$(I + I_a)\ddot{\theta} + b_{55}\dot{\theta} + c_{55}\theta = M(t) \tag{3}$$

$$(m + m_a)\ddot{z} + b_{33}\dot{z} + c_{33}z = F(t), \tag{4}$$

where I and I_a are the moment of inertia and added moment of inertia, respectively; θ is the angular displacement; b and c are the total damping and restoring coefficients, respectively; and $M(t)$ is the pitch wave excitation moment. In Equation (4), m and m_a are the mass and added mass, respectively; z is the vertical displacement; and $F(t)$ is the heave wave excitation force.

The total damping coefficients of pitch and heave were calculated from the free decay test using Equations (5) and (6), respectively.

$$b_{55} = 2\zeta_{55}\sqrt{(I + I_a)c_{55}} \tag{5}$$

$$b_{33} = 2\zeta_{33}\sqrt{(m + m_a)c_{33}}, \tag{6}$$

where ζ is a non-dimensional total damping coefficient estimated by the method of Journée and Massie [9], for which four peaks of motion from the free decay test, as shown in Figure 4, were chosen to evaluate ζ through Equation (7).

$$\zeta = \frac{1}{2\pi} \cdot ln\left(\frac{z_{a1} - z_{a2}}{z_{a3} - z_{a4}} \right) \tag{7}$$

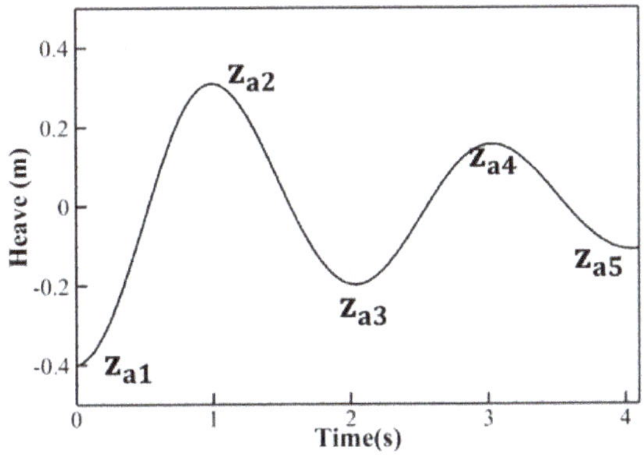

Figure 4. Free decay curve.

The viscous damping coefficients were derived from Equations (8) and (9) proposed by Koh and Cho [18].

$$b_{55,\ vis} = b_{55} - \nu_{55}(\omega_{o,\ 55}) \tag{8}$$

$$b_{33,\ vis} = b_{33} - \nu_{33}(\omega_{o,33}) \tag{9}$$

where $\nu_{55}(\omega_{o,\ 55})$ and $\nu_{33}(\omega_{o,33})$ are the radiation damping coefficients, which were evaluated by the ANSYS-AQWA software in this study, at the undamped natural frequencies of pitch and heave motions defined as $\omega_{o,\ 55} = \sqrt{c_{55}/(I + I_a)}$ and $\omega_{o,\ 33} = \sqrt{c_{33}/(m + m_a)}$, respectively.

3. Numerical Simulations

3.1. Simulation Method

STAR-CCM+ 11.04 was used to simulate the free decay test and for the motion simulations in regular waves. To capture the free surface, the volume of fluid (VOF) method was used. The realizable k-ε model was applied as a turbulent model. In addition, the overset grid and dynamic fluid body interaction techniques were used to handle the motion of the floating body. 3-DOF (surge, heave, and pitch) and 6-DOF simulations were performed for the free decay tests and motion simulations in regular waves, respectively.

With the application of the viscous damping coefficients, which were estimated by CFD simulations, the motion analyses were performed using ANSYS-AQWA, which is based on panel methods.

3.2. Modeling of Lightweight Light Buoys

To improve the motion performance of the recently developed lightweight light buoy, named "Base," two kinds of appendages were conceptually designed and assumed to be installed on the light buoy, as illustrated in Figure 5. The first addition is similar to a heave damping plate for an offshore structure; it is named "Plate" hereafter. The other addition is a conical shape similar to a ship's bilge keel, which is marked as "Cone" in the present paper. From the research of Koh et al. [23], the damping coefficient of a vertical circular cylinder with a porous damping plate is larger than that with a non-porous damping plate. Therefore, the effect of the porosity of the appendage was also evaluated in this study. The models considering the porosity are named "Porous Plate" and "Porous Cone."

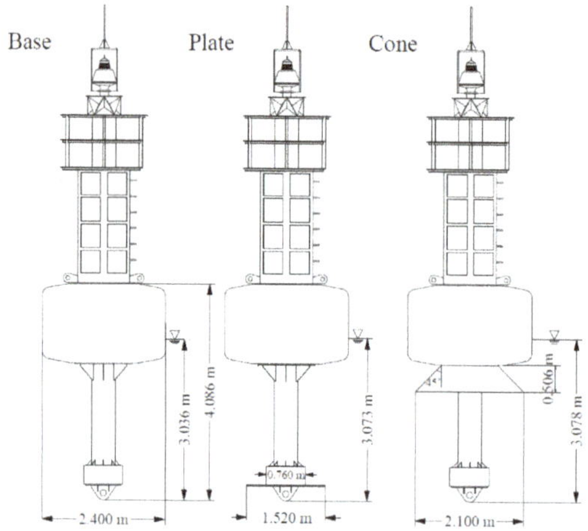

Figure 5. Cross-section views of target lightweight light buoys.

The simplified geometries of the lightweight light buoys with and without the developed appendages for the numerical analysis are shown in Figure 6, where only the major parts affecting the motion of the buoy were modeled considering the total mass and mass moment of inertia of the tower structure. The particulars and hydrostatic properties of the buoys are listed in Table 1. The mass of the light buoy with the appendages is approximately 7% higher than that of the Base model.

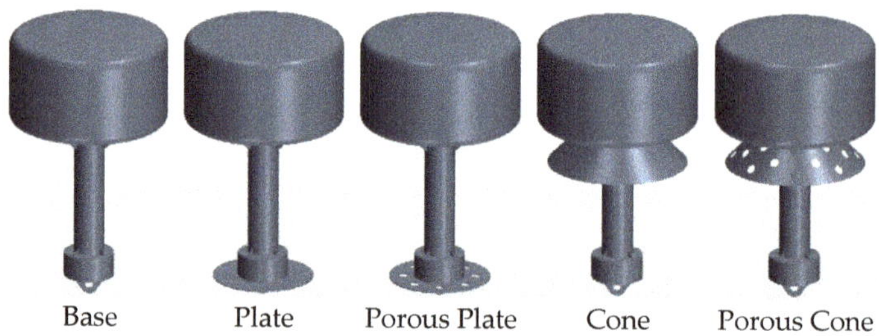

Figure 6. Geometries of lightweight light buoy with and without developed appendages.

Table 1. Main particulars and hydrostatic properties of the lightweight light buoys.

Models	Mass [kg]	Diameter of Buoyant Body [m]	Draft [m]	Center of Gravity [m]	Metacentric Height [m]	Mass Moment of Inertia [kg·m^2]
Base	2423	2.400	3.036	2.136	1.052	10,902
Plate	2596		3.073	2.009	1.144	11,490
Porous Plate	2579		3.069	2.021	1.137	11,435
Cone	2617		3.078	2.150	1.018	10,965
Porous Cone	2597		3.073	2.149	1.021	10,958

3.3. Estimation of Viscous Damping Coefficient Using CFD Simulation

3.3.1. Validation

To confirm the accuracy of the present numerical schemes and methods, CFD simulations of the free pitch and heave decay tests of a circular cylinder were performed under the same conditions as Palm et al. [24], who carried out experimental and numerical tests. A vertically truncated cylinder was tested in a wave tank with a water depth of 0.9m. The mass and diameter of the cylinder were 35.85kg and 0.515m, respectively. The moment of inertia around the center of gravity was 0.9kgm^2 and the center of gravity was placed 0.0758m above the bottom of the buoy along the symmetry z-axis.

Figure 7 shows the comparisons of the pitch and heave time histories obtained from the present CFD simulations with the results of the reference. The results of the present study are in good agreement with those of the reference. The reason for the discrepancy in pitch motion between the numerical and experimental results is thought to be the limitation of a small-scale experiment, in which generally allowable errors in the controlling and measured variables, such as the draft and center of gravity, may result in considerable differences in the motion response, as pointed out by Palm et al. in their work [24].

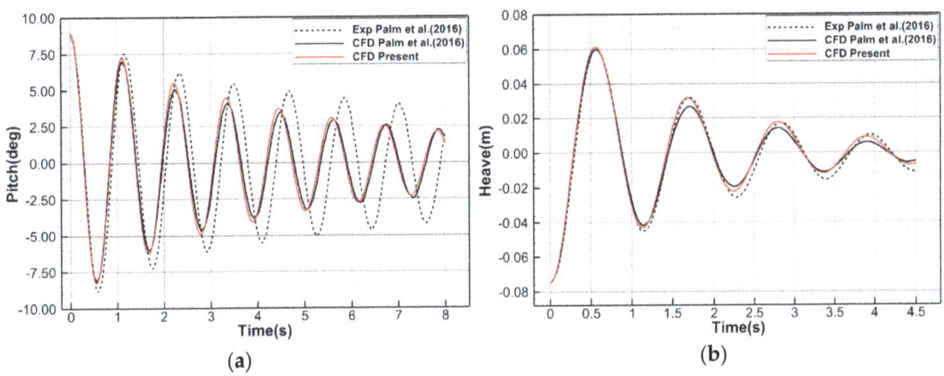

Figure 7. A comparison of the time histories of (**a**) free pitch and (**b**) heave decay curves of a circular cylinder between the present results and those of the reference.

3.3.2. Computational Domain, Boundary Conditions, and Grid System

Figure 8 shows the computational domain and boundary conditions of the free decay test for the light buoys using CFD. The size of the computational domain was set to 25 m in the depth direction below the free surface. The length and width are 30.0 D based on the diameter D (2.40 m) of the light buoy. To suppress the radiated waves from the light buoys, a numerical wave damping scheme is applied at the ends of the side boundaries to approximately one third of the computational domain.

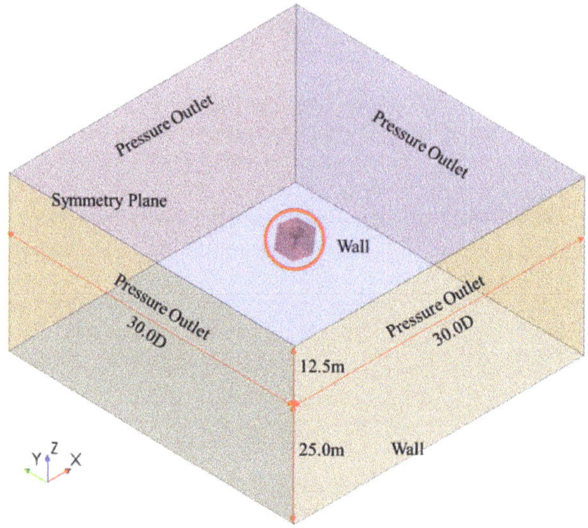

Figure 8. Computational domain and boundary conditions of the free decay test of lightweight light buoys using CFD.

Figure 9 shows the grid system of the free decay test using CFD. The grid system was generated using surface remesher, trimmer mesh, and prism layer mesh in STAR-CCM+. Near the light buoy and free surface, grids are refined to accurately capture the complicated flow around the buoy including the appendage and free surface.

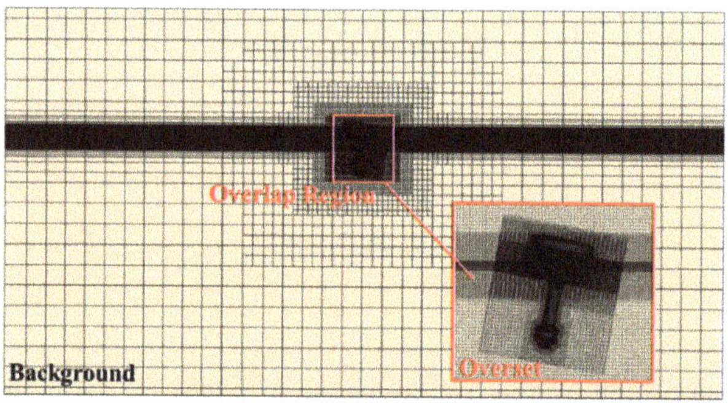

Figure 9. Grid system for the free decay test of lightweight light buoys using CFD.

3.3.3. Results

The free pitch and heave decay curves of the light buoys are shown in Figure 10. As seen in the figure, the decay of the motions becomes faster when the appendages are adopted. The plate-type appendages are more effective than the conical-type ones for the reduction of pitch motions, while the effect of the porous cone appendage is almost the same as that of the plate-type one in reducing heave motions.

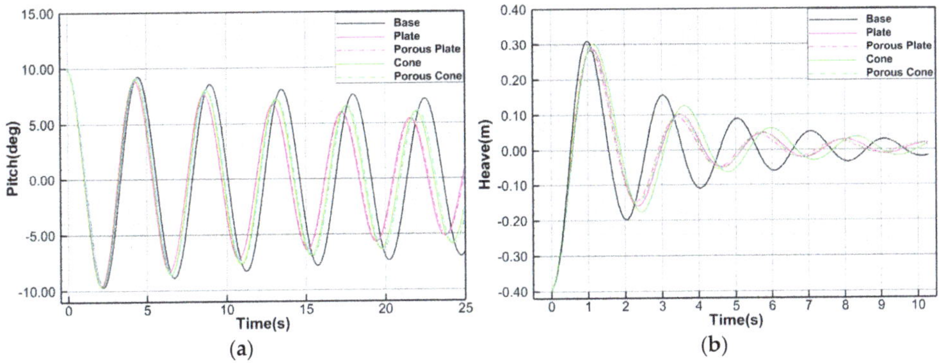

Figure 10. Free pitch (a) and (b) heave decay curves of the lightweight light buoys.

Table 2 shows the estimated non-dimensional damping coefficient and natural frequencies of the light buoys from the free decay tests. With the adopted appendages, the pitch natural frequencies of the buoys became higher than those of the Base model, resulting from the increase in the restoring moment due to the increments in metacentric height (GM) and draft. The reason the heave natural frequencies of the models with appendages become 10%–15% smaller might be related to the change in mass and added mass, and the viscous and radiation damping.

Table 2. Non-dimensional damping coefficients and natural frequencies of lightweight light buoys from free decay tests.

	Pitch		Heave	
	ζ[-]	ω_o [rad/s]	ζ[-]	ω_o [rad/s]
Base	0.0135	1.397	0.1107	3.107
Plate	0.0209	1.452	0.1550	2.748
Porous Plate	0.0219	1.447	0.1660	2.813
Cone	0.0167	1.425	0.1344	2.649
Porous Cone	0.0172	1.418	0.1587	2.795

Figure 11 shows the vorticity distributions at the first and second peaks of motion, which were observed during the free decay simulations. Around the appendages, complicated flows and strong vortexes were observed, which led to high energy dissipations and strong viscous effects.

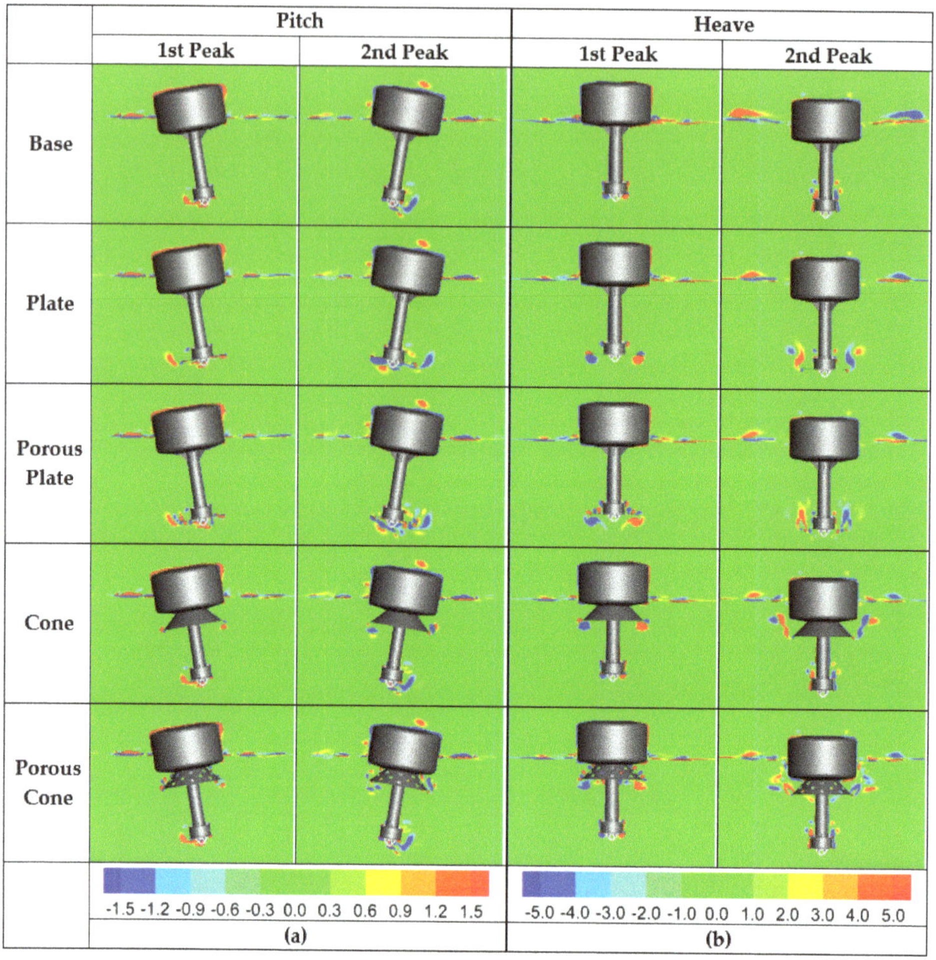

Figure 11. Comparisons of vorticity (j) distributions around the lightweight light buoys during (**a**) free pitch and (**b**) heave decay tests when the first and second peaks of motion occur.

3.4. Potential-Based Motion Analysis

3.4.1. Computational Conditions

Figure 12 shows the computational domain and the panels on the surfaces of the light buoys. Heading waves with frequencies of 0.1 to 6.0 rad/s were considered. The maximum panel size was approximately 1/7 times that of the shortest wavelength. The minimum and maximum numbers of the generated panels were approximately 4000 and 18,300 for the Base and Porous Cone models, respectively. The motion analyses of the light buoys were performed using ANSYS-AQWA with and without the application of the viscous damping coefficients, which were estimated using CFD simulations, as discussed in the previous section.

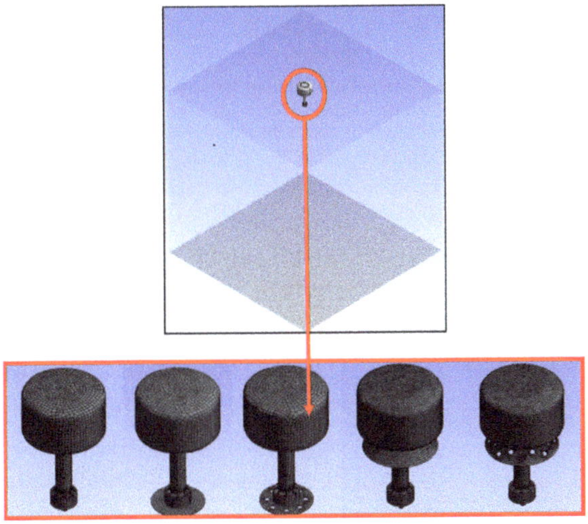

Figure 12. Computational domain and panels of the lightweight light buoys for potential-based simulations.

3.4.2. Results

Table 3 lists the estimated coefficients of the added and total mass moment of inertia, the viscous and radiation damping, and the ratio of viscous damping to total damping for the pitch motion at the natural frequencies of each model from the potential-based motion analysis. As seen in the table, the added mass moment of inertia and viscous damping, which is much larger than the radiation damping, becomes larger with the adopted appendages, which may result in the reduction of motion.

Table 3. Added mass moment of inertia, total mass moment of inertia, coefficients of viscous and radiation damping, and ratio of viscous damping to total damping for a pitch motion at natural frequencies of the models from the potential-based motion analysis.

	I_a [kgm^2]	I+I_a [kgm^2]	$b_{55,\,vis}$ [kgm^2/s]	$\nu(\omega_o)$ [kgm^2/s]	$\frac{b_{55,\,vis}}{b_{55}}$ [-]
Base	1968	12,876	63.359	0.488	0.9924
Plate	2435	13,936	109.952	0.884	0.9920
Porous Plate	2556	14,002	114.261	0.842	0.9927
Cone	2145	13,121	80.067	0.656	0.9919
Porous Cone	2116	13,084	82.690	0.610	0.9927

Figure 13 shows the pitch response amplitude operators (RAOs) of the buoys. Near the natural frequencies of the buoys, the RAOs of all models become very small when the viscous damping is considered. The maximum pitch motions are expected to be reduced by approximately 20%–40% with the adoption of the appendages. Because the radiation damping is small, and the viscous damping was estimated and applied from the CFD simulations of free decay tests, plate-type appendages are more effective than the conical-type ones.

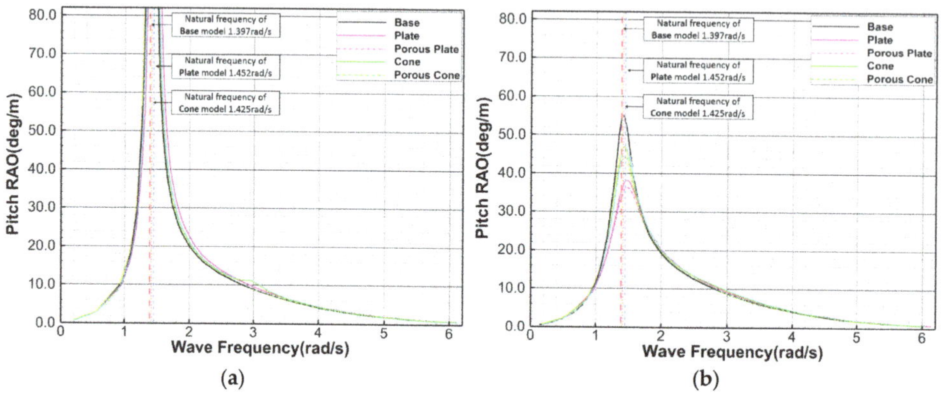

Figure 13. Pitch pitch response amplitude operators (RAOs) estimated from the potential-based motion analysis of the light buoys (**a**) without and (**b**) with the consideration of viscous damping.

Table 4 shows the estimated added mass, total mass, coefficients of viscous and radiation damping, and ratio of viscous damping to total damping for heave motion at the natural frequencies of each model from the potential-based motion analysis. Unlike for pitch motion, radiation damping is greater than viscous damping except for that of the Cone model. The reason for this can be easily deduced, as the vertical translation motion of a cylinder generates more waves than the rotation motion.

Table 4. Added mass, total mass, coefficients of viscous and radiation damping, and ratio of viscous damping to total damping of a lightweight light buoy from the potential-based motion analysis.

	m_a [kg]	$m+m_a$ [kg]	$b_{33,\,vis}$ [kg/s]	$\nu(\omega_o)$ [kg/s]	$\frac{b_{33,\,vis}}{b_{33}}$ [-]
Base	2254	4677	136	3107	0.0419
Plate	2874	5470	2243	2890	0.4368
Porous Plate	3166	5745	2387	2987	0.4441
Cone	2549	5166	2907	1710	0.6295
Porous Cone	2609	5206	2570	2600	0.4970

The heave RAOs of the light buoys without and with considering the viscous damping coefficients are shown in Figure 14. Near the natural frequencies of the buoys, the RAOs of the models with appendages become small when viscous damping is considered, while there is not much difference with the Base model. In the comparison of the maximum heave motions of each model, porous appendages are expected to be more effective than non-porous ones for all frequencies in reducing the heave motion even if the differences are not large.

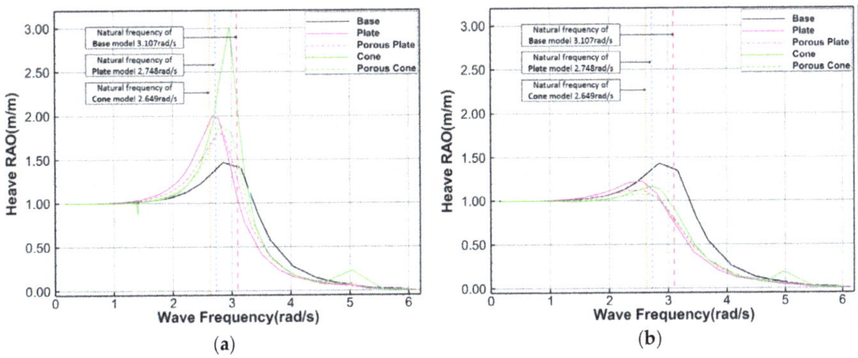

Figure 14. Heave RAOs estimated from the potential-based motion analysis of the light buoys (**a**) without and (**b**) with the consideration of viscous damping.

3.5. Motion Simulation in Regular Waves Using CFD

3.5.1. Computational Domain and Grid System

Figure 15 shows the computational domain and boundary conditions for the motion simulation in regular waves using CFD. To enhance the accuracy of CFD simulations including waves, it is important to minimize the artificial diffusion of generated waves and wave reflections at boundaries. One of the methods to reduce wave reflection and computational cost while maintaining accuracy is the wave forcing method. As an inflow boundary condition, second-order Stokes wave theory was applied as the "velocity inlet" to express waves with the Euler-overlay method (EOM), which is a built-in forcing method of STAR-CCM+ [25,26]. In the overlay zone located from the inlet boundary to 2.4 m ahead of the buoy in the x-direction, analytic solutions and CFD solutions are gradually blended by applying source terms to VOF and momentum equations. No-slip conditions were imposed on the buoy surface and bottom boundary. For outflow boundary, the "pressure outlet" boundary condition was used while adopting grid damping technique [27] to minimize the wave reflection at the boundary. For the side boundaries, the "symmetry plane" boundary condition was used [25,26].

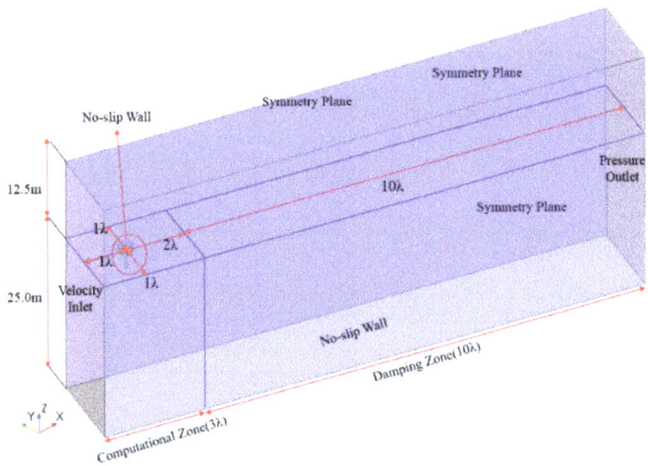

Figure 15. Computational domain and boundary conditions for motion simulations of lightweight light buoys in regular waves using CFD.

Figure 16 shows the grid system for the CFD simulations. The numbers of cells per wavelength and height were 140 and 30, respectively. The minimum and maximum numbers of the total generated cells were approximately 3.0 and 3.74 million for the Base and Porous Cone models, respectively.

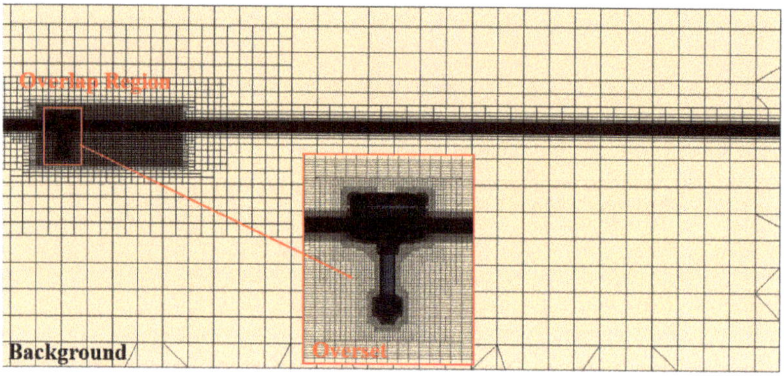

Figure 16. Grid system for motion simulations of lightweight light buoys in regular waves using CFD.

3.5.2. Regular Wave Conditions for Motion Analysis Using CFD

Table 5 shows the wave conditions (i.e., simulation cases) for the CFD motion analysis. Eight waves of different wave frequencies, including the pitch and heave natural frequencies of the Base model, were selected. The amplitude of each wave was determined under the assumption that the wave steepness was 1/40.

Table 5. Regular wave conditions.

Wave Steepness H/λ [-]	Wave Frequency, ω	Wave Period, T [s]	Wave Length, λ [m]	Wave Height, H [m]	Remark
1/40	1.247	5.039	39.638	0.996	
	1.397	**4.498**	**31.583**	**0.794**	**Pitch natural frequency (Base)**
	1.843	3.409	18.147	0.456	
	2.039	3.082	14.826	0.373	
	2.395	2.623	10.746	0.270	
	2.751	2.284	8.145	0.205	
	3.107	**2.022**	**6.385**	**0.160**	**Heave natural frequency (Base)**
	3.500	1.795	5.032	0.126	

3.5.3. Results

Figure 17 shows the time histories of the pitch and heave motions of a light buoy with and without appendages when the frequency of the heading wave was 1.397 rad/s, which is the same as that of the pitch natural frequency of the Base model. The corresponding wave height and length were 0.794 m and 31.583 m, respectively. The effectiveness of the appendages in terms of reducing both motions was confirmed, as shown in the potential-based simulations considering viscous damping. This can also be seen in Figure 18, which shows the snapshots of the free surface and the buoys when the 5th plus peak of the motion was observed. However, unlike the potential-based simulation results, the superiority of the effectiveness of each appendage is difficult to distinguish for the pitch motion, while the plate-type appendages seem to work better for heave motion.

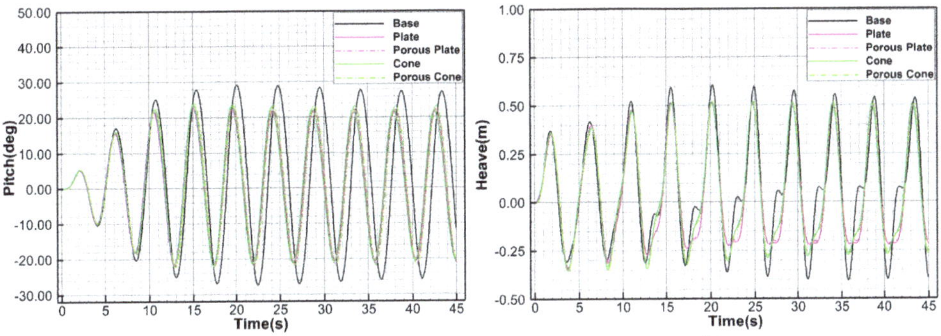

(a) (b)

Figure 17. Time histories of (**a**) pitch and (**b**) heave motions of the buoys in regular waves, for which the frequency is the same as the pitch natural frequency of the Base model, estimated by CFD simulation.

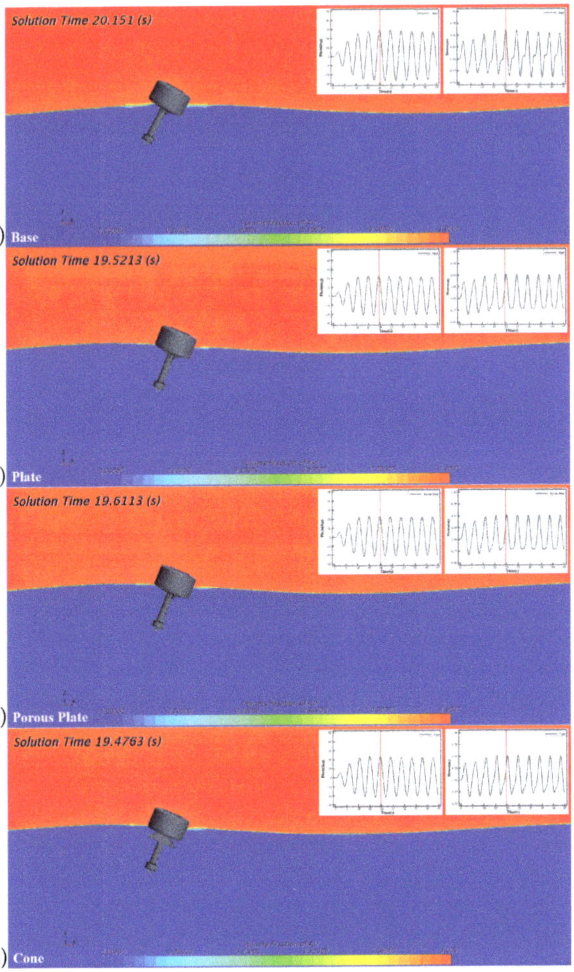

(a) (b) (c) (d)

Figure 18. *Cont.*

(e)
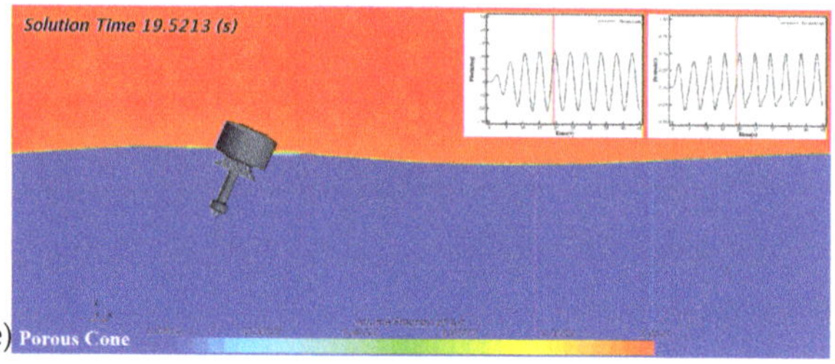

Figure 18. Snapshots of CFD simulations of (**a**) Base, (**b**) Plate, (**c**) Porous Plate, (**d**) Cone and (**e**) Porous Plate models when the 5th plus peak of motion occurs in regular waves, for which the frequency is the same as the pitch natural frequency of the Base model.

Figure 19 shows the time histories of the pitch and heave motion of a light buoy with and without appendages when the frequency of the heading wave was 3.107 rad/s, which is the same as that of the heave natural frequency of the Base model. The corresponding wave height and length were 0.160 m and 6.385 m, respectively. The effectiveness of the appendages in reducing heave motions is clearly seen, while that of the cone-type appendages is trivial for the reduction of pitch motion.

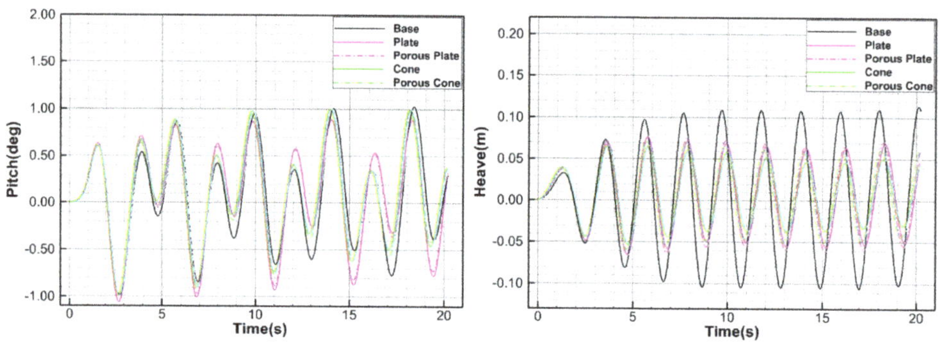

(**a**) (**b**)

Figure 19. Time histories of (**a**) pitch and (**b**) heave motions of the buoys in regular waves, for which the frequency is the same as the heave natural frequency of the Base model, estimated by CFD simulation.

3.6. Comparison of RAOs between the Potential-Based Motion Analysis and CFD Simulations

The comparison of pitch and heave RAOs, and the maximums of each motion between the potential-based motion analysis and CFD simulations are shown and listed in Figure 20 and Table 6, respectively. The pitch RAOs of both results near the natural frequencies of the buoys are expected to be reduced by adopting the appendages. Although the quantitative superiority of the appendages is different, the qualitative superiority is similar between the results of the potential-based and CFD simulations. In addition, if the frequency of the wave is far from that of the natural frequency of the buoys, there is not a large discrepancy in the RAO by both simulations. In the case of heave RAO, discrepancies exit in the low-frequency region, although the effectiveness of the appendages is confirmed near the heave natural frequencies of each model. There may be some reasons for these differences. First, the viscous damping coefficient was evaluated by subtracting the radiation damping coefficient at the natural frequency of the buoy from the total damping coefficient. Second, the wave heights in the low-frequency region are relatively high because the wave steepness is fixed, which may enhance the non-linearity of the motion. The complex interactions among the incoming wave, flows, and vortex around the buoy, and flows near the buoy, may also be causing these differences.

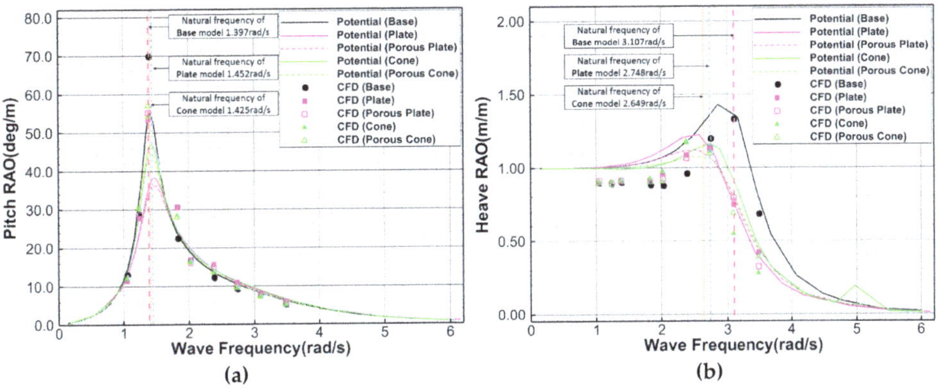

Figure 20. Comparison of (**a**) pitch and (**b**) RAOs between the CFD simulation and potential-based motion analysis.

Table 6. Comparison of maximum RAOs estimated from the potential-based simulation with those from the CFD simulation.

Models	Maximum of pitch RAO [deg/m]			Maximum of Heave RAO [m/m]		
	Potential	CFD	Remark	Potential	CFD	Remark
Base	54.841	69.743	pitch natural frequency of Base model	1.357	1.327	heave natural frequency of Base model
Plate	35.366	53.441		0.760	0.746	
Porous Plate	33.736	55.230		0.790	0.793	
Cone	45.211	54.222		0.920	0.549	
Porous Cone	43.381	57.069		0.815	0.688	

4. Conclusions

In this study, motion analyses for a newly developed lightweight light buoy in waves were performed to predict the buoy's motion performance and check the effect of the developed appendages on the performance. First, free decay tests using CFD including benchmark cases for the validation were carried out to estimate the viscous damping coefficients, which cannot be obtained by potential-based simulations. The results for the validation were in good agreement with those of the reference. Second, potential-based simulations with and without considering the viscous damping coefficients were performed, and the results were compared. From the results, it was confirmed that viscous damping should be considered when the frequency of an incoming wave is close to the natural frequency of the buoy. Finally, motion simulations in regular waves using CFD were carried out to compare the RAOs and maximums of motions with those obtained from a potential-based simulation considering viscous damping. The RAOs of both results near the natural frequencies of the buoys were expected to be reduced by adopting the appendages. Although the quantitative superiority of the appendages was different, the qualitative superiority was similar for both results. The present methods and results would be useful not only for a buoy but for other cylindrical floating structures, such as spars and wave energy converters. Motion simulations in irregular waves using CFD will be carried out to compare its RAO and motion response with those by potential-based simulation.

Author Contributions: Conceptualization, B.-H.S. and S.-M.J.; methodology, S.-M.J. and C.-Y.L..; validation, B.-H.S.; formal analysis, B.-H.S.; investigation, S.-M.J. and C.-Y.L.; resources, S.-M.J.; writing—original draft preparation, S.-M.J.; writing—review and editing, S.-M.J. and C.-Y.L.; visualization, B.-H.S.; supervision, S.-M.J. and C.-Y.L.; funding acquisition, S.-M.J. All authors have read and agreed to the published version of the manuscript.

Funding: This study was supported by research fund from Chosun University (K207177002).

Conflicts of Interest: The authors declare no conflict of interest.

References

1. Jeong, S.M.; Son, B.H.; Ko, S.W.; Park, S.H.; Jeong, D.S. Stability evaluation of light-buoys combined with 7 nautical mile self-contained lantern. In Proceedings of the Oceans-MTS/IEEE Kobe Techno-Oceans (OTO), Kobe, Japan, 28–31 May 2018.
2. Son, B.H.; Ko, S.W.; Yang, J.H.; Jeong, S.M. Motion analysis of light buoys combined with 7 nautical mile self-contained lantern. *J. Korean Soc. Mar. Environ. Saf.* **2018**, *24*, 628–636. [CrossRef]
3. Bhinder, M.A.; Murphy, J. Evaluation of the viscous drag for a domed cylindrical moored wave energy converter. *J. Mar. Sci. Eng.* **2019**, *7*, 120. [CrossRef]
4. Kim, M.G.; Jung, K.H.; Park, S.B.; Suh, S.B.; Park, I.R.; Kim, J.; Kim, K.S. Experimental study on viscous effect in roll and heave motions of a rectangular structure. *Ocean Eng.* **2019**, *171*, 250–258. [CrossRef]
5. Evans, D.V.; Jerey, D.C.; Salter, S.H.; Taylor, J.R.M. Submerged cylinder wave energy device: Theory and experiment. *Appl. Ocean. Res.* **1979**, *1*, 3–12. [CrossRef]
6. Jung, K.H.; Chang, K.A.; Huang, E.T. Two-dimensional flow characteristics of wave interactions with a free-rolling rectangular structure. *Ocean Eng.* **2005**, *32*, 1–20. [CrossRef]
7. Jung, K.H.; Chang, K.A.; Jo, H.J. Viscous effect on the roll motion of a rectangular structure. *J. Eng. Mech.* **2006**, *132*, 190–200. [CrossRef]
8. Bhattacharyya, R. *Dynamics of Marine Vehicles*, 1st ed.; John Wiley & Sons Inc.: Hoboken, NJ, USA, 1978; pp. 75–80.
9. Journée, J.M.J.; Massie, W.W. *Offshore Hydromechanics*; Lecture Notes; Delft University of Technology: Delft, The Netherlands, 2001.
10. Wassermann, S.; Feder, D.F.; Abdel-Maksoud, M. Estimation of ship roll damping—A comparison of the decay and the harmonic excited roll motion technique for a post panamax container ship. *Ocean Eng.* **2016**, *120*, 371–382. [CrossRef]
11. Wilson, R.V.; Carrica, P.M.; Stern, F. Unsteady rans method for ship motions with application to roll for a surface combatant. *Comput. Fluids* **2006**, *35*, 501–524. [CrossRef]
12. Irkal, M.A.R.; Nallayarasu, S.; Bhattacharyya, S.K. Cfd approach to roll damping of ship with bilge keel with experimental validation. *Appl. Ocean Res.* **2016**, *55*, 1–17. [CrossRef]
13. Song, S.S.; Kim, S.H.; Paik, K.J. Determination of linear and nonlinear roll damping coefficients of a ship section using CFD. *Teorija i Praksa Brodogradnje i Pomorske Tehnike* **2019**, *70*, 17–33. [CrossRef]
14. Kianejad, S.S.; Lee, J.S.; Liu, Y.; Enshaei, H. Numerical assessment of roll motion characteristics and damping coefficient of a ship. *J. Mar. Sci. Eng.* **2018**, *6*, 101. [CrossRef]
15. Zhang, X.; Zeng, Q.; Liu, Z. Hydrodynamic performance of rectangular heaving buoys for an integrated floating breakwater. *J. Mar. Sci. Eng.* **2019**, *7*, 239. [CrossRef]
16. Luan, Z.; He, G.; Zhang, Z.; Jing, P.; Jin, R.; Geng, B.; Liu, C. Study on the optimal wave energy absorption power of a float in waves. *J. Mar. Sci. Eng.* **2019**, *7*, 269. [CrossRef]
17. Mohapatra, S.C.; Islam, H.; Guedes Soares, C. Wave diffraction by a floating fixed truncated vertical cylinder based on Boussinesq equations. In *Renewable Energy Offshore*; Taylor & Francis: London, UK, 2019; pp. 281–289.
18. Koh, H.J.; Cho, I.H. Heave motion response of a circular cylinder with the dual damping plates. *Ocean Eng.* **2016**, *125*, 95–102. [CrossRef]
19. Tao, L.; Cai, S. Heave motion suppression of a spar with a heave plate. *Ocean Eng.* **2004**, *31*, 669–692. [CrossRef]
20. Zhu, L.; Lim, H.C. Hydrodynamic characteristics of a separated heave plate mounted at a vertical circular cylinder. *Ocean Eng.* **2017**, *131*, 213–223. [CrossRef]
21. Sudhakar, S.; Nallayarasu, S. Hydrodynamic responses of spar hull with single and double heave plates in random waves. *Int. J. Ocean Syst. Eng.* **2013**, *4*, 188–208. [CrossRef]
22. Sudhakar, S.; Nallayarasu, S. Hydrodynamic response of spar with single and double heave plates in regular waves. *Int. J. Ocean Syst. Eng.* **2014**, *4*, 1–18. [CrossRef]
23. Koh, H.J.; Kim, J.R.; Cho, I.H. Model test for heave motion reduction of a circular cylinder by a damping plate. *J. Ocean. Eng. Tech.* **2013**, *27*, 76–82. [CrossRef]

24. Palm, J.; Eskilsson, C.; Paredes, G.M.; Bergdahl, L. Coupled mooring analysis for floating wave energy converters using cfd: Formulation and validation. *Int. J. Mar. Energy* **2016**, *16*, 83–99. [CrossRef]
25. Lee, Y.G.; Kim, C.H.; Park, J.H.; Kim, H.; Lee, I.; Jin, B. Numerical simulations of added resistance in regular head waves on a container ship. *Brodogradnja* **2019**, *70*, 61–86. [CrossRef]
26. Park, I.R.; Kim, J.I.; Suh, S.B.; Kim, J.; Kim, K.S.; Kim, Y.C. Numerical prediction of the powering performance of a car-Ferry in irregular waves for safe return to port (srtp). *J. Ocean. Eng. Tech.* **2019**, *33*, 1–9. [CrossRef]
27. Kim, S.Y.; Kim, K.M.; Park, J.C.; Jeong, G.M.; Chun, H.H. Numerical simulation of wave and current interaction with a fixed offshore substructure. *Int. J. Nav. Arch. Ocean* **2016**, *8*, 187–197. [CrossRef]

Journal of
Marine Science and Engineering

MDPI

Article

Effect of Top Tension on Vortex-Induced Vibration of Deep-Sea Risers

Jie Zhang [1,2,*], He Guo [1], Yougang Tang [3] and Yulong Li [3]

1 College of Ocean Science and Engineering, Shanghai Maritime University, Shanghai 201306, China; love53531@163.com
2 Department of Civil, Environmental and Geomatic Engineering, University College London, London WC1E 6BT, UK
3 State Key Laboratory of Hydraulic Engineering Simulation and Safety, Tianjin University, Tianjin 300072, China; tangyougang_td@163.com (Y.T.); liylong7@mail.sysu.edu.cn (Y.L.)
* Correspondence: zhangjie@shmtu.edu.cn

Received: 21 January 2020; Accepted: 12 February 2020; Published: 15 February 2020

Abstract: With the increase of water depth, the design and use of the top-tensioned risers (TTR) are facing more and more challenges. This research presents the effect of top tension on dynamic behavior of deep-sea risers by means of numerical simulations and experiments. First, the governing equation of vortex-induced vibration (VIV) of TTR based on Euler-Bernoulli theory and Van der Pol wake-oscillator model was established, and the effect of top tension on natural vibration of TTR was discussed. Then, the dynamic response of TTR in shear current was calculated numerically by finite difference method. The displacement, bending stress and vibration frequency of TTR with the variation of top tension were investigated. Finally, a VIV experiment of a 5 m long flexible top-tensioned model was carried out at the towing tank of Tianjin University. The results show that the vibration displacement of TTR increases and the bending stress decreases as the top tension increases. The dominant frequency of VIV of TTR is controlled by the current velocity and is barely influenced by the top tension. With the increase of top tension, the natural frequency of TTR increases, the lower order modes are excited in the same current.

Keywords: deep-sea riser; top tension; vortex-induced vibration; numerical simulation; experiment

1. Introduction

In recent years, owing to the large demand of crude oil, offshore oil and gas explorations have been shifted to deeper water regions. Risers are one of the basic elements of offshore installations that are usually used for drilling and production. The riser is installed between wellhead at the sea bed and floating platform. Actually, the riser has suffered the strong influences of environmental conditions in unshielded deep-water region, hence it has complicated dynamic behavior under the influence of platform's motions, wind, waves and sea current, as shown in Figure 1. One of the most famous phenomena is vortex-induced vibration (VIV). When the riser moves under the action of sea current, vortices are shed along its surface, resulting in the formation of an unstable wake region behind it. Vortex shedding normally takes place with different frequencies and amplitudes, thus induces a periodically varying transverse force on the riser (i.e., perpendicular to the direction of the current), resulting in a periodically transverse vibration known as the VIV [1–4].

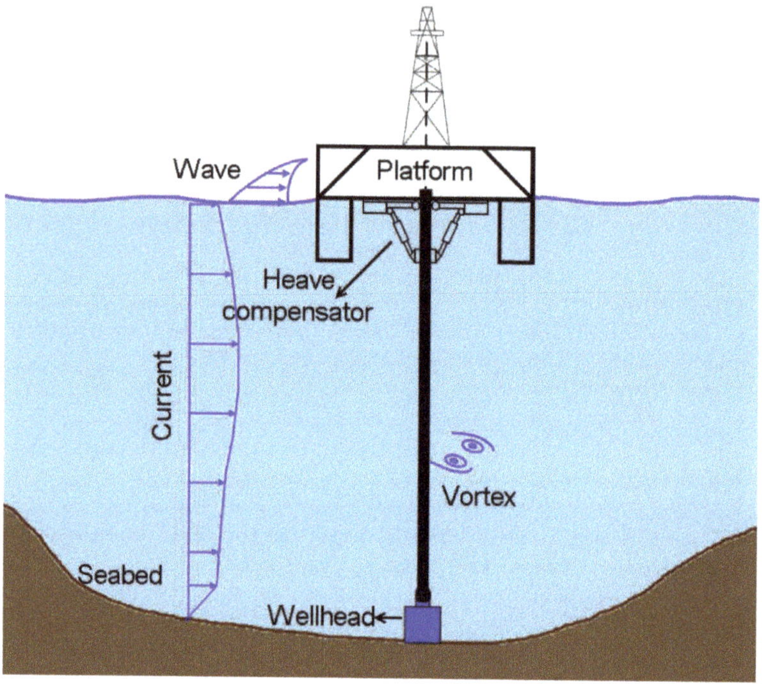

Figure 1. Schematic of a top-tensioned riser in ocean environment.

Drilling and production risers are mostly affected by VIV, which leads to increases in their hydrodynamic loading and reduction in their service life due to fatigue. In the past, to ensure the safety and to enhance the productivity of riser systems, many investigations on VIV of marine risers have been reported. Srinil [5] studied dynamic behavior of a variable tension vertical riser placed in a linearly sheared current by using numerical methods and experiments. He concluded that the current velocity and the variable tension had great influence on VIV response of a deep-sea riser. The prediction of VIV of a deep-sea riser was a challenging task since the incident current was practically non-uniform and the associated fluid-structure interaction (FSI) phenomena was highly complex. Wang and Xiao [6] presented an investigation on the VIV of the riser in uniform and linearly sheared current by means of a fully 3D FSI simulation. They pointed out that the dominating modes of the riser in uniform and linearly sheared current with the same surface velocity had no great difference. However, the lock-in frequency range of the riser was wider in the sheared current, and the maximum stress was smaller than in the uniform current. The contribution from in-line vibration has been noticed by many researchers [7–11]. The coupled model of In-Line (IL) and Cross-Flow (CF) VIV was established, and the results showed that IL VIV would affect the mode of the vortex and the amplitude of the CF VIV. As the water depth increases, the IL-VIV was more easily excited than CF VIV, and IL VIV had a significant contribution to fatigue on the deep-sea riser for the low-order modes. Duan et al. [12] presented a numerical study on VIV of a riser with internal flow. Guo and Lou [13] performed an experiment in water with a pipe made of rubber to investigate the effect of internal flow on VIV. The results indicated that the internal flow would reduce the natural frequency of the riser system, and the response amplitude increases while the vibration frequency decreases. Chen et al. [14] studied the dynamic response of the riser suffering both the floating top-end and VIV. An integrated system with both platform and riser was established by means of finite element numerical simulations. The result showed that the displacement of the riser was multiple times larger than that of the case without moving top-end. Wang et al. and Yin et al. [15,16] performed experiments in towing tank for the riser which was excited at the top by platform surge motion, and the results showed that the platform surge had a great influence on VIV of the riser.

The top-tensioned risers (TTR) are widely deployed by FPSOs, Semi-submersible platforms, Spar or Tensioned-leg platforms for drilling and production. The bottom end of TTR is connected with the wellhead by ball joint, and the top end is connected with the platform by heave compensator (or tensioner, see Figure 1). The compensator provides a large static tensile tension at the top end of TTR (known as "top tension" for short in this paper), which can keep TTR vertical and avoid buckling due to its large length. TTR is very sensitive to heave movements due to waves and to VIV caused by sea current, since the rotation at the top and bottom connections is limited. The heave movement also requires top tension equipment to compensate for the lack of tension. Therefore, the compensator has the buffering function as a spring on the riser to reduce the heave motion effect [17,18]. The top tension is generally 1.1~1.6 times of the submerged weight of TTR. Insufficient top tension might cause buckling failure of the riser, while excessive top tension might result in stress concentration at the top or bring about failure of the lower flexible joint. Yang et al. [19] pointed out that the stress and displacement of TTR was controlled by the tensioners which provided constant axial tension to the riser. Moghiseh et al. [20] concluded that increasing the pretension in the middle of the riser could effectively reduce the displacement of the riser. Fujiwara et al. [21] performed a large-scale experiment in deep-sea basin with a Teflon tube of 28.5 m long that was tensioned at the top end, and they found out that the displacement of VIV of TTR was increased as the top tension increased.

Unlike the shallow sea risers, the deep-sea riser has a large slenderness ratio (around 10^4). In order to study how VIV response of TTR changes with the change of top tension, this research presents the effect of top tension on VIV of deep-sea risers by means of numerical simulations and experiments.

2. Governing Equations

In this research, TTR is modeled as simply-supported beam, as shown in Figure 2. The sea current is in y direction and simplified as shear current. The influence of IL VIV is neglected, since the IL VIV is an order of magnitude smaller than the CF VIV [22].

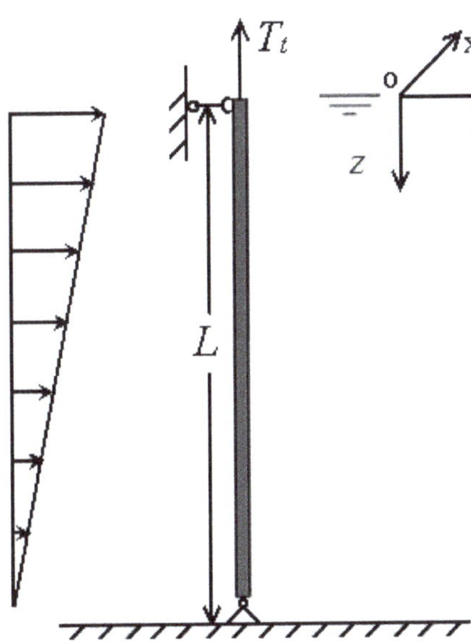

Figure 2. Simply supported TTR.

The Euler-Bernoulli theory is applicable for description of TTR dynamic bending. The motion equation governing the transverse displacement $x(z,t)$ of TTR from its straight vertical equilibrium as a function of depth z and time t can be written as:

$$EI\frac{\partial^4 x(z,t)}{\partial z^4} - \frac{\partial}{\partial z}\left[(T_t - wz)\frac{\partial x(z,t)}{\partial z}\right] + \overline{m}\frac{\partial^2 x(z,t)}{\partial t^2} = f_x(z,t) \tag{1}$$

where EI is the bending stiffness. $\overline{m}$ is the mass per unit length of the system (riser with internal fluid and added mass). w is the submerged weight of TTR per unit length. T_t is the top tension of TTR.

$$T_t = f_{top} \cdot w \cdot L \tag{2}$$

where f_{top} is a dimensionless pretension factor which generally varies between 1.1 and 1.6. L is the length of TTR.

$f_x(z,t)$ is the hydrodynamic force per unit length of the riser. Generally, $f_x(z,t)$ is calculated by Morison's Equation:

$$f_x(z,t) = \frac{1}{2}\rho_w D C_L U^2 - \frac{1}{2}\rho_w D C_d |\dot{x}|\dot{x} \tag{3}$$

where ρ_w is the density of seawater, D is the outer diameter of the riser, C_L is the lift force coefficient. U is the current velocity in any depth, $U = (1 - \frac{z}{L})U_0$, U_0 is the current velocity at sea level. C_d is the drag coefficient.

A forced Van der Pol oscillator equation is used to describe the dynamics of the riser's wake [23]:

$$\ddot{q} + \varepsilon\Omega_f(q^2 - 1)\dot{q} + \Omega_f^2 q = F \tag{4}$$

where the variable q is defined as the local fluctuating lift coefficient, $q = 2C_L/C_{L0}$, the reference lift coefficient C_{L0} is that observed on a fixed structure subjected to vortex shedding. Ω_f is the vortex-shedding frequency, which is related to the non-dimensional Strouhal number S_t, $\Omega_f = 2\pi S_t U/D$. It is common practice to assume $S_t = 0.2$ in the sub-critical range, $300 < R_e < 1.5 \times 10^5$. The reference lift coefficient C_{L0} being usually taken as $C_{L0} = 0.3$ in the large range of R_e. The right-hand side forcing term F models the effect of the cylinder motion on the near wake, $F = A\ddot{x}/D$. ε and A are given parameters, and the values are estimated through experimental data on free and forced vortex shedding behind cylinders, $\varepsilon = 0.3$, $A = 12$ [23].

The boundary conditions at the ends of the riser are given as

$$x(z,t) = 0, \quad \frac{\partial^2 x(z,t)}{\partial z^2} = 0 \quad (z = 0, z = L) \tag{5}$$

The bending stress is dependent on the second derivative of displacement of TTR:

$$\sigma_b(z) = \frac{EDx''(z,t)}{2} \tag{6}$$

3. Effect of Top Tension on Natural Vibration of TTR

3.1. Natural Frequencies and Mode Shapes

The natural vibration characteristics of TTR mainly include natural frequencies and mode shapes, which are the basis of complex vibration analysis. The natural frequency directly affects the excitation of vortex-induced resonance for deep-sea risers, and the mode shape can be used to analyze the vibration response.

For the deep-sea riser, an accepted conclusion is that the low order natural frequencies and mode shapes can be expressed by simplified formulas which depend on the tension in the riser [24,25].

$$\omega_n = \frac{n\pi}{2L\sqrt{m}}(\sqrt{T_t} + \sqrt{T_b}) \tag{7}$$

$$\phi_n(z) = \left(\frac{T_b}{T_t - wz}\right)^{1/4} \sin\left(\frac{n\pi\left(\sqrt{T_t - wz} - \sqrt{T_b}\right)}{\sqrt{T_t} - \sqrt{T_b}}\right) \tag{8}$$

where ω_n is the nth natural frequency of TTR. $\phi_n(z)$ is the nth mode function. T_b is the tension at the bottom end of TTR, $T_b = (f_{top} - 1)wL$.

It shows that the deep-sea riser is like a cable owing to the large slenderness ratio, the lower order modes are mainly affected by tension and unrelated to the bending stiffness. The natural frequency of TTR will increase with the increase of top tension, it is equivalent to increasing the stiffness of TTR. The nth natural frequency is about n times of the first frequency (i.e., $\omega_n = n\omega_1$), this is typical phenomenon of deep-sea risers. Meanwhile, the axial tension in riser gradually decreases with the increase of water depth due to the self-weight; the mode shape is no longer a standard sine function.

The mode curvature is obtained by finding the second derivative of the variable z in Equation (8):

$$\varphi_n(z) = \phi_n''(z) \tag{9}$$

3.2. System Parameters and Basic Calculations

The adopted values of the system parameters are shown in Table 1. It should be pointed out that the data in Table 1 are from the Truss Spar (named Horn Mountain) servicing in the Gulf of Mexico [26].

The natural frequencies of TTR are shown in Table 2. It shows that the low order natural frequencies of TTR are small. The current velocity at sea level generally varies between 0.1 m/s and 1 m/s without considering influence of waves in South China Sea [27], and the range of vortex-shedding frequencies will be concentrated in the low order mode frequencies of TTR.

Table 1. Base set of system parameters.

No.	Property	Symbol	Value	Unit
1	Riser length	L	1500	m
2	Outer diameter	D	0.3048	m
3	Wall thickness	t	0.0136	m
4	Young's modulus	E	2.1E11	Pa
5	Density of steel	ρ_s	7850	kg/m^3
6	Density of seawater	ρ_w	1025	kg/m^3
7	Density of internal fluid	ρ_f	800	kg/m^3
8	Added mass coefficient	C_a	1.0	-
9	Drag coefficient	C_d	1.0	-

Table 2. Natural frequencies of TTR (rad/s).

	ω_1	ω_2	ω_3	ω_4	ω_5
$f_{top} = 1.1$	0.1075	0.2149	0.3224	0.4298	0.5373
$f_{top} = 1.3$	0.1329	0.2658	0.3986	0.5315	0.6644
$f_{top} = 1.5$	0.1521	0.3042	0.4563	0.6083	0.7604

The 4th mode shape and mode curvature of TTR are shown in Figures 3 and 4. Under the same condition, with the increase of top tension, the nodes and anti-nodes of the mode shape move up, and the mode shape more conforms to the standard sine function. The amplitude of mode shape decreases as the top tension decreases. However, the smaller the top tension, the larger amplitude of mode curvature at the bottom of TTR.

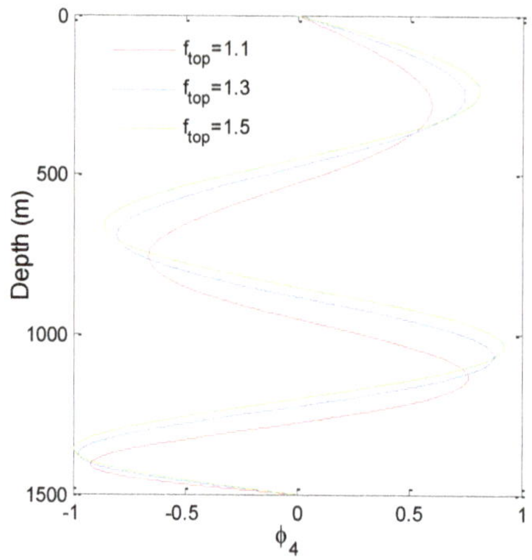

Figure 3. The 4th mode shape of TTR.

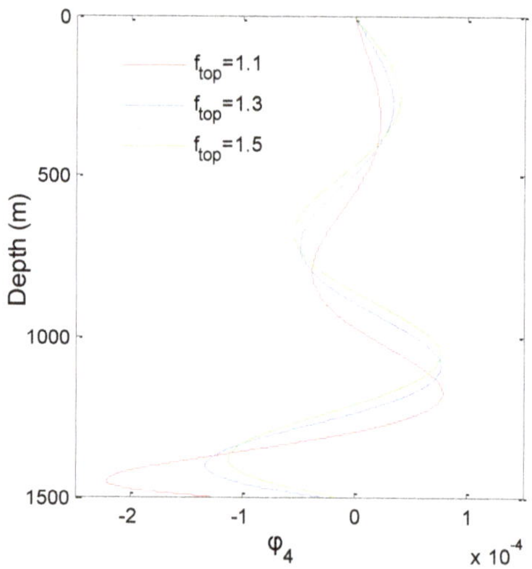

Figure 4. The 4th mode shape curvature of TTR.

4. Effect of Top Tension on VIV of TTR

The VIV response of TTR can be calculated numerically by finite difference method (FDM) according to Equations (1)–(6). The time history of VIV of TTR at each depth are obtained. Figure 5 shows the displacement and bending stress of TTR at midpoint ($z = 750$ m). It can be seen that the change trend of displacement is not obvious under different top tensions, while the bending stress decreases significantly with the increase of top tension.

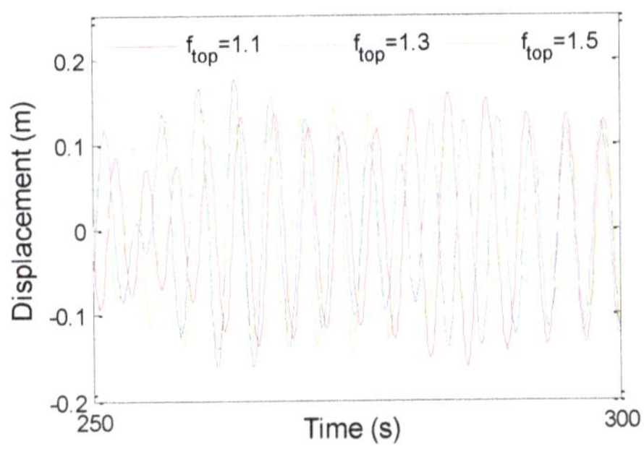

(**a**)

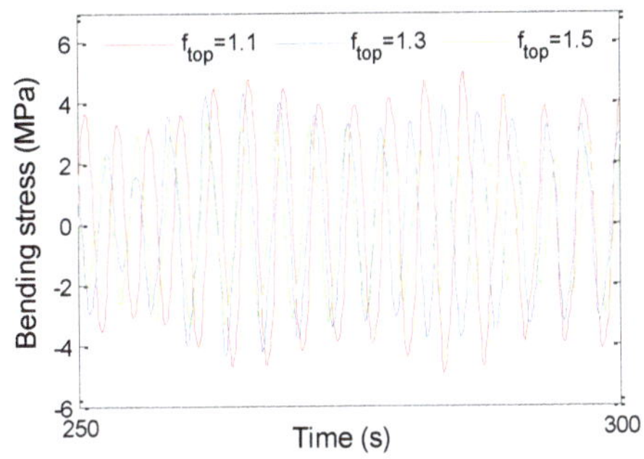

(**b**)

Figure 5. Dynamic response of TTR (U_0 = 0.5 m/s, z = 750 m). (**a**) Time history of displacement. (**b**) Time history of bending stress.

The frequency spectrum of TTR is shown in Figure 6. The dominant frequency of TTR decreases slightly with the increase of top tension. It indicates that the dominant frequency is mainly controlled by vortex-shedding frequency, which is related to the current velocity and the riser's outer diameter.

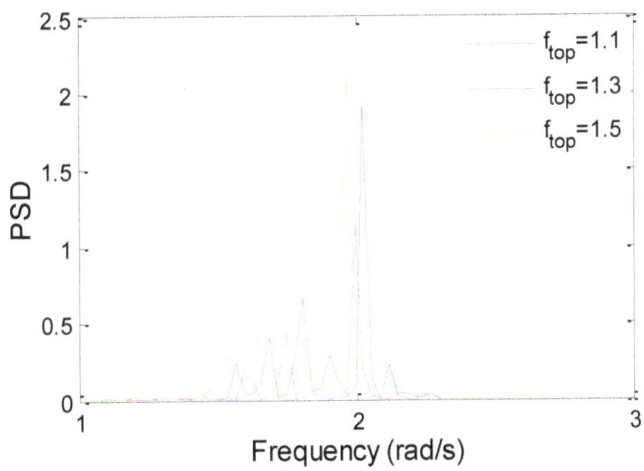

Figure 6. Frequency spectrum of TTR (U_0 = 0.5 m/s).

According to the time history of displacement and bending stress of each depth, the maximum displacement and the maximum bending stress of TTR are obtained, as shown in Figures 7 and 8.

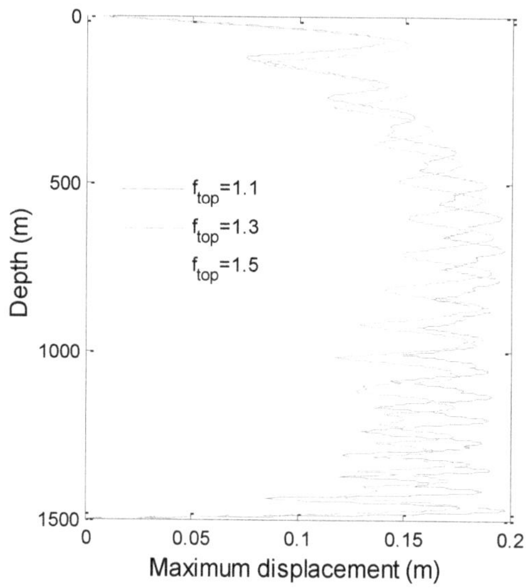

Figure 7. The maximum displacement of TTR(U_0 = 0.5 m/s).

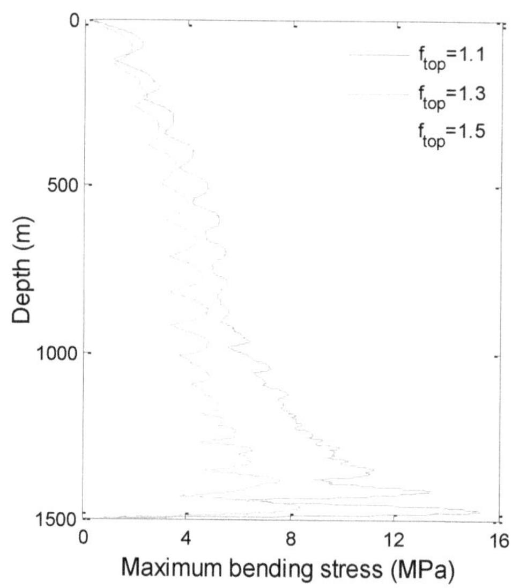

Figure 8. The maximum bending stress of TTR (U_0 = 0.5 m/s).

It shows that, with the increase of top tension, the maximum displacement increases slightly and the bending stress decreases significantly. This phenomenon is especially apparent at the bottom of TTR, since the axial tension in a riser decreases with the increase of water depth, the effective tension at the bottom of TTR is small, and the top pretension has a greater influence on the effective tension at the bottom of TTR.

This is a typical phenomenon of deep-sea risers, which is different from the conventional beam or shallow sea riser. The reason for this phenomenon is that: increasing the top tension is equivalent to increasing the stiffness of TTR, leading to the increase of the riser's natural frequency. However, the vortex-shedding frequency does not change under the same conditions, resulting in the lower modes of TTR becoming excited (or it can be equivalent to the stiffness staying the same but the current velocity decreasing). When the low-order modes of TTR are excited, the displacement is generally large but the stress is small.

5. VIV Experiment

5.1. Test Introduction

This experiment was carried out at the towing tank of Tianjin University (137 × 7 × 3 m). The TTR model was placed horizontally due to the limitation of the depth of tank, and the riser was hung at the bottom and moved at a constant speed in the towing tank. The top tension is presented by hanging weights at the top end of TTR, as shown in Figure 9. The stress of TTR at the midpoint was monitored by use of strain gauges.

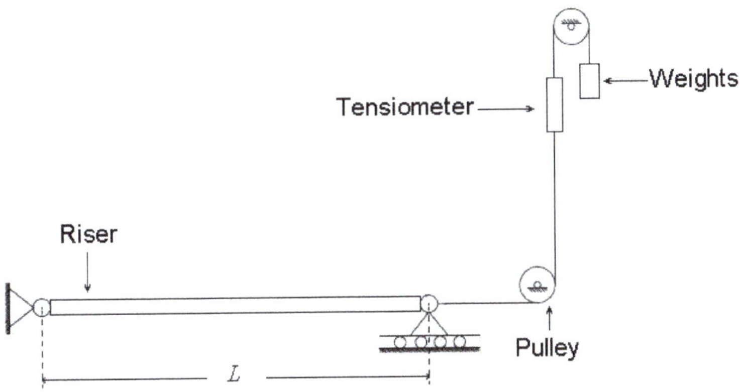

Figure 9. VIV test of TTR.

The TTR model was made of Teflon. The parameters of the model are shown in Table 3. The test conditions are shown in Table 4.

Table 3. Specifications of riser model.

No.	Property	Value	Unit
1	Length	5	m
2	Outer diameter	0.016	m
3	Wall thickness	0.001	m
4	Young's modulus	7.38×10^8	Pa
5	Material	Teflon	-
6	Density of Teflon	2178	kg/m^3
7	Density of seawater	1000	kg/m^3
8	Density of internal fluid	1000	kg/m^3

Table 4. Test conditions.

No.	Current Velocity (U)	Pretension (T_t)
1	0.3 m/s	80 N
2	0.3 m/s	100 N
3	0.3 m/s	120 N
4	0.5 m/s	80 N
5	0.5 m/s	100 N
6	0.5 m/s	120 N

The first five natural frequencies of the riser model under different pretension are shown in Table 5.

Table 5. Natural frequencies of TTR test model (Hz).

	ω_1	ω_2	ω_3	ω_4	ω_5
T = 80 N	1.3273	2.6738	4.0579	5.4976	7.0093
T = 100 N	1.4858	2.9887	4.5253	6.1119	7.7637
T = 120 N	1.6198	3.2552	4.9217	6.6342	8.4069

5.2. Verification for Numerical Simulation

The bending stress, frequency and displacement of VIV of TTR are obtained by means of numerical simulations and tests. The comparisons are shown in Figures 10–12. It should be pointed out that the displacement data of TTR were from Fujiwara et al. [21], since the displacement was not measured in this experiment.

After a lot of comparisons between the numerical simulations and experiments, it is proved that the prediction model of VIV of TTR in this paper is reasonable, and the calculating program is reliable.

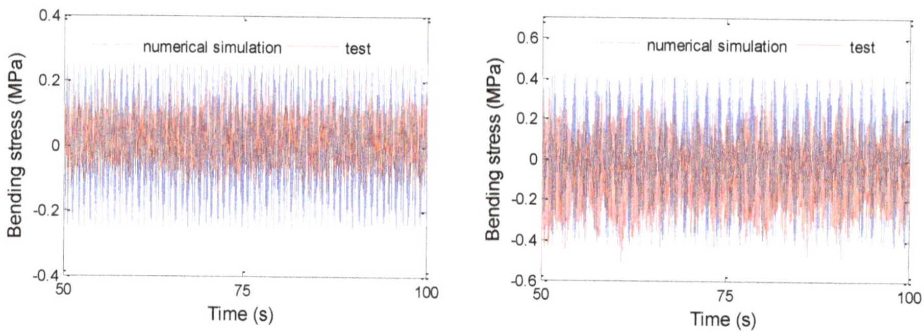

(**a**) (**b**)

Figure 10. Bending stress time history of riser at midpoint. (**a**) U = 0.3 m/s, T_t = 100 N. (**b**) U = 0.5 m/s, T_t = 100 N.

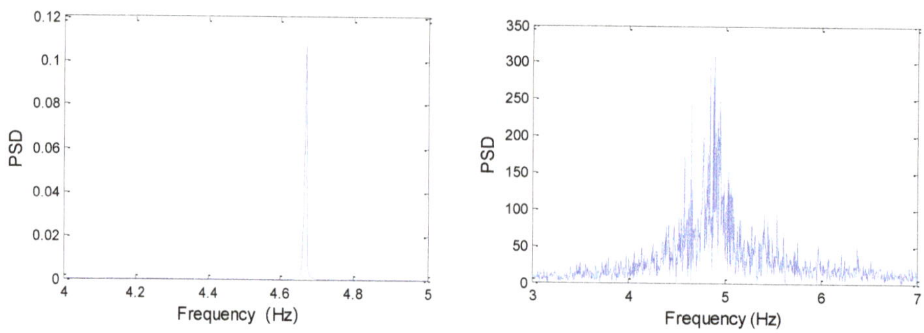

(**a**) (**b**)

Figure 11. Frequency spectrum of TTR (U = 0.5 m/s). (**a**) numerical simulation. (**b**) experiment.

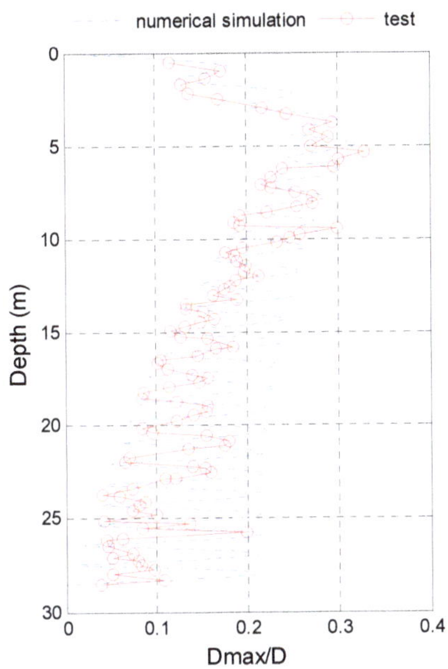

(a)

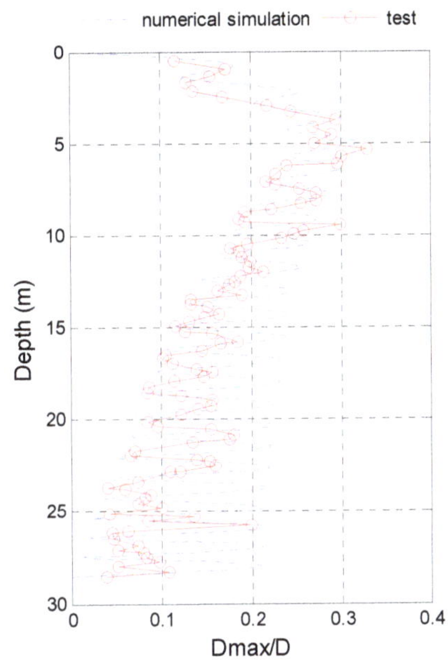

(b)

Figure 12. The maximum displacement of TTR (test data were from [21]). (**a**) Self-weight balanced in water. (**b**) Add-tens./Self-weight = 0.58.

5.3. Effect of Top Tension on VIV of TTR

The bending stress time history of TTR at midpoint and the frequency spectrum are shown in Figures 13 and 14. It can be found that the larger the top tension, the smaller the bending stress. However, increasing the top tension has little influence on the dominant frequency of VIV of TTR. This indicates that the VIV frequency of TTR is controlled by the current velocity, but it is also related to the natural frequency of TTR. In general, the low order modes of TTR are more easily excited than the high order modes.

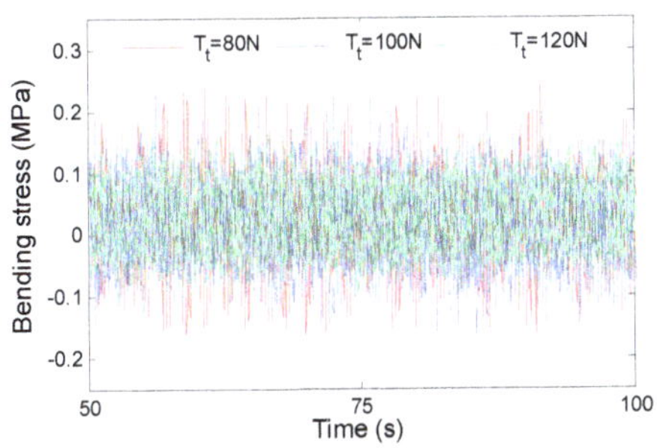

(a)

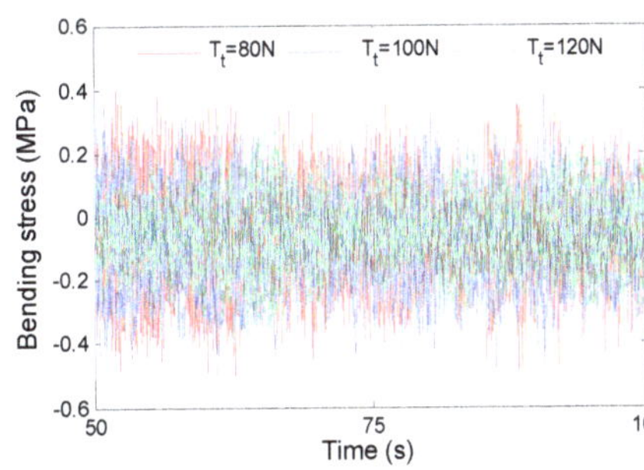

(b)

Figure 13. Bending stress time history of TTR at midpoint. (**a**) U = 0.3 m/s. (**b**) U = 0.5 m/s.

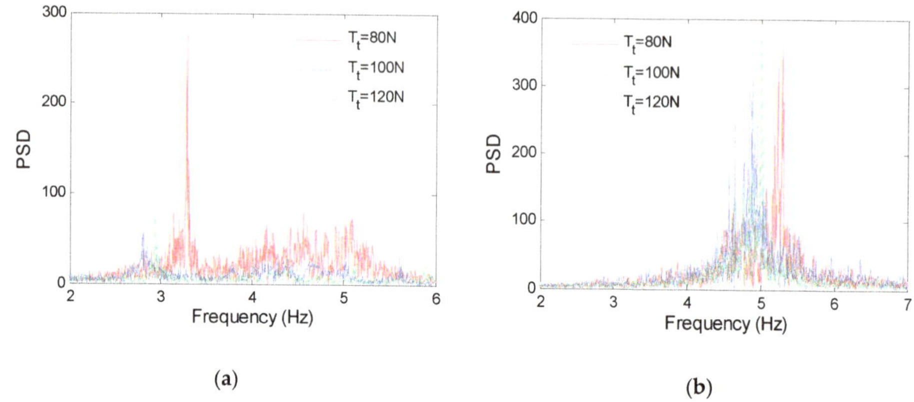

Figure 14. Frequency spectrum of TTR. (**a**) U = 0.3 m/s. (**b**) U = 0.5 m/s.

The maximum displacement of VIV of TTR is shown in Figure 15. The higher the top tension, the larger the displacement of TTR. It indicates that the natural frequency of TTR increases with the increase of the top tension, and the lower order modes of TTR are excited at the same current velocity. Therefore, the displacement is larger but the stress is smaller when the top tension is increasing.

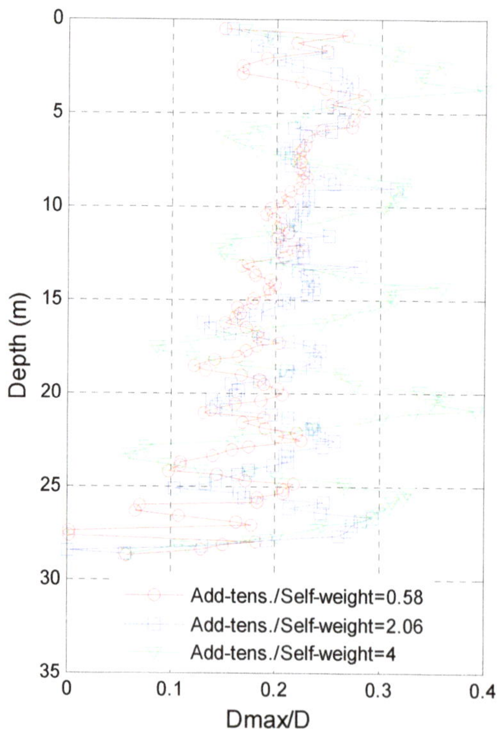

Figure 15. The maximum displacement of TTR (test data were from [21]).

6. Conclusions

Unlike the shallow sea risers, there will be some new dynamic phenomena of deep-sea risers under the complex ocean environment. In order to study the effect of top tension on VIV of TTR, the governing equation of VIV of TTR is established, and the dynamic response of TTR is calculated numerically by FDM based on Van der Pol wake-oscillator model. In addition, a validation experiment in the towing tank of Tianjin University is presented. The following conclusions can be drawn:

The natural frequency of TTR increases as the top tension increases, and it is equivalent to increasing the stiffness of TTR.

With the increase of top tension, the lower order modes are excited in the same current, and the vibration displacement of TTR increases and the bending stress decreases.

The dominant frequency of VIV of TTR is controlled by the current velocity and is barely influenced by the top tension.

Author Contributions: All four authors contributed equally to forming the objectives of this paper and discussing the numerical simulations, tests, and results. J.Z. performed the simulations and tests; J.Z. and H.G. analyzed the data, and wrote the paper under the supervision of Y.T. and Y.L. All authors have read and agreed to the published version of the manuscript.

Funding: This research is supported by the National Natural Science Foundation of China (11601323, 41906143) and the State Key Laboratory of Hydraulic Engineering Simulation and Safety (Tianjin University-HESS1904).

Conflicts of Interest: The authors declare no conflict of interest.

References

1. Griffin, O.M.; Ramberg, S.E. Some recent studies of vortex shedding with application to marine tubulars and risers. *J. Energy Res. Technol.* **1982**, *104*, 2–13. [CrossRef]
2. Gabbai, R.D.; Benaroya, H. An overview of modeling and experiments of vortex-induced vibration of circular cylinders. *J. Sounds Vib.* **2005**, *282*, 575–616. [CrossRef]
3. Wu, X.D.; Ge, F.; Hong, Y.S. A review of recent studies on vortex-induced vibrations of long slender cylinders. *J. Fluids Struct.* **2012**, *28*, 292–308. [CrossRef]
4. Hong, K.S.; Shah, U.H. Vortex-induced vibrations and control of marine risers: A review. *Ocean Eng.* **2018**, *152*, 300–315. [CrossRef]
5. Srinil, N. Analysis and prediction of vortex-induced vibrations of variable-tension vertical risers in linearly sheared currents. *Appl. Ocean Res.* **2011**, *33*, 41–53. [CrossRef]
6. Wang, E.H.; Xiao, Q. Numerical simulation of vortex-induced vibration of a vertical riser in uniform and linearly sheared currents. *Ocean Eng.* **2016**, *121*, 492–515. [CrossRef]
7. Baarholm, G.S.; Larsen, C.M.; Lie, H. On fatigue damage accumulation from in-line and cross-flow vortex induced vibration on risers. *J. Fluids Struct.* **2006**, *22*, 109–127. [CrossRef]
8. Bourguet, R.; Karniadakis, G.E.; Triantafyllou, M.S. Phasing mechanisms between the in-line and cross-flow vortex-induced vibrations of along tensioned beam in shear flow. *Comput. Struct.* **2013**, *122*, 155–163. [CrossRef]
9. Thorsen, M.J.; Sævik, S.; Larsen, C.M. Fatigue damage from time domain simulation of combined in-line and cross-flow vortex-induced vibrations. *Mar. Struct.* **2015**, *41*, 200–222. [CrossRef]
10. Yuan, Y.C.; Xue, H.X.; Tang, W.Y. An improved time domain coupled model of Cross-Flow and In-Line Vortex-Induced Vibration for flexible risers. *Ocean Eng.* **2017**, *136*, 117–128. [CrossRef]
11. Feng, Y.L.; Li, S.W.; Chen, D.Y.; Xiao, Q. Predictions for combined In-Line and Cross-Flow VIV responses with a novel model for estimation of tension. *Ocean Eng.* **2019**, *191*, 106531. [CrossRef]
12. Duan, J.L.; Chen, K.; You, Y.X.; Li, J.L. Numerical investigation of vortex-induced vibration of a riser with internal flow. *Appl. Ocean Res.* **2018**, *72*, 110–121. [CrossRef]
13. Guo, H.Y.; Lou, M. Effect of internal flow on vortex induced vibration of risers. *J. Fluids Struct.* **2008**, *24*, 496–504. [CrossRef]
14. Chen, W.M.; Li, M.; Zheng, Z.Q.; Guo, S.X.; Gan, K. Impacts of top-end vessel sway on vortex-induced vibration of the submarine riser for a floating platform in deep water. *Ocean Eng.* **2015**, *99*, 1–8. [CrossRef]

15. Wang, J.G.; Xiang, S.; Fu, S.X.; Cao, P.M.; Yang, J.M.; He, J.X. Experimental investigation on the dynamic responses of a free-hanging water intake riser under vessel motion. *Mar. Struct.* **2016**, *50*, 1–19. [CrossRef]
16. Yin, D.C.; Passano, E.; Lie, H.; Grytøyrb, G.; Aronsenb, K.; Tognarellic, Mi.; Kebadze, E.B. Experimental and numerical study of a top tensioned riser subjected to vessel motion. *Ocean Eng.* **2019**, *171*, 565–574. [CrossRef]
17. Bai, Y.; Bai, Q. *Subsea Pipelines and Risers*; Elsevier: Amsterdam, The Netherlands, 2005; pp. 423–436.
18. Kuiper, G.L.; Brugmans, J.; Metrikine, A.V. Destabilization of deep-water risers by a heaving platform. *J. Sound Vib.* **2008**, *310*, 541–557. [CrossRef]
19. Yang, J.; Meng, W.; Yao, M.B.; Gao, D.L.; Zhou, B.; Xu, Y.J. Calculation method of riser top tension in deep water drilling. *Petroleum Explor. Dev.* **2015**, *42*, 119–122. [CrossRef]
20. Moghiseh, A.; Chaloshtory, H.R.; Rahi, A. Effect of middle tension on dynamic behaviour of marine risers. *J. Marit. Res.* **2012**, *9*, 63–70.
21. Fujiwara, T.; Uto, S.; Kanada, S. An Experimental Study of the Effects that Change the Vibration Mode of Riser VIV. In Proceedings of the 30th International Conference on Offshore Mechanics and Arctic Engineering, Rotterdam, The Netherlands, 19–24 June 2011.
22. Le Cunff, C.; Biolley, F.; Fontaine, E.; Etienne, S.; Facchinetti, M.L. Vortex-induced vibrations of risers: Theoretical, numerical and experimental investigation. *Oil Gas Sci. Technol.* **2002**, *57*, 59–69. [CrossRef]
23. Facchinetti, M.L.; Langre, E.D.; Biolley, F. Coupling of structure and wake oscillators in vortex-induced vibrations. *J. Fluids Struct.* **2004**, *19*, 123–140. [CrossRef]
24. Senjanovic, I.; Ljustina, A.M.; Parunov, J. Natural vibration analysis of tensioned risers by segmentation method. *Oil Gas Sci. Technol.* **2006**, *61*, 647–659. [CrossRef]
25. Zhang, J.; Tang, Y.G. Further analysis on natural vibration of deep-water risers. *J. Ship Mech.* **2014**, *18*, 165–171. (In Chinese)
26. Basil, T.P. Numerical Simulation of the Truss Spar 'Horn Moutain' Using Couple. Master's Thesis, Texas A&M University, College Station, TX, USA, May 2006.
27. Gao, Y.; Zong, Z.; Sun, L. Numerical prediction of fatigue damage in steel catenary riser due to vortex-induced vibration. *J. Hydrodyn.* **2011**, *23*, 154–163. [CrossRef]

Journal of
Marine Science and Engineering

MDPI

Article

An Experimental and Theoretical Study of Wave Damping due to the Elastic Coating of the Sea Surface

Igor Shugan [1,2,*], Ray-Yeng Yang [3] and Yang-Yih Chen [1]

1 Department of Marine Environment and Engineering, National Sun Yat-Sen University, Kaoshiung 804, Taiwan; yichen@mail.nsysu.edu.tw
2 Laboratory of shelf and sea coasts, Shirshov Institute of Oceanology, Russian Academy of Sciences, Moscow 117997, Russia
3 Department of Hydraulic and Ocean Engineering, National Cheng Kung University, Tainan 701, Taiwan; ryyang@mail.ncku.edu.tw
* Correspondence: ishugan@rambler.ru

Received: 15 June 2020; Accepted: 27 July 2020; Published: 29 July 2020

Abstract: Flexible plates or membranes located on the sea surface can be effective for attenuation waves approaching the beach. The most efficient structures should be found through comprehensive research using developed experiments and theory. Our experimental work was focused on the wave propagation and attenuation passing through floating elastic structures. The experiments were conducted at the wave flume of Tainan Hydraulics Laboratory, National Cheng Kung University, Taiwan. The experiment mainly analyzes the reflection coefficient, transmission coefficient and energy loss of the regular wave of intermediate water depth after passing through the elastic structure under different wave steepness and other different wave conditions. Our experiments also explore the comparison of energy dissipation effects and the differences in motion characteristics between different elastic plates and different plate fixing methods. Three elastic materials were tested in the experiments: Latex, cool cotton and polyvinyl chloride (PVC). A model of a thin elastic plate covering the sea surface was used to evaluate the effectiveness of the structure of the wave barrier. The results of experiments carried out in the wave flume were compared with theoretical predictions in a wide range of generated waves.

Keywords: surface waves; wave breaker; elastic plate

---

1. Introduction

Taiwan is densely populated and surrounded by the sea. Not only the continuous development of inland lands, but also the development of areas around the coast is quite frequent. Therefore, how to protect yourself from coastal erosion and other natural disasters has always been the subject of the efforts of scientists and experts in coastal engineering. However, traditional protection methods mainly use rigid structures such as breakwaters, sea embankments and jetty to protect the coast, block water flow, prevent seawater intrusion and land loss.

However, such rigid construction methods can cause problems such as increased reflection and premature erosion or structural damage, and can even cut off sand downstream drift and cause loss on the beach.

Consequently, the use of coastal space and coastal hydrophilicity has also become one of development considerations. Widely proposed new methods and concepts, mainly based on the idea of not harming the beach area. Even in addition to protecting the coast, they can also serve as ecological services to the landscape while relaxing. Therefore, in recent years, flexible construction methods have gradually replaced rigid construction methods and become a tendency to protect the coast. In flexible construction methods, various construction methods have their own characteristics and effects. Among them, the design of highly adaptable floating breakwater is widely used in marine engineering, because various types of breakwaters may have different interactions with waves to achieve the effect of energy dissipation. Therefore, various floating structures are constantly being proposed and discussed.

There are many relevant literatures about different floating structures. Theoretical analysis of the transmission coefficient of incoming waves after passing through a fixed plate and a floating plate is performed in [1]. The main result is that the value of the transmission coefficient is mainly affected by the obtained wavelength. The transition and reflection of waves interacting with floating or submerging objects causing radiation problems in two or three directions of fluctuation was under discussion in [2,3].

A π-type floating breakwater was first introduced in [4]. An extremely simple formula is proposed for the transmittance, which depends only on the length of the incoming wave, the geometry of the breakwater, and the depth of the water. The model underestimates the transmission of short waves and overestimates the transmission of long waves, but at the same time, the model was a good starting point for further research and improvement. A number of works are devoted to the further development of a π-type floating breakwater—a review and new approaches are presented in [5]. There are still many problems for analysis: Complex hydrodynamics with vortices and turbulence, a variety of mooring systems, scaling issues, etc.

The reflection and transmission of waves by submerged horizontal rigid plate was studied in [6,7]. It was shown, that plate can reflect significant amount of wave energy for specific wavelengths.

The transmittance of water waves passing through a floating plate was experimentally investigated in [8]. The authors concluded that when the length of the floating horizontal plane is several times the length of the incoming wave, the attenuation of the waves is stronger.

Experimental study of the transmission and reflection coefficient of waves of intermediate water depth through a lattice-shaped perforated horizontal plate was presented in [9]. The experimental results suggest that as the length and wavelength ratio of the structure and the steepness of the wave increase, the transmission coefficient decreases. It was also found that when the depth of the structure is increased, the reflection coefficient increases.

The effect of immersed or surface-mounted horizontal flexible membranes on the dynamics of waves was studied by the methods of linear hydro elastic theory in [10]. The numerical results for the reflection and attenuation of the incident waves were reasonably consistent with existing laboratory experiments.

A freely floating porous box was theoretically investigated in [11]. The small drift forces of the porous bodies gave the box an advantage for use as a floating breakwater.

The spar buoy floating breakwater is presented in [12]. An investigation into the wave reflection and transmission properties was carried out with a study on mooring tension induced by the waves.

A hydraulic model was used in [13] to study the transmission coefficient of waves after passing through floating structures with different properties and forms. It was found that the effect of installing a grid under the flat panel was the best, and showed that the steepness of the waves increased, and the transmission coefficient also decreased.

A new structure of a floating breakwater, consisting of several connected perforated diamond-shaped blocks, was proposed in [14]. It is experimentally shown that the proposed wave breaking system can dissipate the energy of the incoming wave more efficiently than already known. Possible mooring forces are discussed.

The interaction between the floating thin elastic plate and water surface waves was investigated in [15]. Authors suggested different oscillation modes of a two-phase structure in dependence of the elasticity properties and geometrical scales of the cover, characteristics of incoming waves and depth of fluid.

In this paper, the performance of an elastic thin plate floating breakwater is experimentally and theoretically examined. Wave transmission and attenuation in dependence of its steepness, elastic properties of the membrane, its length, and water depth are investigated. The theoretical model is constructed in the long wave approximation which is usually applicable in the beach zone of the Ocean for the relatively long and energetic incoming swell. We consider the case of fix edges of the elastic plate boundary conditions as the most effective configuration for transmitted wave suppression [7].

The structure of the article is as follows.

Section 2 describes the experimental setup. We present laboratory equipment and a detailed test configuration. There is a full and detailed description of the laboratory tank, measuring instruments, the characteristics of a set of elastic plates and the organization of the experiment itself.

Section 3 of this article presents experimental results. The reflection, transmittance and energy loss corresponding to different generated wave steepness, the ratios of the wave length and length of the plate are discussed. Corresponding reflections, transmittance and energy loss under different elastic plates are also analyzed.

The theoretical model of the surface wave's interaction with the surface-mounted elastic plate is presented in Section 4. Theoretical predictions of the model are presented in Section 5. Some particularities of the theoretical solution are discussed. Conclusions are made in Section 6.

2. Experimental Setup

The experiments were carried out in a laboratory wave flume (Figure 1) located in the Tainan Hydraulics Laboratory (THL), National Cheng Kung University (NCKU), Taiwan. The dimensions of the flume are 0.5 m wide, 0.7 m deep and 20 m long.

Figure 1. Laboratory tank.

It is mainly used to measure the attenuation height of the wave in the water tank after passing through the elastic membrane covering the water surface. Therefore, the equipment used is a test tank, wave maker, wave elimination section, elastic plate, CCD camera, wave gauges, power supply and signal amplifier. Both sides of the tank are laid with reinforced transparent glass to observe the movement in the flume. A flat push wave maker is set at the left end of the tank. A schematic representation of the experimental setup is shown in Figure 2. A piston-type wave maker was used, controlled by software developed in the Hydraulic Research Institute. This wave-making system can produce periodic regular waves, irregular waves and solitary waves. The regular wave range created by the experimental wave maker has period 0.65–1.45 s wave height 0–6 cm.

A sloping beach was constructed at the end of the wave flume, which can reduce the wave reflection to 5%. The wave-cutting section was a 1:11.7 acrylic slope, located at the other end of the tank. It is 0.48m high and 5.65m long. The bottom is supported by a steel frame to avoid deformation of the acrylic plate caused by wave water pressure and wave energy.

This experiment uses the capacitive wave gauges developed by the National Cheng Kung University Hydraulic Engineering Institute. The output signal is 0–5 V, the acquisition frequency can reach 50 Hz, and the wave height measurement accuracy is ±0.5 mm.

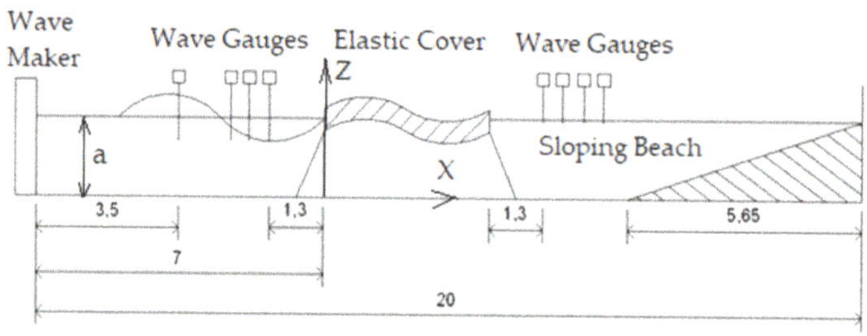

Figure 2. Schematic view of the experimental setup.

A total of eight capacitive wave gauges were used in this test. The first wave gauge was placed at a distance of 3.5 m from the wave generator to check the incident wave height. The second, third, and fourth wave gauges are placed at the front of the plate used to measure and analyze the height of incoming and reflected waves. The height of the fifth wave probe is 1.3 m behind the plate, used to measure transmission wave height and attenuation after passing through the cover. The sixth, seventh and eighth wave gauges are in front of the wave-elimination band, and are used to observe the wave height after the transmission wave is stabilized, and the wave reflectivity of the wave-elimination band.

The shooting range of CCD camera is 25 cm × 25 cm. 1.2 m away from the tank. Shooting at 30 frames per second for 30 s. It is used to observe the movement of the plate and fluid.

Anchor has the four points on the front and rear ends of the membrane are provided with 10 kg iron pieces for fixing the components according to different fixing methods.

Three different elastic plates where tested. They are the latex membrane with better elasticity (Figure 3), the foam component with the second best elasticity (Figure 4) and a flat rigid plastic component—Figure 5 (Table 1).

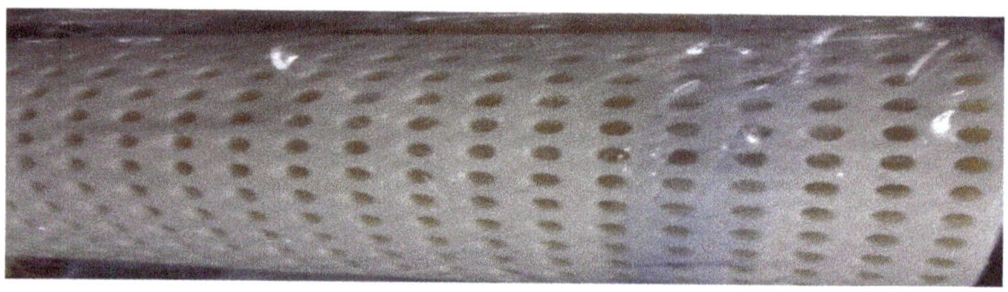

Figure 3. Latex component (Latex): length 1.9 m and 5.1 m; density 147.37 (kg/m^3); weight 7 kg, thickness 0.02 m.

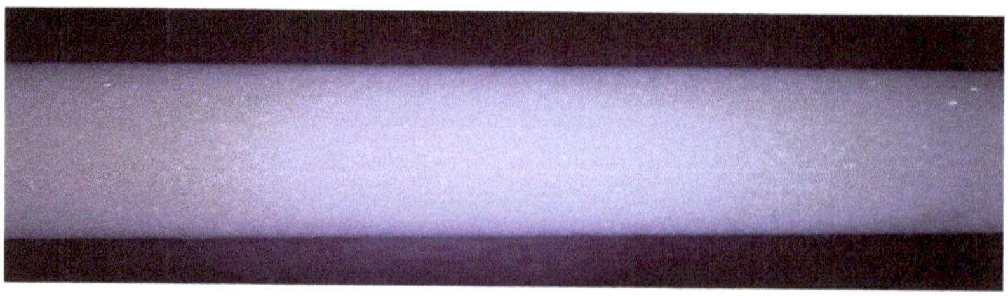

Figure 4. Foam component (Steep cotton): length 1.9 m; density 63.18 (kg/m^3); weight 3 kg, thickness 0.02 m.

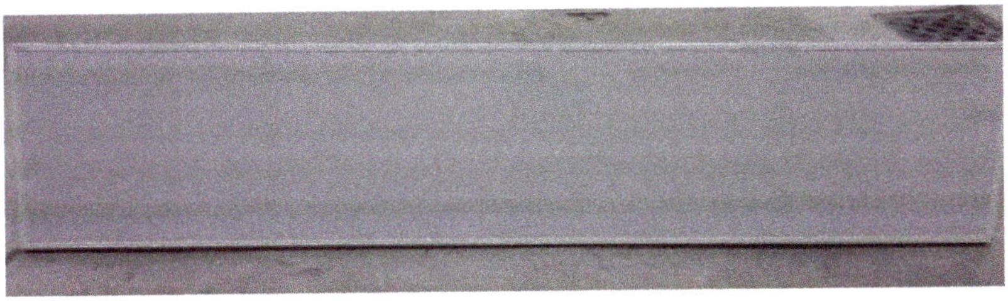

Figure 5. Rigid member (PVC): length 1.9 m; density 181.05 (kg/m^3); weight 8.6 kg, thickness 0.02 m.

Table 1. Properties of elastic plates and generated waves.

Materials	Length L (m)	Density (kg/m^3)	Elastic Modulus (GPa)	Depth of Fluid a (m)	Period of Waves (s)	Height of Waves H (cm)
Latex	1.9 and 5.1	147.37	0.01	0.25 0.3 0.35	0.65 0.73 0.81 1 1.45	2 4 6
Steep cotton	1.9	63.18	0.07	0.25 0.3 0.35	0.65 0.73 0.81 1 1.45	2 4 6
PVC	1.9	181.05	3.5	0.25 0.3 0.35	0.65 0.73 0.81 1 1.45	2 4 6

In order to compare their efficiencies on attenuating surface waves under different wave and water depth conditions, monochromatic waves were generated in the range: the wave height is 2–6 cm, and the period is 0.65–1.45 s. The test conditions are as follows (see Table 1):

The characteristics of the generated waves with different time periods were determined from the dispersion relation for the intermediate water depth and are presented in Tables 2–4.

Table 2. Characteristics of generating waves (depth of water a = 0.25 m).

H (m)	T (s)	λ (m)	(H/λ)	k	ka	a/λ
0.020	0.650	0.649	0.031	9.686	2.422	0.385
0.020	0.730	0.799	0.025	7.862	1.966	0.313
0.020	0.810	0.951	0.021	6.608	1.652	0.263
0.020	1.000	1.303	0.015	4.822	1.206	0.192
0.020	1.450	2.088	0.010	3.010	0.753	0.120
0.040	0.650	0.649	0.062	9.686	2.422	0.385
0.040	0.730	0.799	0.050	7.862	1.966	0.313
0.040	0.810	0.951	0.042	6.608	1.652	0.263
0.040	1.000	1.303	0.031	4.822	1.206	0.192
0.040	1.450	2.088	0.019	3.010	0.753	0.120
0.060	0.650	0.649	0.092	9.686	2.422	0.385
0.060	0.730	0.799	0.075	7.862	1.966	0.313
0.060	0.810	0.951	0.063	6.608	1.652	0.263
0.060	1.000	1.303	0.046	4.822	1.206	0.192
0.060	1.450	2.088	0.029	3.010	0.753	0.120

The test preparation was based on the following plan. The wave maker and signal amplifier was fully warmed up before the experiment to ensure the stability of the wave maker operation, and then the wave making and data collection can be performed according to the test conditions. The wave making time of each set of data is 45 s, the wave gauge records at 60 Hz per second for 60 s and the CCD camera shoots at 30 frames per second for 30 s. The interval of each test group is 10 min to ensure that the disturbance of the previous group of tests tends to be stable. After completing the entire set of tests to be carried out, the fixing method and the elastic plate can be replaced, and the above steps are repeated until all tests are complete.

Table 3. Characteristics of generating waves (depth of water $a = 0.3$ m).

H (m)	T (s)	λ (m)	(H/λ)	k	ka	a/λ
0.020	0.650	0.655	0.031	9.595	2.879	0.458
0.020	0.730	0.815	0.025	7.709	2.313	0.368
0.020	0.810	0.980	0.020	6.408	1.922	0.306
0.020	1.000	1.372	0.015	4.580	1.374	0.219
0.020	1.450	2.247	0.009	2.796	0.839	0.134
0.040	0.650	0.655	0.061	9.595	2.879	0.458
0.040	0.730	0.815	0.049	7.709	2.313	0.368
0.040	0.810	0.980	0.041	6.408	1.922	0.306
0.040	1.000	1.372	0.029	4.580	1.374	0.219
0.040	1.450	2.247	0.018	2.796	0.839	0.134
0.060	0.650	0.655	0.092	9.595	2.879	0.458
0.060	0.730	0.815	0.074	7.709	2.313	0.368
0.060	0.810	0.980	0.061	6.408	1.922	0.306
0.060	1.000	1.372	0.044	4.580	1.374	0.219
0.060	1.450	2.247	0.027	2.796	0. 839	0.134

Table 4. Characteristics of generating waves (depth of water $a = 0.35$ m).

H (m)	T (s)	λ (m)	(H/λ)	k	ka	a/λ
0.020	0.650	0.657	0.030	9.558	3.345	0.533
0.020	0.730	0.823	0.024	7.632	2.671	0.425
0.020	0.810	0.999	0.020	6.292	2.202	0.350
0.020	1.000	1.424	0.014	4.413	1.545	0.246
0.020	1.450	2.384	0.008	2.635	0.922	0.147
0.040	0.650	0.657	0.061	9.558	3.345	0.533
0.040	0.730	0.823	0.049	7.632	2.671	0.425
0.040	0.810	0.999	0.040	6.292	2.202	0.350
0.040	1.000	1.424	0.028	4.413	1.545	0.246
0.040	1.450	2.384	0.017	2.635	0.922	0.147
0.060	0.650	0.657	0.091	9.558	3.345	0.533
0.060	0.730	0.823	0.073	7.632	2.671	0.425
0.060	0.810	0.999	0.060	6.292	2.202	0.350
0.060	1.000	1.424	0.042	4.413	1.545	0.246
0.060	1.450	2.384	0.025	2.635	0.922	0.147

Several methods of separation for incident and reflected waves exist [16–18]. The most common procedure is to install a number of wave gauges at relatively short distances from each other and to make the simultaneous measurements of the waves at all wave gauges. We used three wave gauges and the least squares method for separating incident and reflected waves [17]. The second, third and fourth wave gauges are placed at the front of the plate used to measure and analyze the height of incoming and reflected waves. The wave reflection coefficient and incident amplitude are estimated from wave heights measured at three fixed wave gauges with unequal spacing. The spacings between wave gauges must not be an integer that is a multiple of a half wave length. The optimal spacing between wave gauges is still under discussion [18].

The distance between the wave gauges on both sides of the structure must be at least three times the water depth to reduce the impact of dissipated waves. In this paper, the mature wave height after 15 s is taken as the analysis; and the blank test shows that the reflectivity of the elimination band is less than 0.1.

According to the theory of potential current, without energy loss, the ratio of the sum of the square of the transmitted wave height and the reflected wave height is 1, that is, the square of reflectance plus the square of transmittance:

$$E = \frac{H_R^2 + H_T^2}{H_I^2}$$

Then the wave energy loss coefficient can be expressed as:

$$Eloss = \sqrt{1 - K_r^2 - K_t^2} \quad (1)$$

Then the wave energy loss coefficient can be regarded as the energy coefficient reduced by the action of the elastic plate body.

3. Experimental Results and Discussion

When the membrane is not fixed and the front end is fixed, the movement of the plate's surge, sway, and heave is more intense than when the two ends are fixed, and the disturbance of the dissipation wave may have a greater impact on the test. Therefore, we were concentrated on the effect of wave steepness on reflection and transmission coefficients, energy loss coefficient and attenuation of the wave height when different elastic plates are used and both ends of the floating cover are fixed.

Figure 6 shows the transmission coefficient K_t for latex plate and different steepness of the incoming wave, when the wave height is 2 cm, 4 cm and 6 cm. The graph clearly shows that for any wave height, the greater the steepness of the wave, the lower the transmission coefficient. From this figure, it can be seen that the elastic plate has a certain effect of reducing waves. However, when the wave steepness is very small, the transmission coefficient is close to 1 (that is, the wave is almost completely transmitted). Reflection coefficients of the latex plate are presented in Figure 7.

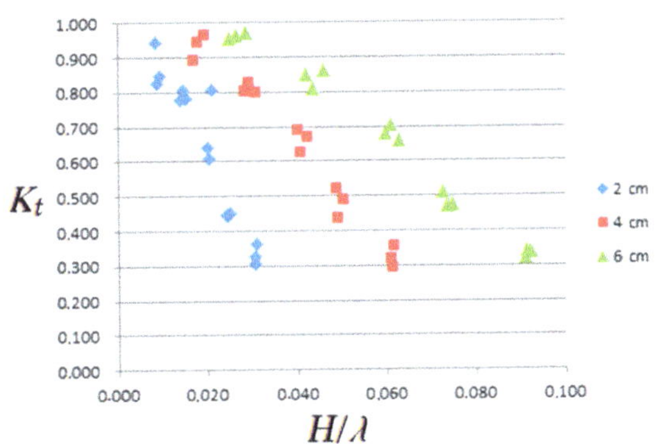

Figure 6. The relationship between the steepness of the wave and the transmittance of the latex plate fixed at both edges, water depth a = 0.3 m.

The graph shows that at different wave heights, the reflection coefficient has 1–2 peaks in the middle. The trend line is closer to the quadratic curve. It is speculated that this phenomenon is related to the period (wavelength) of the wave. That is, under different periods or wavelengths, the wave reflections caused by its components have different characteristics.

Figure 8 shows the energy loss (energy reduction) coefficients corresponding to different wave steepness when the wave height is 2 cm, 4 cm and 6 cm. First, it can be observed from the figure that when the steepness of the wave is very small (when the wavelength is longer), the trend of loss caused by the steepness change is not obvious. When the loss coefficient is greater than 0.5, the greater the steepness of the wave corresponds to a larger energy loss coefficient. Secondly, from this figure it can also be found that the energy dissipation effect of the elastic plate is significant, and the energy loss coefficient can reach about 0.9.

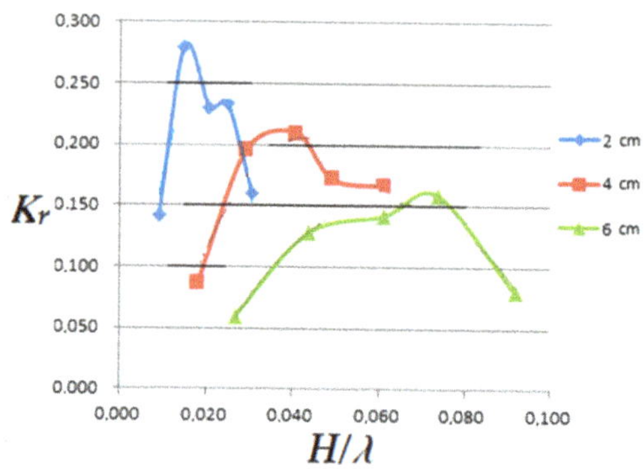

Figure 7. The ratio between the steepness of the incoming wave and the reflection coefficients of the latex plate at a wave height of 2 cm, 4 cm and 6 cm, water depth a = 0.3 m.

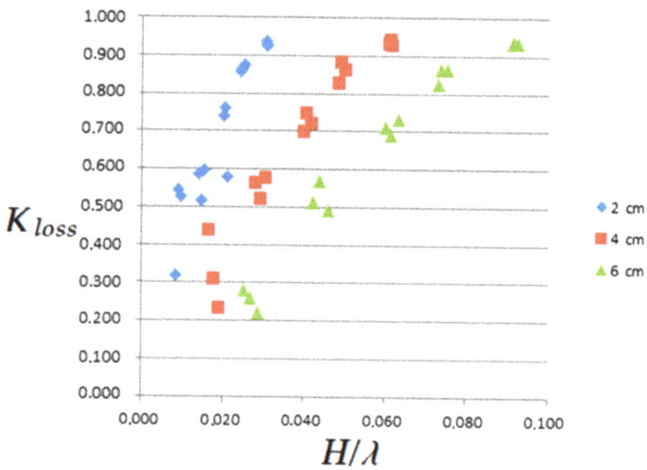

Figure 8. Relationship between wave steepness and energy loss coefficient, water depth a = 0.3 m.

It can be seen from the above test results that the steepness of the wave has a significant effect on the interaction between the elastic plate and the wave, and it is also seen that the damping effect of the elastic membrane is strong. The period and wavelength should be the main parameters that affect the motion characteristics.

Next, we analyzed the properties of transmission and reflection of waves depending on the depth of water a, which are presented in Figures 9 and 10.

It can be found that at different steepness of the waves, the transmission and reflection coefficients corresponding to different water depths do not differ much. The reason should be that all test conditions relate to the intermediate water depth, and the range of water depth changes has a smaller effect on this component than other parameters.

Figure 11 shows the relationship between the ratio of the plate length and the wavelength of the incident wave (L/λ) and the corresponding transmission coefficient. It can be seen that the greater the ratio (L/λ), the smaller the transmission coefficient, that is, the effect of energy dissipation is stronger.

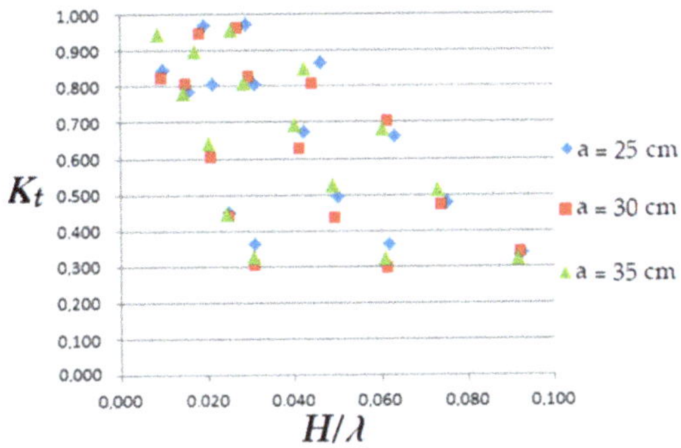

Figure 9. The relationship between wave steepness and transmission coefficient at different water depths.

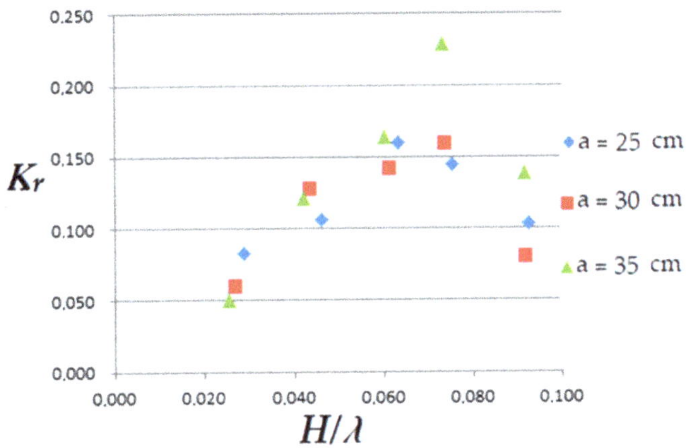

Figure 10. The relationship between wave steepness and reflection coefficient at different water depths.

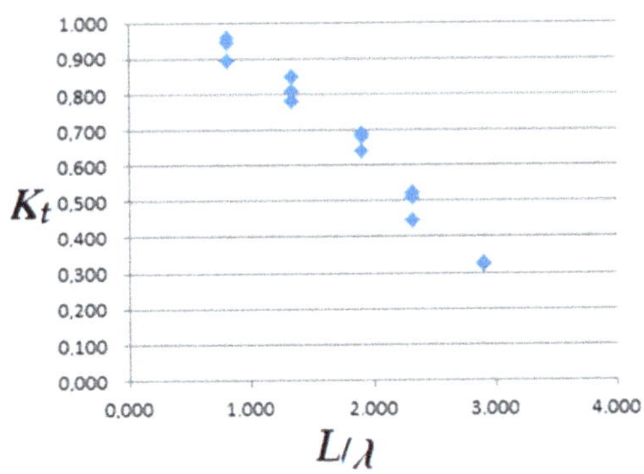

Figure 11. Relationship between wave steepness and transmission coefficient at different plate lengths.

When the wavelength of the incident wave is greater than the length of the plate, that is, the ratio (L/λ) is less than 1, the transmission coefficient is about 0.9, and the wave elimination effect is not effective.

Figure 12 shows the transmittances corresponding to the steepness of the waves at different plate lengths. It can be clearly seen that under each wave condition, the transmission coefficient of the long plate is smaller than that of the short plate.

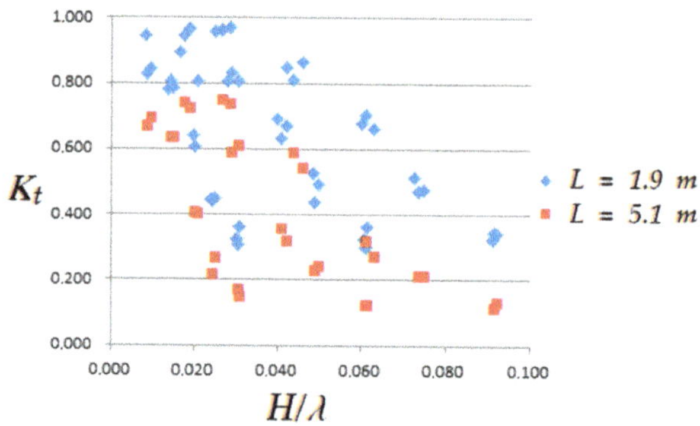

Figure 12. Relationship between wave steepness and transmission coefficient at different plate lengths.

Different elastic plates may have different oscillation modes and different interactions with waves. Therefore, we analyzed the motion characteristics of regular waves acting on different elastic membrans (latex E = 0.01 gpa, foam E = 0.07 gpa and rigid member E = 3.5 gpa) when both ends are fixed.

Figure 13 shows the relationship between the wave steepness and transmission coefficient of different elastic plates when the wave height is 2, 4 and 6 cm at a fixed water depth 0.3 m. It shows that the transmission coefficient of the rigid plate is slightly lower than that of the other two more soft plates. The transmission coefficient of the foam pate is the largest.

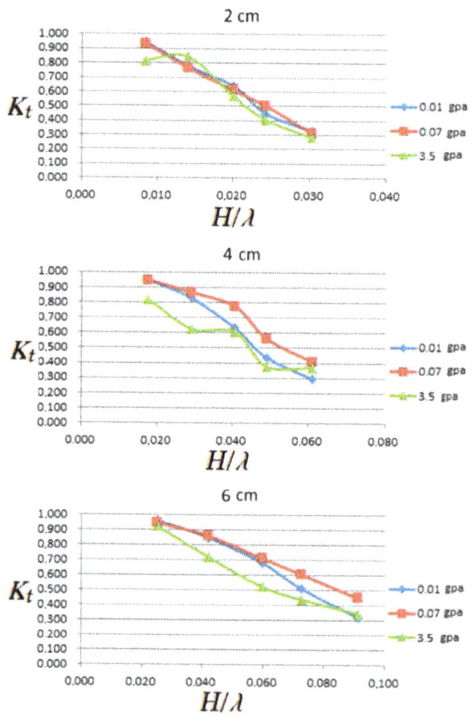

Figure 13. The relationship between the wave steepness and transmission coefficient of different elastic plates when the wave height is 2, 4 and 6 cm at a fixed water depth 0.3 m.

4. Analytical Model

We consider two-dimensional interaction (2D) of a surface-mounted elastic plate of length L and a thickness $H_w = 2$ cm with a monochromatic incident surface wave of height H_I and wavelength λ. Orthogonal system of coordinate is chosen with X-axis directed horizontally and Z-axis directed upward (see Figure 2). We assume also an ideal incompressible fluid. Taking in mind that breakwaters should be located in coastal areas and effective ones should have a length comparable to the length of incoming waves, we consider the long wave approximation model: the characteristic wave length λ and the scale of the wave breaker L are much larger than the water depth a: $L \gg a; \lambda \gg a$.

That is why we will use the linearized standard shallow water model for the open sea water $(X < 0, X > L)$:

$$\begin{aligned} \frac{\partial H}{\partial T} &= \frac{\partial (aU)}{\partial X}; \\ \frac{\partial U}{\partial T} &= -\frac{1}{\rho_f}\frac{\partial (P)}{\partial X}; \\ P - P_A &= -\rho_f g (Z - H), \end{aligned} \tag{2}$$

where $H = H(X,T)$ is the sea surface elevation, U—depth integrated velocity of fluid, P, P_A are pressure in fluid and atmospheric pressure, respectively, g is the gravity acceleration, ρ_f—density of fluid.

The dynamics of the elastic membrane on the sea surface $(0 < X < L)$ can be described within the thin elastic plate approximation model [19,20]:

$$P - P_A = 2H_w\rho_w\frac{\partial^2 H}{\partial T^2} + \frac{2H_w^3 E_0}{3(1-\nu_w^2)}\frac{\partial^4 H}{\partial X^4} \tag{3}$$

where the terms in the left side of the equation express the external load acting on the surface of the plate in the transverse direction, P is the pressure in fluid on the internal plate surface $Z = H(X,T)_{(-)}$, constant atmospheric pressure P_A is assumed on the external plate surface $Z = H(X,T)_{(+)}$; E_0 is these Young modulus, $(\nu_w = 0.3)$ is the Poisson coefficient and ρ_w is the density of the plate material.

The fluid motion under the elastic membrane $(0 < X < L)$ in the long wave approximation can be also described by the shallow water model:

$$\begin{aligned} \frac{\partial^2 H}{\partial T^2} &= \frac{a}{\rho_f}\frac{\partial^2 (P_0)}{\partial X^2}; \\ P &= -\rho_f g Z + P_0(X,T). \end{aligned} \tag{4}$$

The solutions of the Equations (3)–(5) in the regions $X < 0, 0 < X < L, X > L$ should be connected by the mass and momentum conservation laws at the boundaries:

$$\begin{aligned} \lim(U|_{X=0^+}) &= \lim(U|_{X=0^-}); \\ \lim P|_{X=0^+} &= \lim P|_{X=0^-}; \\ \lim U|_{X=L^+} &= \lim U|_{X=L^-}; \\ \lim P|_{X=L^+} &= \lim P|_{X=L^-}, \end{aligned} \tag{5}$$

where superscripts (+) and (−) correspond to the limiting values of functions at the different sides of the boundary cross sections.

Boundary conditions for the elastic membrane oscillations include zero displacements and transverse forces at the edges of plate [19,20]:

$$\begin{aligned} &H(X=0) - a = H(X=L) - a = 0, \\ &\frac{\partial^2 H}{\partial X^2}(X=0) = \frac{\partial^2 H}{\partial X^2}(X=L) = 0 \end{aligned} \tag{6}$$

We introduce dimensionless variables in the following form:

$$\begin{array}{l} t = \frac{\sqrt{ga}}{L}T, x = \frac{X}{L}, z = \frac{Z}{a}, \\ u = \frac{U}{\sqrt{ga}}, p = \frac{P}{\rho_f ga}, h = \frac{H}{a}. \end{array} \tag{7}$$

Then, the equations of the two-phase motion and corresponding boundary conditions (1)–(5) can be written in the dimensionless form

For $(x < 0, x > 1)$

$$\begin{array}{l} \frac{\partial h}{\partial t} = \frac{\partial(u)}{\partial x}; \frac{\partial u}{\partial t} = -\frac{\partial(p)}{\partial x}; \\ p - p_A = -(z-1), \end{array} \tag{8}$$

For $(0 < x < 1)$

$$\frac{\partial^2 h}{\partial t^2} = \frac{\partial^2(p_0)}{\partial x^2}; p = -z + p_0(x,t); \tag{9}$$

$$-h + p_0(x,t) - p_A = \overline{\gamma}\frac{\partial^2 h}{\partial t^2} + \overline{\beta}\frac{\partial^4 h}{\partial x^4} \tag{10}$$

where $\overline{\gamma} = 2\frac{\rho_w H_w}{\rho_f a}\left(\frac{a}{L}\right)^2; \overline{\beta} = \frac{2}{3}\frac{E_0}{(1-v_w{}^2)H_w\rho_w g}\left(\frac{\rho_w}{\rho_f}\right)\left(\frac{H_w}{L}\right)^4$.

Boundary conditions for fluid ($x = 0$, $x = 1$)

$$\begin{array}{l} \lim u|_{x=0^+} = \lim u|_{x=0^-}; \\ \lim p|_{x=0^+} = \lim p|_{x=0^-}; \\ \lim u|_{x=1^+} = \lim u|_{x=1^-}; \\ \lim p|_{x=1^+} = \lim p|_{x=1^-}, \end{array} \tag{11}$$

Boundary conditions for elastic plate ($x = 0$, $x = 1$)

$$\begin{array}{l} h(x=0) - 1 = h(x=1) - 1 = 0, \\ \frac{\partial^2 h}{\partial x^2}(x=0) = \frac{\partial^2 h}{\partial x^2}(x=1) = 0. \end{array} \tag{12}$$

We will investigate solutions for transmitting and scattering by the flexible membrane of the initially monochromatic wave in the open water:

$$h = 1 + \mathrm{Re}\left(h_I e^{-ikx+i\omega t}\right), x < 0. \tag{13}$$

Dispersion relation for the elastic membrane oscillations can be found from the system of Equations (10) and (11):

$$\omega^2 + k^2(1 - \overline{\gamma}\omega^2) + \overline{\beta}k^6 = 0 \tag{14}$$

Different modes of oscillations of the membrane around a constant level have the form:

$$h = \mathrm{Re}\left(h_i e^{k_i x + i\omega t}\right). \tag{15}$$

Solution of Equation (15) defines six own modes of membrane oscillations h_i on the water surface with frequency ω:

$$\begin{array}{l} k_{1.2} = \pm\sqrt{(A+B)}, \\ k_{3,4} = \pm\sqrt{-(A+B)/2 + i\sqrt{3}/2(A-B)}; \\ k_{5,6} = \pm\sqrt{-(A+B)/2 - i\sqrt{3}/2(A-B)}; \end{array} \tag{16}$$

where

$$A = \left(-\frac{\omega^2}{2\overline{\beta}} + \sqrt{(1-\gamma\omega^2)/(27\overline{\beta}^3) + \omega^4/(4\overline{\beta}^2)}\right)^{1/3};$$
$$B = \left(-\frac{\omega^2}{2\overline{\beta}} - \sqrt{(1-\gamma\omega^2)/(27\overline{\beta}^3) + \omega^4/(4\overline{\beta}^2)}\right)^{1/3}. \quad (17)$$

Boundary conditions (13) for the plate oscillations define four relationships for six wave amplitudes:

$$\sum_{i=1}^{6} h_i = \sum_{i=1}^{6} k_i^2 h_i = 0;$$
$$\sum_{i=1}^{6} e^{k_i} h_i = \sum_{i=1}^{6} k_i^2 e^{k_i} h_i = 0. \quad (18)$$

The other four relations for h_i together with amplitudes for reflected $h_r e^{ikx+i\omega t}(x < 0)$ and transmitted $h_t e^{-ikx+i\omega t}(x > 1)$ waves are followed from the boundary conditions (12):

$$h_I - h_r = -ik \sum_{i=1}^{6} h_i/k_i;$$
$$h_I + h_r = -k^2 \sum_{i=1}^{6} h_i/k_i^2;$$
$$h_t = -ik \sum_{i=1}^{6} e^{k_i} h_i/k_i = -k^2 \sum_{i=1}^{6} e^{k_i} h_i/k_i{}^2. \quad (19)$$

5. Theoretical Results and Discussion

Floating plate wave barriers have been known and analyzed for decades [5]. Their basic properties are also well known: the length of the plate should be at least comparable to the length of the incoming waves in order to have some chance of suppressing their amplitude and, accordingly, be effective. Too long plates, several times longer than typical energy waves coming to the beach, are not practical and are not very suitable for use. The idea of this study is to consider floating elastic plates with lengths of the same order as the incoming waves. The elastic plate interacts with the surrounding water and has its own life—a set of natural modes of oscillation, depending on the bending stiffness, geometric dimensions, mooring, etc. Our modeling analysis of the coupled fluid-elastic plate system gives the dispersion relation (15) of oscillations of the coating plate, which is characterized by pair of dimensionless parameters $\overline{\beta}$ and $\overline{\gamma}$. The parameter $\overline{\beta}$ determines the elasticity property of the plate and also depends on the length of the plate and its thickness. Another parameter $\overline{\gamma}$ defines the relative role of the depth of fluid on oscillation of the plate.

The main problem that we first analyzed was the following: can bending plate stiffness significantly affect the transmission characteristics of waves. We consider the most intriguing case when the length of incoming wave λ is not far away from length of the plate L = 1.9 m. The results of our numeral simulations are presented in Figure 14a. We fix all other parameters and so only the Young modulus E_0 was widely changed.

Results were quite surprising: dependence of wave transmittance is not monotonic—it has the same minimum for all considered range of waves: $\overline{\beta} \sim 10^{-3}$. Results of simulations for another length of plate L = 5.1 m are presented in Figure 14b. As one can see transmitting coefficient has the same minimum $\overline{\beta} \sim 10^{-3}$. Our experimental results qualitatively correspond to numerical simulations, but all they are in the region of high transmittance and so, not so interesting. The conclusion we made from our two simulations that probably not just specific elasticity itself provides the essential attenuation of waves, but the specific region of the parameter $\overline{\beta} \sim 10^{-3}$. Such a hypothesis needs a further research.

The coefficient of transmission is almost independent from the value of parameter $\overline{\gamma}$ which expresses the influence of water depth. This conclusion is confirmed by experiments.

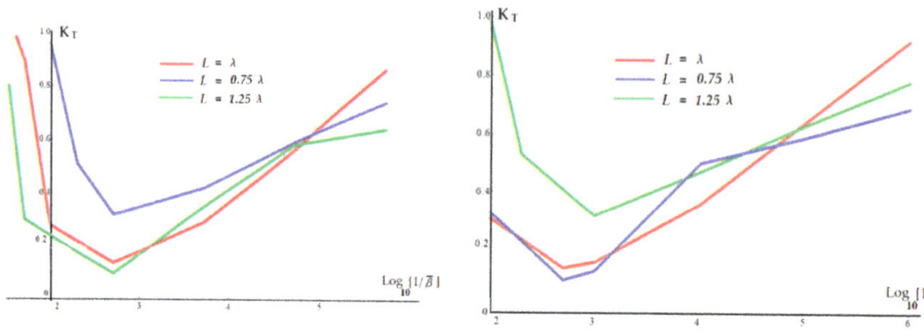

(a) (b)

Figure 14. Wave transmission coefficient K_T in dependence of plate elasticity coefficient $\overline{\beta}$ for different ratios (L/λ) (**a**) length of plate L = 1.9 m, (**b**) length of plate L = 5.1 m. For all cases depth of water a = 0.3 m, thickness of plate H_w = 0.02 m.

Three elastic materials were tested in the experiments: latex, cool cotton and polyvinyl chloride (PVC) with a very different stiffness and two lengths of plates: 1.9 m and 5.1 m. The results of experiments and numerical calculations of transmission coefficient $K_T = H_T/H_I$ in dependence of the relative length of incoming waves for all cases are presented in Figure 15a–d. Red lines and triangles (experimental points) correspond to experiments, blue lines—to results of numerical simulations of the model.

One can see in Figure 15a,b,d some spikes where transmittance suddenly jumps to unit-absolute transmittance. This fact has the following physical explanation: elastic plate with fixed edges (zero boundary conditions) has its own system of eigenvalue modes for discrete set of frequencies. If one of these frequencies coincides with the frequency of incoming wave we will have the situation of total transparency and coefficient of transmittance jumps to unit. This fact was fully confirmed by our numerical simulations. This is a local effect just in one point of frequency spectrum and seems to be more formal than practically observable.

As can be seen from the experiments, a very common property of wave propagation is that its amplitude decreases with increasing plate length. Wave attenuation is ineffective for short elastic plates whose length is much less than the length λ of the incoming wave. Coefficient of wave transmission sharply decreases only when the length of the plate is comparable or more than λ.

The wave transfer coefficient decreases sharply only when the plate length is comparable or greater than λ.

Plate rigidity increases the damping of transmitted waves (see Figure 15a–c) and relative water depth (parameter $\overline{\gamma}$) has no essential influence on wave propagation. Wave attenuation increases with rigidity of the elastic plate (see Figure 16).

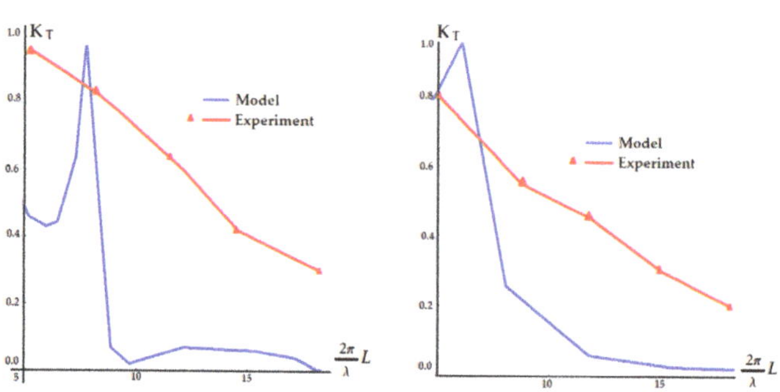

(a) (b)

Figure 15. *Cont.*

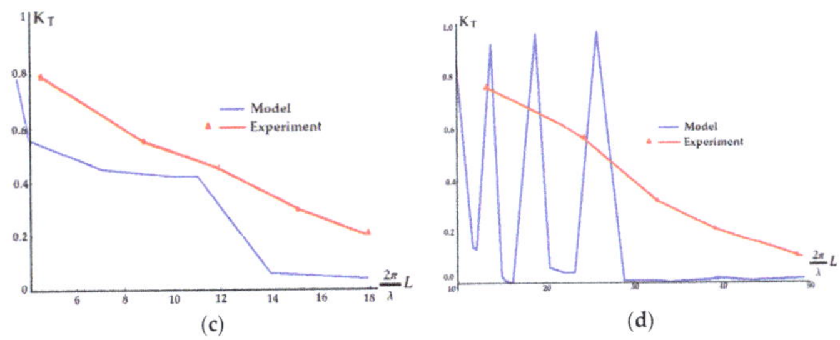

Figure 15. Wave transmission coefficient $K_T = H_T/H_I$ in dependence of $(2\pi L/\lambda)$ ratio for different plate materials (**a**) latex, length of plate L = 1.9 m, $\overline{\beta} \sim 5.6 \times 10^{-6}$. (**b**) polyvinyl chloride (PVC), length of plate L = 1.9 m, $\overline{\beta} \sim 0.02$ (**c**) steep cotton, length of plate L = 1.9 m, $\overline{\beta} \sim 5 \times 10^{-3}$. (**d**) latex, length of plate L = 5.1 m. For all cases depth of water a = 0.3 m, thickness of plate H_w = 0.02 m. Red lines and triangles (experimental points) correspond to experiments, blue lines–to results of numerical simulations of the model.

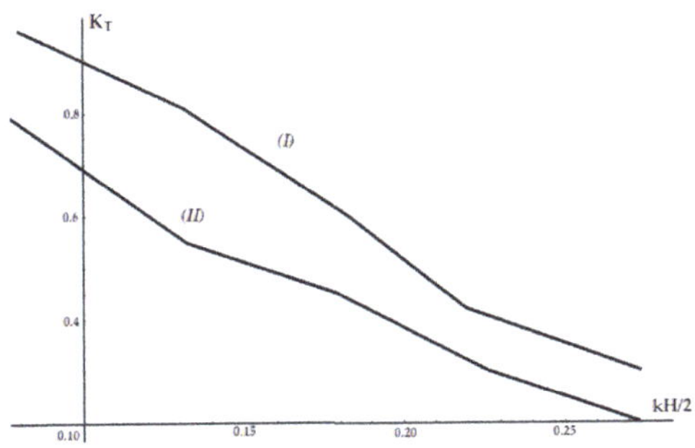

Figure 16. Wave transmission coefficient $K_T = H_T/H_I$ in dependence of incoming wave steepness $kH/2$. Curve (I)—latex, curve (II)—polyvinyl chloride (PVC).

6. Conclusions

From the results of the experiments, it is clear that the elastic membrane covering the surface of the water can have a significant effect of suppressing and reducing waves.

The transmission coefficient is mainly related to the steepness of waves, and the reflection is related to the fixing method and the elastic material of the plate.

The steepness of incoming waves has an essential effect on the transmission, reflection and energy loss.

Different water depths have a relatively small effect on the transmission coefficient. (Intermediate water depth)—the greater the relative water depth generates a slightly more energy loss.

Long plates have a large reflection coefficient and a small transmission coefficient. The longer plate gives a less transmission compared to the short plate. The energy absorption of the whole plate is still higher than that of the short plate.

When both ends are fixed, the transmission coefficient is smaller than other methods, and the reflection coefficient is larger.

The reflection coefficient of rigid components is significantly greater than the other two components, while the transmission coefficients are similar. And the characteristics of the reflection coefficient corresponding to the wave steepness of each plate are also different.

The main difference between elastic and rigid components is reflecting in front of the structure. Since the incident wave energy of the rigid plate is converted into reflection, it may cause the structure to be washed in advance.

Various mounting methods: when both ends are fixed, the transmission coefficient is less than other methods, and the reflection coefficient is greater, the influence caused by the cable force or angle can be further explored.

Our theoretical analysis revealed the region of minimum transmission of waves defined by the value of parameter $\overline{\beta} \sim 10^{-3}$ depending from the elasticity, length and other properties of elastic plate. The region of applicability of such a hypothesis needs further experimental and theoretical efforts.

The simulation showed the existence of a discrete set of frequencies of incoming waves with a full transmittance depending on the characteristics of the elastic plate.

Our theoretical estimates are somewhat inconsistent with the experimental results, perhaps for several reasons:

(1) The model is idealized, does not take into account viscosity, nonlinearity, wave overlap, etc.
(2) On the other hand, the experimental conditions are also not fully realized, for example, the conditions of the fixed edges of the plate, overlapping was observed, and drift and rocking, etc.

Nevertheless, on the basis of the conducted studies, it can be argued that a properly designed wave barrier in the form of a horizontal elastic membrane on the water surface can become an effective shore protection structure.

Author Contributions: Conceptualization, I.S., R.-Y.Y., Y.-Y.C.; methodology I.S., R.-Y.Y., Y.-Y.C.; software, I.S., R.-Y.Y.; validation, I.S., R.-Y.Y.; formal analysis, I.S.; investigation, I.S.; resources, R.-Y.Y.; writing—original draft preparation, I.S.; writing—review and editing I.S., R.-Y.Y., Y.-Y.C.; visualization, R.-Y.Y.; supervision, Y.-Y.C.; project administration, Y.-Y.C.; funding acquisition, R.-Y.Y.; All authors have read and agreed the publication of the manuscript.

Funding: The reported study was funding by RFBR and TUBITAK according to the research project 20-55-46005. The study was also funded by the Ministry of Science and Higher Education of the Russian Federation, theme no. 0149-2019-0005.

Acknowledgments: The authors acknowledge the financial support from Ministry of Science and Technology of Taiwan, under Grant Number MOST 106-2221-E-110-036-MY3 for this study.

Conflicts of Interest: The authors declare no conflict of interest. The funders had no role in the design of the study; in the collection, analyses, or interpretation of data; in the writing of the manuscript, or in the decision to publish the results.

References

1. John, F. On the Motion of Floating Bodies I. *Commun. Pure Appl. Math.* **1949**, *2*, 13–57. [CrossRef]
2. Black, J.L.; Mei, C.C. Scattering and radiation of surface waves by rectangular obstacles in water of finite depth. *J. Fluid Mech.* **1969**, *38*, 499–511.
3. Black, J.L.; Mei, C.C.; Bray, M.C.G. Radiation and scattering of water waves by rigid bodies. *J. Fluid Mech.* **1971**, *46*, 151–164. [CrossRef]
4. Macagno, E.O. Houle dans un canal presentant un passage en charge. *La Houille Blanche* **1954**, *1*, 31–41. [CrossRef]
5. Ruol, P.; Martinelli, L.; Pezzutto, P. Formula to Predict Transmission for -Type Floating Breakwaters. *J. Waterw. Port Coast. Ocean Eng.* **2013**, *139*, 1–8. [CrossRef]
6. Siew, P.; Hurley, D. Long surface waves incident on a submerged horizontal plate. *J. Fluid Mech.* **1977**, *83*, 141–151. [CrossRef]
7. McIver, M. Diffraction of water waves by a moored horizontal, flat plate. *J. Eng. Math.* **1985**, *19*, 297–319. [CrossRef]
8. Sendil, U.; Graf, W. Transmission of regular waves past floating plates. In Proceedings of the International Conference on Coastal Engineering, Copenhagen, Denmark, 24–28 June 1974; p. 14.
9. Arunachalam, V.M.; Raman, H. Experimental studies on a perforated horizontal floating plate breakwater. *Ocean Eng.* **1982**, *9*, 35–45. [CrossRef]
10. Cho, I.; Kim, M. Interactions of a horizontal flexible membrane with oblique incident waves. *J. Fluid Mech.* **1998**, *367*, 139–161. [CrossRef]

11. Stiassnie, M.; Drimer, N. On a Freely Floating Porous Box in Shallow Water Waves. *Appl. Ocean Res.* **2003**, *25*, 263–268. [CrossRef]
12. Liang, N.; Huang, J.; Li, C. A study of spar buoy floating breakwater. *Ocean Eng.* **2004**, *31*, 43–60. [CrossRef]
13. Dong, G.H.; Zheng, Y.N.; Li, Y.C.; Teng, B.; Guan, C.T.; Lin, D.F. Experiments on wave transmission coefficients of floating breakwaters. *Ocean Eng.* **2008**, *35*, 931–938. [CrossRef]
14. Wang, H.Y.; Sun, Z.C. Experimental study of a porous floating breakwater. *Ocean Eng.* **2010**, *37*, 520–527. [CrossRef]
15. Shugan, I.V.; Hwung, H.-H.; Yang, R.-Y.; Hsu, W.-Y. Elastic plate as floating wave breaker in a beach zone. *Phys. Wavew Phenom.* **2012**, *20*, 199–203. [CrossRef]
16. Goda, Y.; Suzuki, T. Estimation of incident and reflected waves in random wave experiments. In Proceedings of the 15th Coastal Engineering Conference, Honolulu, HI, USA, 11–17 July 1976; pp. 828–845.
17. Mansard, E.P.D.; Funke, E.R. The measurement of incident and reflected spectra using a list squares method. In Proceedings of the 17th International Conference on Coastal Engineering, ASCE, New York, NY, USA, 23–28 March 1980; pp. 154–172.
18. Zelt, J.; Skjelbreia, J. Estimating incident and reflected wave fields using an arbitrary number of wave gauges. In Proceedings of the 23rd ICCE, Venice, Italy, 4–9 October 1992; pp. 777–789.
19. Marchenko, A.; Semenov, A. Edge waves of shallow liquid under an elastic plate with a crack. *Izv. Ross. Akad. Nauk.* **1994**, *4*, 185–189.
20. Marchenko, A. Resonance interactions of waves in an ice channel. *J. Appl. Math Mech.* **1997**, *61*, 931–940. [CrossRef]

Journal of
Marine Science
and Engineering

Article

Dynamic Amplification of Gust-Induced Aerodynamic Loads Acting on a Wind Turbine during Typhoons in Taiwan

Tsung-Yueh Lin [1], Chun-Yu Yang [2,*], Shiu-Wu Chau [3] and Jen-Shiang Kouh [3]

1 Research Department, CR Classification Society, Taipei City 104707, Taiwan; tylin@crclass.org
2 Renewable Energy Department, CR Classification Society, Taipei City 104707, Taiwan
3 Department of Engineering Science and Ocean Engineering, National Taiwan University, Taipei 10617, Taiwan; chausw@ntu.edu.tw (S.-W.C.); kouhjsh@ntu.edu.tw (J.-S.K.)
* Correspondence: cyyang@crclass.org; Tel.: +886-2-2506-2711 (ext. 905)

Abstract: Typhoons, such as Soudelor, which caused the collapse of several onshore wind turbines in 2015, pose a considerable challenge to Taiwan's wind energy industry. In this study the characteristics of the aerodynamic loads acting on a wind turbine due to wind gusts in a typhoon are studied with a view to providing a proper definition of the S-Class wind turbine proposed in International Electrotechnical Commission (IEC) 61400-1. Furthermore, based on analysis of wind data during typhoons, as obtained from the meteorological mast in the Zhangbin coastal area, an extreme wind speed and gust model corresponding to the typhoon wind conditions in Taiwan are herein proposed. Finally, the flow fields around a parked wind turbine experiencing both an unsteady gust and a steady extreme wind were simulated by a numerical approach. Numerical results show that the aerodynamic shear force and overturning moment acting on the target wind turbine in a steady wind are significantly lower than those under an unsteady gust. The gust-induced amplification factors for aerodynamic loadings are then deduced from numerical simulations of extreme wind conditions.

Keywords: typhoon; gust; extreme wind; aerodynamic load; numerical simulation

Citation: Lin, T.-Y.; Yang, C.-Y.; Chau, S.-W.; Kouh, J.-S. Dynamic Amplification of Gust-Induced Aerodynamic Loads Acting on a Wind Turbine during Typhoons in Taiwan. *J. Mar. Sci. Eng.* **2021**, *9*, 352. https://doi.org/10.3390/jmse9040352

Academic Editors: Dong-Sheng Jeng and Raúl Guanche García

Received: 23 February 2021
Accepted: 22 March 2021
Published: 24 March 2021

1. Introduction

Taiwan, benefitted by its unique geographical location, has rich wind resources particularly in the region of the Taiwan Strait. This geographical advantage clearly favors large-scale wind farm development, but the frequent incidence of typhoons in Taiwan is a real threat to wind farm safety, having already caused significant damage to installed wind turbines in recent years. For example, Typhoon Soudelor in 2015 collapsed six wind turbines and seriously damaged the blades of a seventh wind turbine in the Taichung wind farm. The maximum wind speed at a nearby meteorological mast was reported as 62.6 m/s by Liu and Chen [1]. This event indicates that a wind turbine may not intactly survive the extreme winds of a typhoon if this scenario is not well-defined. Although both steady and turbulent wind conditions are considered in the Extreme Wind Speed Model (EWM) of the International Electrotechnical Commission (IEC) 61400-1 [2] standard, the corresponding wind characteristics and induced aerodynamic loads on wind turbines under such extreme conditions still require further investigations and inspired this study to address the dynamic effect of gusts during extreme typhoons.

In order to define the proper design requirements for wind turbines operating in typhoon-prone areas, several studies have, in recent decades, investigated the characteristic wind conditions of typhoons, such as wind profile, turbulence intensity, and gust factor, based on measured data. Ishizaki [3] proposed relationships between turbulence intensity, mean wind speed, ground height, and gust factor from a statistical analysis of typhoon measurements. Cao et al. [4] analyzed the wind conditions of Typhoon Maemi through the wind speed samples measured by nine vane-type and seven sonic-type anemometers at a height of 15 m. They found that the turbulence intensity decreases with increasing wind speed and remains almost constant at high wind speeds, with a reported gust factor of 1.6. Clausen et al. [5] proposed a method to characterize tropical cyclones and to derive the structural design wind speed at a given site based on the existing and publicly available cyclone data. Garciano and Koike [6] employed an extreme wind speed delivered by generalized extreme value distribution to estimate the buckling strength by a buckling capacity model recommended by the ISO. They further assessed the probability of buckling failure of a wind turbine considering extreme wind speed distributions from both typhoon-prone and non-typhoon-prone areas, as well as the buckling resistance of a tower as a function of wind speed. They proposed a reference wind speed for wind turbines in typhoon-prone areas based on the annual extreme wind speeds measured at fifty weather stations around the Philippines. Their reference wind speed is about 16% higher than that suggested for wind turbines of Class I.

Several studies have focused on the damage inflicted to wind turbines by typhoon-induced wind loads, with a specific focus on the fluid-structure interactions. Ishihara et al. [7] analyzed the damage to the wind turbines on Miyakojima Island after the onslaught of typhoon Maemi, where the estimated maximum gusts were over 70 m/s. They employed a finite element method (FEM) to forecast the displacement at the tower top together with the bending moment at the turbine foundation. Uchida et al. [8] investigated the cause of blade damage to wind turbines in southern Honshu when Typhoon Melor struck Japan in 2009. The WRF-ARW meteorological model was employed to predict the mesoscale flow behavior, followed by a large eddy simulation (LES) model, i.e., RIAM-COMPACT, to describe the near-field flow features, and, finally, a Reynolds-Averaged Navier–Stokes (RANS) model was coupled with an FEM model to determine the flow field around the blades and the resulting stress on the blades. In a recent study [9], the aerodynamic loads on a 5-MW wind turbine during an extreme gust, modelled in accordance with the IEC 61400-1 standard, were assessed via an unsteady RANS method, and the increase of blade loading was quantified and verified. The dynamic response of this 5-MW wind turbine on a floating platform in irregular seas was investigated via an aero-servo-elastic modeling approach [10]. From the perspective of dynamic analysis, Amaechi et al. [11] analyzed submarine hoses attached to a mooring buoy and suggested a dynamic amplification factor (DAF) of 2 on the environmental loads. Haddadin et al. [12] defined the DAF of a lattice structure as the ratio between the peak total response and the peak quasi-static response, which is the definition adopted in the present study for the aerodynamic response of a wind turbine.

Unfortunately, no local wind characteristics during typhoons in Taiwan are explicitly disclosed in the aforementioned references. For this reason, the present study first applied a generalized extreme value analysis to local wind measurements during typhoons to propose an extreme wind speed and a gust model for the Taiwan region. Figure 1 illustrates the framework of this paper. A measurement-based approach to describe the extreme wind conditions of typhoons is given in Section 2. Then, Section 3 uses this extreme condition to conduct steady and transient RANS simulations, and examines the aerodynamic loads on a parked wind turbine. Because of a lack of in-situ load measurements of a wind turbine in such an extreme typhoon in Taiwan, a benchmark study of the target wind turbine under its rated condition was performed to ensure the reliability of the RANS results. Finally, Section 4 discusses the load amplification factors due to gust effects based on the comparison of the steady and unsteady simulation results. Note that the amplification induced by the turbulent wind in EWM is not performed in the present study, so only the partial dynamic behaviour of this wind turbine is resolved by the simulations.

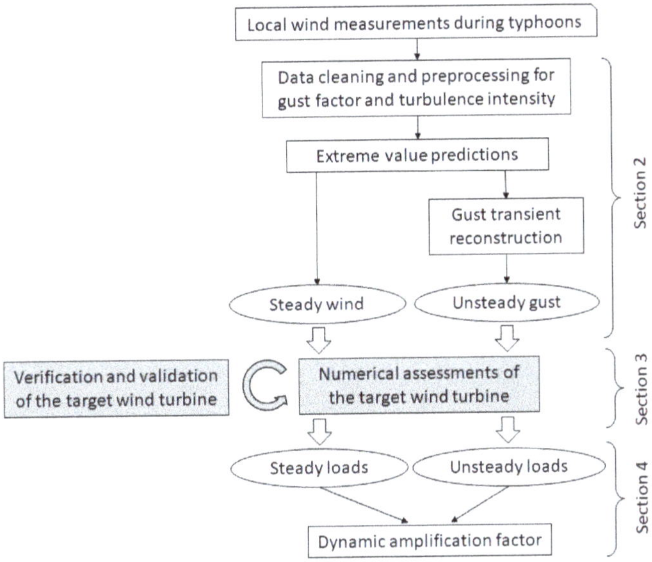

Figure 1. Framework of this study.

2. Typhoon Wind Condition

2.1. Wind Speed Measurement

Wind speed data measured by a meteorological mast located in the Zhangbin coastal area (24°06′ N 120°23′ E), from 2007 to 2015, were analyzed to determine the wind characteristics on the Taiwanese west coast. Vane-type anemometers, with sampling rates of 10 Hz and data logging intervals of one minute, were installed at heights of 10 m, 30 m, 50 m, and 70 m above ground, as shown in Figure 2. The accessible wind speed data consisted of ten-minute average V_{600}, one-minute average V_{60}, the maximum of 3-s averages taken over a minute V_{max}, and the 10-min standard deviation.

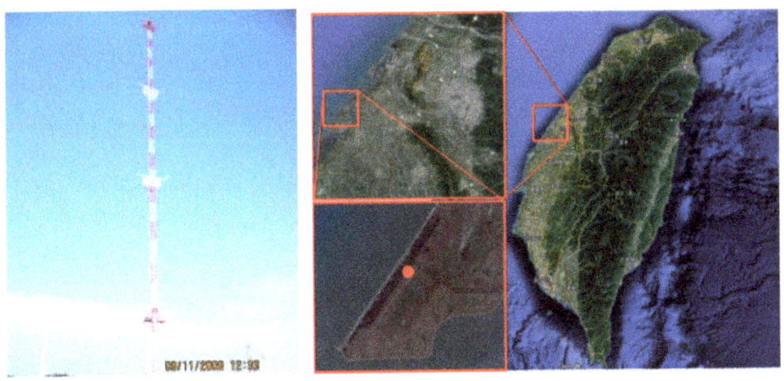

Figure 2. Meteorological mast on the Zhangbin coastal area.

In this study, the turbulence, wind shear, and gust during typhoons were considered. Turbulence is the phenomenon of random fluctuations of flow. The turbulence intensity describes the strength of the turbulence and is defined as the ratio of the wind speed standard deviation to the corresponding average wind speed. In IEC 61400-1, the wind shear is defined as the variation of wind speed with height above ground and is modelled by the power law. Equation (1) [2] and Figure 3 give the mathematical and graphical description of wind shear, respectively, where V is the wind speed, z is the height above the ground, z_{ref} is the reference height, and V_{ref} is the velocity at the reference height.

The power law exponent of the meteorological mast α is deduced as 0.3, according to the regression fit to the anemometers at different heights [13].

$$V(z) = V_{\text{ref}}\left(\frac{z}{z_{\text{ref}}}\right)^{\alpha} \tag{1}$$

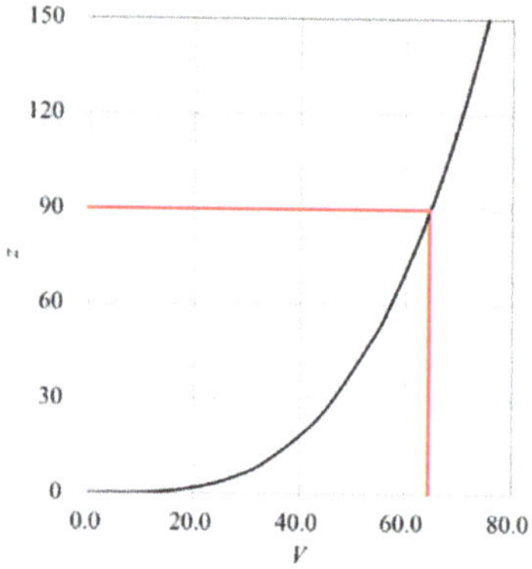

Figure 3. Wind shear profile.

A gust is meteorologically defined as the wind condition where the instantaneous wind speed is at least 5 m/s faster than the corresponding average wind speed. The World Meteorological Organization (WMO) [14] describes gust magnitude in terms of the gust factor GF, which is defined as the ratio of the peak τ-second average wind speed V_τ to the corresponding T_0-second average wind speed V_{T_0}, as shown in Equation (2), where T_0 is a base reference observation period. To comply with the specifications of the data logger in the device, τ was set as 3 s and T_0 was taken as 60 s to compute GF.

$$GF(\tau, T_0) \equiv \frac{V_\tau}{V_{T_0}} \tag{2}$$

Following from the local wind measurements, the gust factor GF and the turbulence intensity I of the recorded data from the meteorological mast in the Zhangbin area during typhoons from 2007 to 2014 are shown in Figure 4a,b. According to Equation (2), the gust factors were calculated from the maximum 3-s average wind speeds and 1-min average wind speeds, and the turbulence intensities were calculated from the wind speed 10-min standard deviations and the 10-min average wind speeds. Since GF and I refer to different averaging durations of wind speed, a correlation between V_{60} and V_{600}, Figure 4c, is further required so that we can convert I into terms of V_{60}, so as to have the same argument as GF.

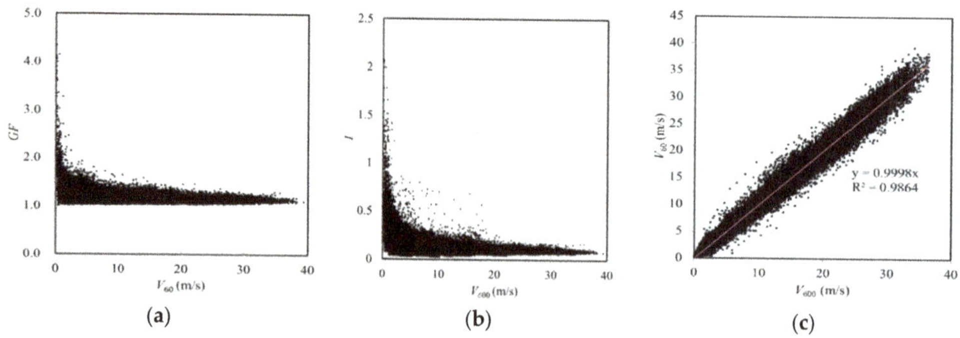

Figure 4. Wind measurements. (**a**) Gust factor; (**b**) turbulence intensity; (**c**) wind speed correlation.

2.2. Extreme Wind Condition

In order to find the extreme condition of the gust factor and turbulence intensity, the gust factor and the turbulence intensity of the recorded data during typhoons were processed by the following steps, as shown in the flowchart of Figure 5:

1. Filter 1: The raw data were firstly filtered by removing zeros and empty fields. High-frequency noise was trimmed and low gusts that did not meet this definition were removed.
2. Grouping: The recorded data were divided into 1 m/s bins, employing the 1-min average wind speeds for gust factors and 10-min average wind speeds for turbulence intensities.
3. Filter 2: The data in each interval were assumed to fit a normal distribution. Unreasonable values and outliers of each interval were filtered by box-whisker plot. This filter performed the calculation of the first (Q1) and third (Q3) quartiles of the gust factor and turbulence intensity, and then the interquartile range (IQR) was computed. Data values that exceeded Q3 by three times IQR or were less than Q1 by three times IQR were identified as outliers and removed.
4. Chi-squared test: The maximum cumulative probability of all the intervals filtered by chi-squared test was applied to estimate the extreme value of each interval by the assumption of normal distribution with a confidence interval of 95%.
5. Fitting: The fitting curve of the estimated extreme values of each wind speed group was calculated by an exponential function with a least-squares fitting algorithm.

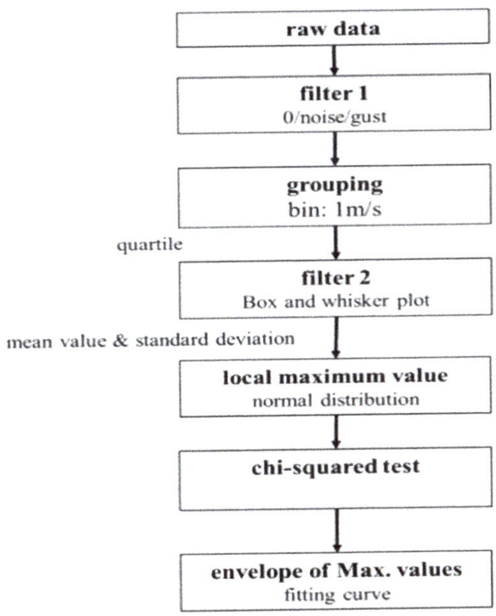

Figure 5. Data processing procedure for wind measurements.

The extreme values of the gust factor and the turbulence intensity during typhoons are shown in Figure 6. The fitting curve for the gust factor at 99.999% extreme values is given by Equation (3), while Equation (4) expresses the fitting curve for 99.99999% extreme values of turbulence intensity.

$$GF = 1 + 8.4177 \cdot {V_{60}}^{-0.9702} \tag{3}$$

$$I = 1.0231 \cdot {V_{600}}^{-0.5715} \tag{4}$$

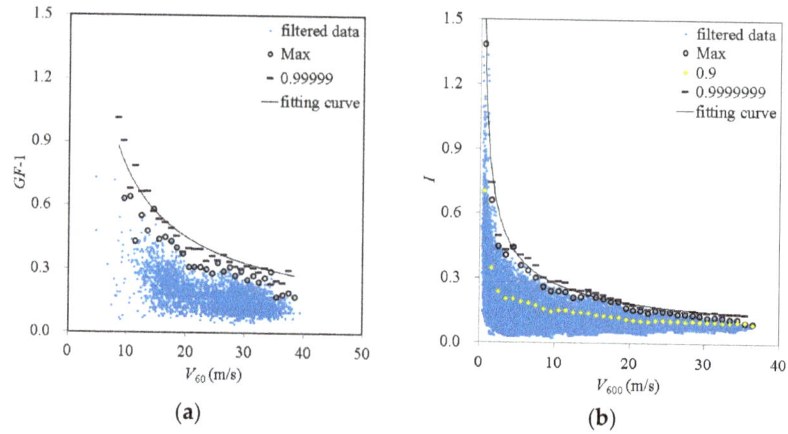

Figure 6. Extreme value analysis during typhoons. (**a**) Gust factor; (**b**) turbulence intensity.

Equations (3) and (4) are the calibrated curves based on local measurement in the Zhangbin area, and the corresponding comparison to the IEC 61400-1 design standards for gust factor and turbulence intensity are shown in Figure 7. One can easily see that the maximum wind speeds during typhoons in the Zhangbin area clearly exceed those suggested in the IEC standards. The present gust model predicts about 20 m/s higher maximum wind speeds. The stronger gust fluctuations might impose higher wind loads on the turbine, and this is the main reason we proceeded further with numerical simulations based on Equation (3), instead of the value defined in IEC standards. Additionally, the turbulence intensities at hub height also differ from values adopted in IEC standards. During typhoons, the turbulence intensity is found higher than any of the design turbulence types at low wind speeds below 30 m/s, but it decreases quickly with increasing wind speed. Taking a wind speed of 36 m/s, for example, the corresponding turbulence intensity is 0.13, which is close to the type B defined in the IEC standards.

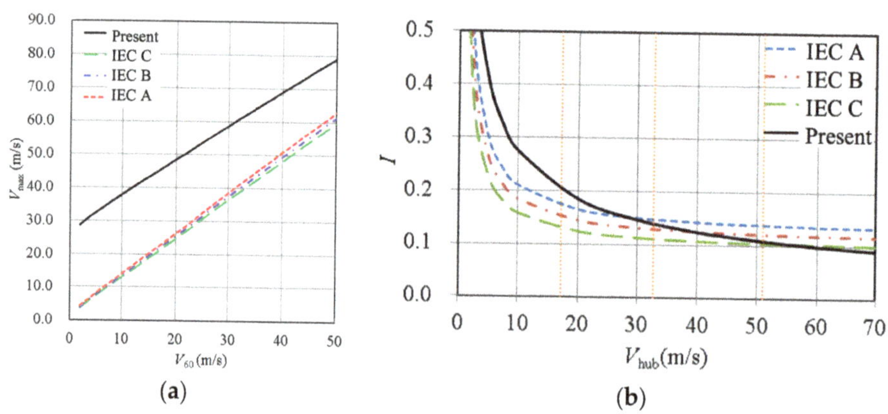

Figure 7. Wind condition comparison between local statistics and design standard. (**a**) Gust factor; (**b**) turbulence intensity.

2.3. Reconstruction of Transient Gust

In the IEC standards, the extreme wind speed is defined as the maximum value of 3-s averages taken over a minute, which was 45.36 m/s as obtained from the typhoon wind measurements in the Zhangbin area. The local gust duration time T, suggested by a related study [15], was 6 s. By utilizing the extreme wind conditions, i.e., Equations (2) and (3), V_{60} was iteratively solved and the corresponding gust factor GF was subsequently determined as 1.26.

There is a difficulty in reconstructing this extreme GF to a transient gust time series, which is explicitly indicated by the extreme wind speed model (EWM) in the IEC 61400-1 standard that unsteadiness should be considered. The anemometer data logger did not record the short transient state of a gust. Furthermore, the extreme operating gust model (EOG) in IEC 61400-1 [2] is only applied to describe the gust for a wind turbine in operating condition, but not in the parked condition, such as during typhoons. To overcome the mismatch, the transient profile of the EOG model, $g(t)$, was adopted, and the model constant K was calibrated by the local measurements of GF obtained in the Zhangbin area, as per Equation (5), where t is time, T is the duration of gust, and V_0 is the initial wind speed.

$$V(t) = \begin{cases} V_0[1 - K \cdot g(t)],\ 0 \le t < T \\ V_0\ ,\ t \ge T \end{cases},\ \text{where } g(t) = \sin\frac{3\pi}{T}t\left(1 - \cos\frac{2\pi}{T}t\right) \tag{5}$$

Then, in order to express K in terms of GF, the peak average wind speed V_τ, and the average wind speed over gust duration, V_T, are subsequently expressed by Equations (6) and (7). Additionally, V_T has to be transformed into the base reference observation period time V_{T_0}, such that it is the same as in the definition of GF, as per Equation (8). Substituting V_τ and V_{T_0} (Equations (6) and (8)) into Equation (2), and expressing the model constant K explicitly yields Equation (9). Finally, using the extreme GF in the Zhangbin area as calculated in the previous section, 1.26, K was calculated by Equation (9) as 0.4.

$$\begin{gathered} V_\tau = \frac{1}{\tau}\int_{\frac{T-\tau}{2}}^{\frac{T+\tau}{2}} V(t)dt = V_0\left(1 - K\frac{G(\tau)}{\tau}\right),\ \text{where} \\ G(\tau) = \int_{\frac{T-\tau}{2}}^{\frac{T+\tau}{2}} g(t)dt = -\frac{T}{\pi}\left(\sin\frac{\pi}{2}\frac{\tau}{T} + \frac{2}{3}\sin\frac{3\pi}{2}\frac{\tau}{T} + \frac{1}{5}\sin\frac{5\pi}{2}\frac{\tau}{T}\right) \end{gathered} \tag{6}$$

$$V_T = \frac{1}{T}\int_0^T V(t)dt = V_0\left(1 + \frac{8}{15\pi}K\right) \tag{7}$$

$$V_{T_0} = \frac{V_0(T_0 - T) + \overline{V}_T T}{T_0} = V_0\left(1 + K\frac{8}{15\pi}\frac{T}{T_0}\right) \tag{8}$$

$$K = -\frac{GF(\tau, T_0) - 1}{\frac{G(\tau)}{\tau} + \frac{8}{15\pi}\frac{T}{T_0}GF(\tau, T_0)} \tag{9}$$

After all the required parameters for Equation (5) were determined, the extreme gust transient profile during typhoons in the Zhangbin area was illustrated, as in Figure 8. The related wind speeds following this extreme gust profile are summarized in Table 1, where V_{max} was used in the steady simulation, i.e., the extreme wind condition. The measured maximum 1-s average wind speed of the Soudelor typhoon in 2015 was reported as 59.4 m/s [1], which agrees well with the present gust model with an error of 3%. With the help of Equation (1), the wind speeds from the meteorological mast height of 70 m were able to be extrapolated to the hub height of 90 m. So, the spatial and temporal distributions of wind speed were known for computing aerodynamic loads on a wind turbine. The turbulence intensity during a typhoon in the Zhangbin area was estimated as 0.13 from Equation (4) as input to the following simulations.

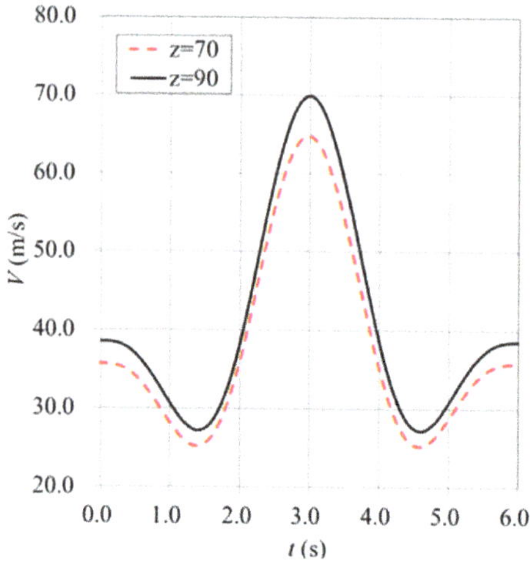

Figure 8. Time history of wind speed in a gust period.

Table 1. Wind speed parameters.

Wind Speed (m/s)	z = 70 m (Met. Height)	z = 90 m (Hub Height)
V_0	35.75	38.55
V_{60}	36.00	38.82
V_3	45.36	48.91
V_1	61.41	66.22
V_{max}	64.85	69.93

3. Wind Load Assessment

3.1. Wind Turbine Model

The target wind turbine is a 5-MW horizontal-axis wind turbine proposed by the National Renewable Energy Laboratory (NREL) [16], and its geometric and operational parameters are given in Table 2. The tilt angle is defined as the angle between the rotor's rotation axis and the ground level (Figure 9a) and the pitch angle is defined as the angle between the rotor's rotation plane and the chord line of the blade tip (Figure 9b). The detailed section profiles for each radial position of the blade are given in [16]. According to the blade geometry and operating conditions of the target wind turbine, the Mach number is smaller than 0.3 and the Reynolds number ranges between 5×10^6 and 1.6×10^7, which justifies the turbulent flow simulation employed in this study.

During typhoons, the wind turbine is designated as "parked". The rotor is then stationary with a blade pitch angle of 90° and the yaw system keeps the wind turbine facing into the wind to reduce the induced aerodynamic loads. The investigated aerodynamic loads acting on the wind turbine are given in Figure 9c, where (F_x,F_y,F_z) denote the resultant forces acting on the wind turbine along three coordinate directions, (Q_x,Q_y,Q_z) denote the resultant moments with respect to three coordinate axes, Q_A denotes the rotor moment with respect to the rotor axis, and (Q_{p1},Q_{p2},Q_{p3}) denote three blade pitch moments with respect to the blade axis.

Owing to a nontrivial tilt angle, the rotor axis is principally not parallel to the incoming wind, so that the angle of attack at each blade section is not constant about the rotor axis. Figure 10 illustrates the section profile in the A-A plane, together with the relative velocity vector triangle, of a blade with azimuthal position θ_a. The change of the angle of attack due to the tilt angle is described by Equation (10) that indicates the change of the angle of attack of the blade section decreasing with an azimuthal angle between 0° and 180° (on the right-hand side of sweeping disk), and increasing with an azimuthal angle between 180° and 360°. Therefore, the distribution of the rotor moment Q_A and pitch moment of the blade (Q_{p1},Q_{p2},Q_{p3}) are expected to be similar to a sine function. The peak loads at certain azimuthal angles are then obtained in the steady simulations.

$$\Delta\alpha = \tan^{-1}\left(\frac{\sin\theta_t \sin\theta_a}{\cos\theta_t}\right) \tag{10}$$

Table 2. Main particulars of the target wind turbine.

Rotor type	Upwind, 3 blades
Diameter (D)	126 m
Hub height (H_{hub})	90 m
Hub diameter (D_{hub})	3 m
Tilt angle (θ_t)	5°
Pitch angle (θ_p)	Operating: 0°~23.5°; Parked: 90°
Rated wind speed	11.4 m/s
Rated rotor speed	12.1 rpm

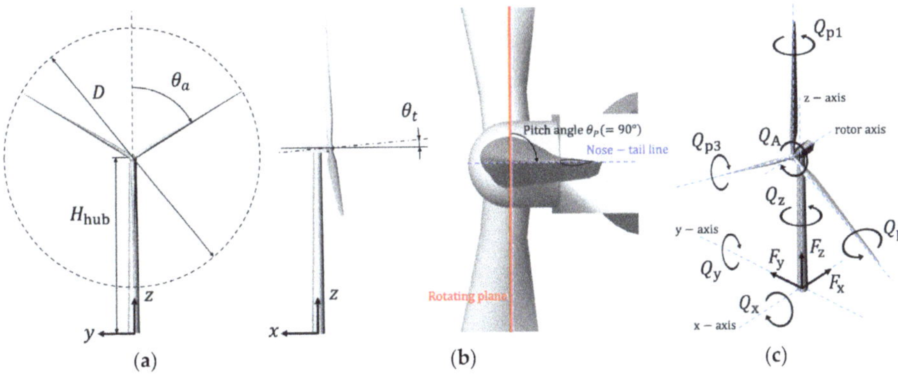

Figure 9. National Renewable Energy Laboratory (NREL) 5-MW wind turbine. (**a**) Tilt angle; (**b**) blade pitch angle; (**c**) aerodynamic loads.

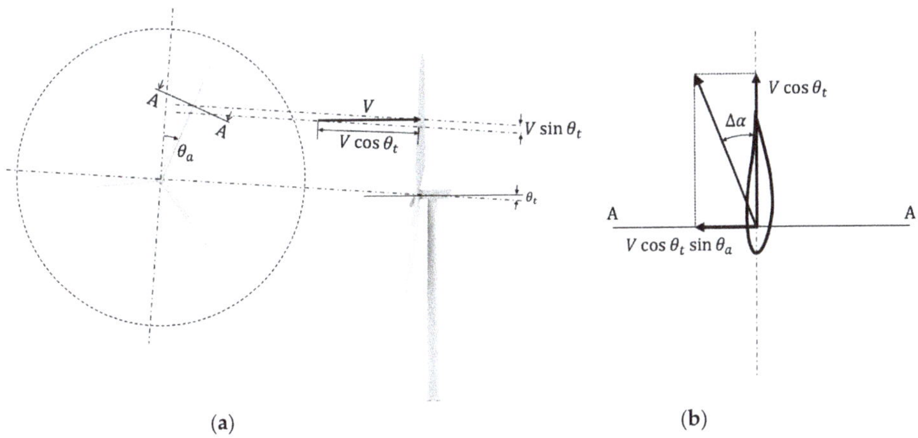

Figure 10. Change of angle of attack due to tilt in a blade rotation. (**a**) Rotor view; (**b**) section view.

3.2. Numerical Method

The characteristics of the aerodynamic loads acting on a wind turbine during typhoons were studied by numerical simulations. The flow solver ANSYS Fluent [17], a general-purpose finite volume and RANS code, was applied to simulate the flow field around the target wind turbine. The wind conditions during a typhoon were described by the results of the statistical analysis of the recorded data from the meteorological mast in the Zhangbin area. The flow around wind turbines is considered an incompressible and turbulent flow. The corresponding governing equations for conservation of mass and momentum are, respectively, as per Equations (11) and (12), where u_i is the mean velocity component in the x_i direction, p is the pressure, ρ and μ are the density and viscosity of air, respectively, $\rho\overline{u_i'u_j'}$ is the Reynolds stress, and ρb_i is the gravitational force component in the x_i direction. The SST k-ω model, as per Equations (13) and (14), has been widely accepted for simulating flow past airfoils [9] and was applied in this study to compute the Reynolds stress term in Equation (12), where k is the turbulent kinetic energy, ω is the dissipation rate of k, τ_{ij} is the shear stress, μ_t is the turbulent viscosity, and (β^*, σ_k, γ, β, σ_ω, δ, $\sigma_{\omega 2}$) are the equation constants.

$$\frac{\partial u_i}{\partial x_i} = 0 \tag{11}$$

$$\frac{\partial u_i}{\partial t} + \frac{\partial}{\partial x_j}(u_i u_j) = -\frac{1}{\rho}\frac{\partial p}{\partial x_i} + \frac{1}{\rho}\frac{\partial}{\partial x_j}\left[\mu\left(\frac{\partial u_i}{\partial x_j} + \frac{\partial u_j}{\partial x_i}\right)\right] - \frac{\partial\left(\overline{u_i'u_j'}\right)}{\partial x_j} + b_i \tag{12}$$

$$\frac{\partial k}{\partial t} + \frac{\partial}{\partial x_j}(u_j k) = \frac{1}{\rho}\tau_{ij}\frac{\partial u_i}{\partial x_j} - \beta^*\omega k + \frac{1}{\rho}\frac{\partial}{\partial x_j}\left[(\mu + \sigma_k\mu_t)\frac{\partial k}{\partial x_j}\right] \tag{13}$$

$$\frac{\partial \omega}{\partial t} + \frac{\partial}{\partial x_j}(u_j\omega) = \frac{\gamma}{\mu_t}\tau_{ij}\frac{\partial u_i}{\partial x_j} - \beta\omega^2 + \frac{1}{\rho}\frac{\partial}{\partial x_j}\left[(\mu + \sigma_\omega\mu_t)\frac{\partial \omega}{\partial x_j}\right] + 2(1-\delta)\sigma_{\omega 2}\frac{1}{\omega}\frac{\partial k}{\partial x_j}\frac{\partial \omega}{\partial x_j} \tag{14}$$

The computational domain of the numerical model is shown in Figure 11a. There is a distance of two times the rotor diameter between the upstream boundary and the wind turbine, six times the rotor diameter between the wind turbine and the downstream boundary, 1.5 times the rotor diameter between the wind turbine and each of the side boundaries, and 3 times the rotor diameter between the bottom and top boundaries, as listed in Table 3. Figure 11b illustrates the locations for the adopted boundary conditions. The bottom boundary and the wind turbine surface are defined as no-slip walls. The upstream, top, and side faces are defined as inlet, where the inflow condition is specified according to the calibrated gust condition. Assuming that the distance between the wind turbine and the downstream boundary is far enough, a vanishing velocity gradient is employed at the outlet. The boundary conditions for velocity are summarized in Table 4.

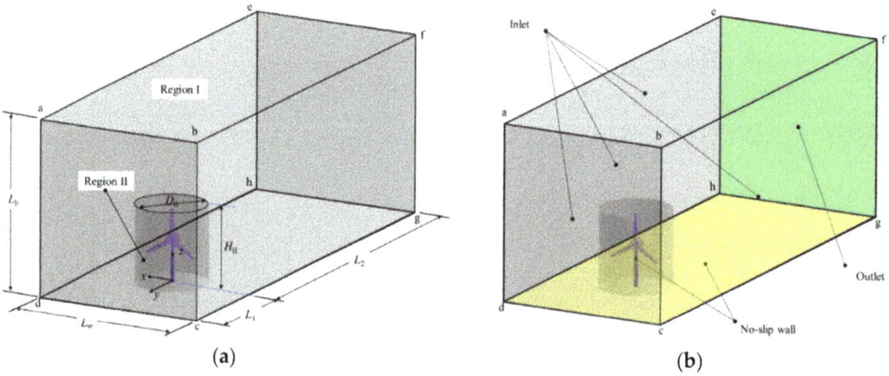

Figure 11. (**a**) Computational domain; (**b**) boundary conditions.

Table 3. Geometrical parameters of computational domain.

L_h	L_w	L_1	L_2	D_{II}	H_{II}
$3D$	$3D$	$2D$	$6D$	$1.27D$	$1.31D$

Table 4. Boundary conditions.

Boundary	Type	Velocity Condition
abcd (upstream)	Inlet	Velocity specified
abfe, *aehd*, *bcgf* (far field)	Inlet	Velocity specified
ghef (downstream)	Outlet	Zero velocity gradient
cdhg (ground)	No-slip wall	Zero velocity
Wind Turbine Surface	No-slip wall	Zero velocity

3.3. Verification and Validation

As the study utilizes a computational model, the verification process is essential to mitigate modelling errors within tolerance. For the purpose of verification, a steady wind speed of 60 m/s at hub height with the exponential profile $\alpha = 0.3$ was used. The second-order upwind scheme was used for spatial discretization. Since the real geometry of the wind turbine is considered, a multi-block unstructured body-fitted grid arrangement was employed. The grid distribution is shown in Figure 12, where a refined cylindrical region was generated around the turbine. Several prismatic layers were fit around the body surfaces, and the thickness of the first layer was tuned to achieve y^+ less than 10 for the requirement of the selected SST turbulence model. A grid dependence test, consisting of five levels of systematic grid refinement, was conducted to verify that the solution is numerically convergent. The aerodynamic moment Q_y about the tower base was selected as the target function in terms of grid number. Figure 13a displays the dependence of Q_y on grid number, where Q_y approached a constant value with an increasing grid number. The discretization error E for a method of second-order accuracy is estimated through Richardson's extrapolation, where dx denotes the dimensionless grid size. Figure 13b indicates that the model with grid number over 10 million is capable of reaching a discretization error less than 2%.

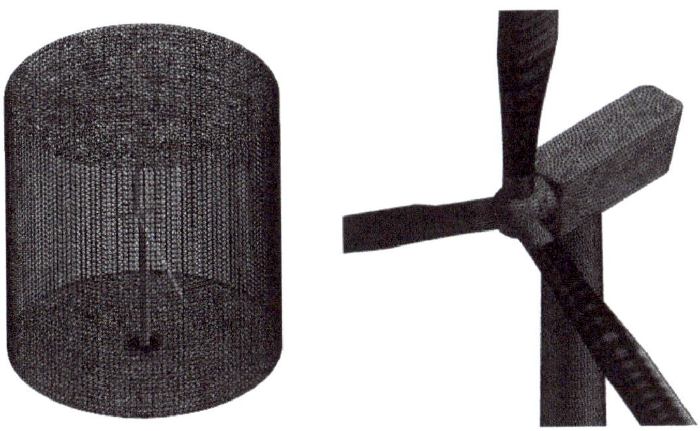

(**a**) (**b**)

Figure 12. Body-fitted grid for steady wind case. (**a**) Refined cylindrical region; (**b**) surface grid.

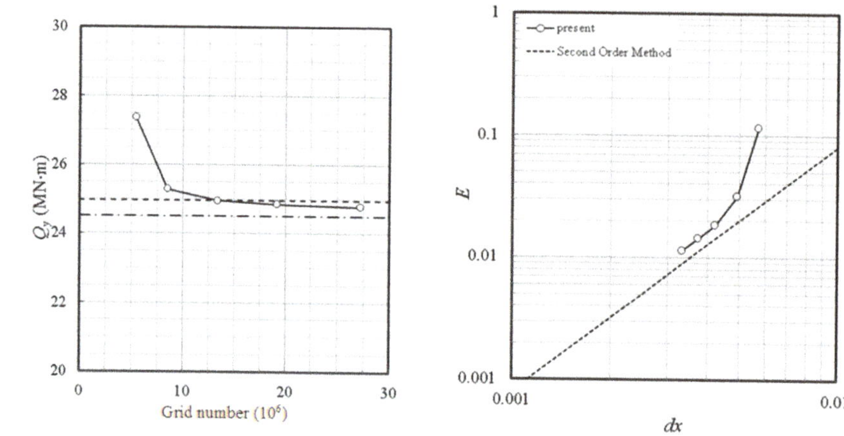

(a) (b)

Figure 13. (**a**) Dependence of aerodynamic load on grid number; (**b**) discretization error.

The proposed numerical model is then validated for the rotor torque Q_A at rated condition, namely wind speed 11.4 m/s, rotor speed 12.1 rpm, and zero blade pitch angle where an unsteady simulation was performed. The time step was 5 ms and it corresponded to 1000 steps per revolution or a blade rotation of 0.36° per time step. The time step was small enough to accurately resolve the temporal variation of wind turbine aerodynamics. A sliding mesh approach was applied to model the rotating rotor disk to cope with the relative motion between the rotor and tower in the unsteady flow simulation. Figure 14a depicts the evolution of Q_A with respect to time. After a simulation time of 40 s, i.e., 10 turns of the rotor, the rotor torque converges. The mean value of Q_A over the 11th revolution of the rotor was 4.09 MN-m, which was about 2% less than the benchmark value of 4.18 MN-m. Figure 14b shows the velocity contours and vectors on the midplane between the rotor plane and the tower as well as the pressure contours on the blades for the blade configuration of 0°–120°–240°. The axial wake region induced by the rotor is clearly seen and a radial wake expansion is implied by outward vectors across the rotor disk boundary.

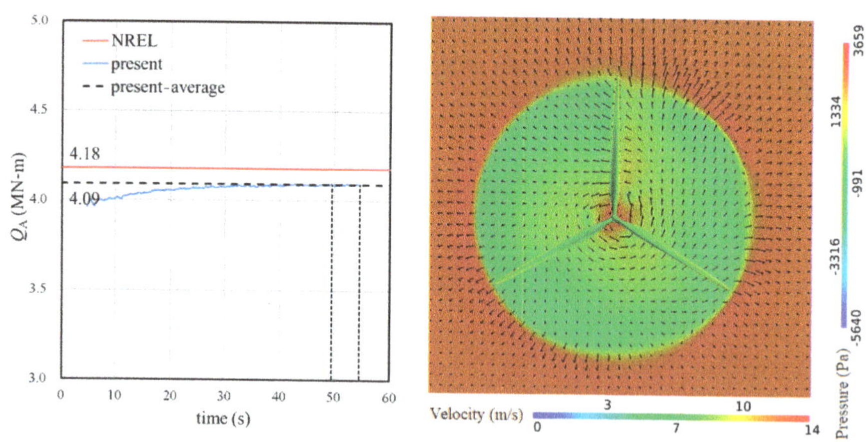

(a) (b)

Figure 14. (**a**) Rotor torque at rated condition; (**b**) axial velocity contour and in-plane vector on the midplane between the rotor plane and the tower, and pressure contour on the blade surface for the blade configuration of 0°–120°–240°.

4. Results and Discussion

To investigate the characteristics of the aerodynamic loads acting on the wind turbine, several cases with different wind conditions were simulated. The predicted cases were separated into two categories: in the first category, the flow field around the parked wind turbine was simulated with constant extreme wind speed. These simulations were conducted in steady mode since the rotor was not moving and a mean wake was of main interest. Flow separation may occur behind the tower, but it is expected to be insignificant, especially in high wind conditions. In the other category, the flow field around the parked target wind turbine was simulated with a gust in transient mode. To enhance the numerical stability of the transient simulations, a precursor calculation of constant wind speed was first conducted. Then, one gust period with a varying wind speed was simulated using the precursor flow field as an initial condition. Four blade configurations were studied in the steady simulation, while only one blade configuration was investigated in the gust simulation.

4.1. Steady Aerodynamic Loadings

The simulation results of steady extreme wind speed are summarized in Table 5. The drag force F_x acting on the tower as well as rotor is insignificantly impacted by the blade configuration, except the case of 60°–180°–300° when the tower is directly shadowed by a blade. In this case the rotor shows a lower aerodynamic loading because of a small mean blade height implying a low incoming wind speed along with zero angle of attack, whereas the tower loading slightly grows due to a local flow acceleration due to the blockage effect of the blade. However, the total drag of the tower and rotor delivers a similar magnitude. The rotor torque Q_A is relatively sensitive to the blade configuration and fluctuates in the range between 1.2 MN-m and 5.0 MN-m where the blade configuration of 60°–180°–300° clearly shows the smallest value benefitted from its advantage in mean blade height and effective angle of attack. The blade pitch moment Q_P is approximately −0.167 MN-m in average, where the minus sign indicates aerodynamic loading tending to decrease the blade pitch angle, i.e., to lead to an unfavorable blade position to acquire higher blade pitch moment. In the range of $0^\circ \leq \theta_a \leq 180^\circ$ the blade pitch moment generally declines with the blade azimuthal angle, while the blade pitch moment grows with the blade azimuthal angle for $180^\circ \leq \theta_a \leq 360^\circ$. As expected, the blade pitch moment at $\theta_a = 180^\circ$ is the smallest among all studied blade positions due to a small local wind speed as well as angle of attack. The wind turbine suffers from a higher overturning risk along the y axis than along the x axis because Q_y is much higher than its counterpart Q_x that constantly changes its direction in a rotor rotation period. Additionally, Q_y varies little among different blade configurations and has a mean value of 32.3 MN-m. The overturning moment along the y axis is mainly contributed by the rotor and tower where they deliver a comparative importance in Q_y. The twisting moment Q_z is mostly governed by the rotor but its magnitude is about one-order of magnitude smaller than Q_x and approximately two-order of magnitude less than Q_y.

4.2. Dynamic Amplification

The gust simulation results for the blade configuration of 0°–120°–240° are shown in Table 6, which lists the aerodynamic loading at the time of peak wind speed, i.e., t = 3 s, as well as the time instance of the maximum loading during the unsteady gust. Wind profile is depicted in Figure 15, which also shows the low speed wake behind the blades and tower. The pressure contour at t = 2 s is plotted on the right-hand side of Figure 15 to illustrate a positive pressure gradient along the wind direction during the acceleration phase of the gust.

Table 5. Wind loads under steady extreme wind.

Load (kN/MN-m)	Part	Blade Configuration			
		0°–120°–240°	30°–150°–270°	60°–180°–300°	90°–210°–330°
	Tower	276.0	280.3	306.0	284.3
F_x	Rotor	172.5	175.7	146.6	177.3
	Total	451.6	458.9	454.5	465.5
Q_A	Rotor	3.80	5.00	1.20	2.50
	Blade 1	−0.20	−0.23	−0.23	−0.21
Q_P	Blade 2	−0.17	−0.17	−0.10	−0.12
	Blade 3	−0.13	−0.12	−0.13	−0.17
	Tower	−3.7	−1.6	−10.8	−1.2
Q_x	Rotor	−1.0	5.2	1.8	4.0
	Total	−5.0	3.7	−9.1	3.1
	Rotor	18.1	18.2	15.8	17.8
Q_y	Tower	13.8	13.9	15.9	14.2
	Total	32.3	32.5	31.9	32.5
	Rotor	0.60	1.30	−0.09	0.50
Q_z	Tower			<0.001	
	Total	0.60	1.30	−0.06	0.60

Table 6. Wind loads under unsteady extreme gust for the blade configuration of 0°–120°–240°.

Load (kN/MN-m)	Part	At t = 3 s	Maximum	t_{max} (s)
	Tower	400.3	458.5	2.70
F_x	Rotor	209.8	223.0	2.86
	Total	621.3	693.8	2.74
Q_A	Rotor	5.11	5.31	2.87
	Blade 1	−0.19	−0.19	3.01
Q_P	Blade 2	−0.19	−0.19	3.00
	Blade 3	−0.16	−0.16	2.93
Q_x	Total	0.76	−1.78	2.28
Q_y	Total	39.16	42.04	2.80
Q_z	Total	0.80	0.84	3.06

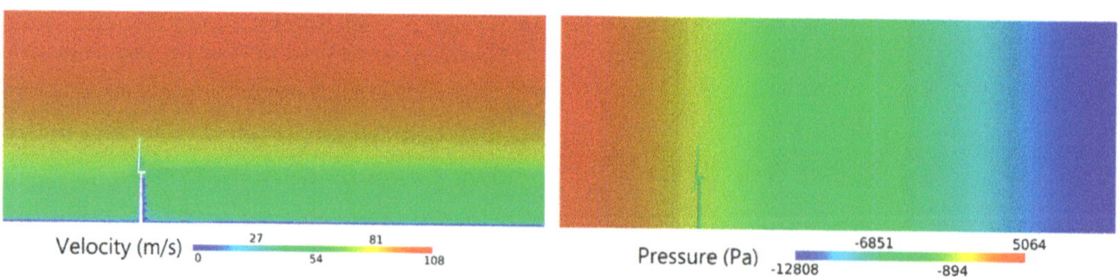

Figure 15. Velocity and pressure contours on the centerplane of wind field at t = 2.0 s.

In order to investigate the correlation between the aerodynamic loads and wind speed in the gust cases, the amplification factor, DAF, is defined as follows [12]:

$$DAF = \frac{\phi}{|\phi|_{max}} \tag{15}$$

where ϕ denotes any physical quantity, and $|\phi|_{max}$ is the maximum absolute value of the physical quantity ϕ. Both the rotor and tower drags reach their peak values somewhat in advance of the 3-s peak. Figure 16a compares the time history of different drag components in a gust period where top represents the contribution of rotor-nacelle assembly. The aerodynamic loading is found to principally mimic the wind speed variation. The tower drag force plays a major role in the total drag while the contribution of nacelle is relatively trivial. The rotor torque Q_A behaves similar to the rotor drag force and Q_A precedes the gust peak by 0.13 s. Different from the rotor torque, the peak blade pitch moments echo the arrival of the top wind speed. Figure 16b displays the blade pitch moment of the blade configuration of 0°–120°–240°. The wind unsteadiness interestingly leads to a more uniform peak blade pitch moment among the three blades accompanied by a slight increase in the mean value. The overturning moment Q_y and Q_x show a leading phase of the gust peak while the phase of Q_z lags the gust peak by 0.16 s. The phase inconsistency between the maximum aerodynamic loadings and the gust peak is a contribution of time terms in the momentum equations. In the unsteady flow the flow velocity as well as its time derivative contributes to the stagnation pressure whereas the contribution of time derivative vanishes in the steady case. Figure 17 depicts the normalized profiles of wind velocity, the time derivative of wind velocity and the drag force experienced by the wind turbine in a gust period where the superscript * represents the physical quantity normalized by its maximum value in a gust period. Figure 17 indicates that the wind velocity just experiences a sharp deceleration process as the flow velocity slowly reaches the gust peak. The combining effect of a gradually rising wind velocity and a steep velocity gradient in time clearly leads to a peak-load offset away from the time instance of gust peak, especially the aerodynamic load is governed by the stagnation pressure. For the aerodynamic load mainly stemming from viscous shear, such as the blade pitch moment and the twisting moment, this offset behavior is less unnoticeable.

Table 7 summarizes the amplification factor of aerodynamic loads given in Tables 5 and 6. The ratio of the peak tower drag in gust to the steady tower drag is approximately proportional to the square of the gust factor, GF, but the corresponding ratio for rotor drag varies quite linearly with the gust factor. This is explained by the blunt shape of the tower and the streamlined geometry of the rotor blade where the former is governed by stagnation pressure characterized by the square of flow velocity and the latter is dominated by viscous shear governed by the velocity gradient. The rotor torque Q_A is mainly contributed by the lateral force exerted on the blades where pressure and shear components are both important in the head wind condition. This accounts for an amplification factor of 1.4 falling between GF (shear dominant) and GF^2 (pressure dominant). The unsteady amplification of blade pitch moment is not very significant where only 10% increase is found for this blade configuration (0°–120°–240° or λ-shaped). The gust exhibits a positive impact on Q_x with a peak reduction by 64% in a λ-shaped blade configuration. Contrary to Q_x, the extreme moments along other two axes, i.e., Q_y and Q_z, are governed by the shear effect where the corresponding amplification factor is close to the gust factor.

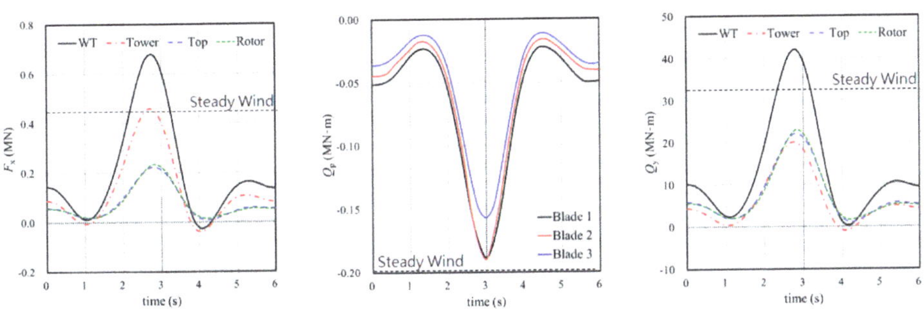

(**a**) (**b**) (**c**)

Figure 16. Time history of wind load for the blade configuration of 0°–120°–240°: (**a**) F_x; (**b**) Q_p; (**c**) Q_y.

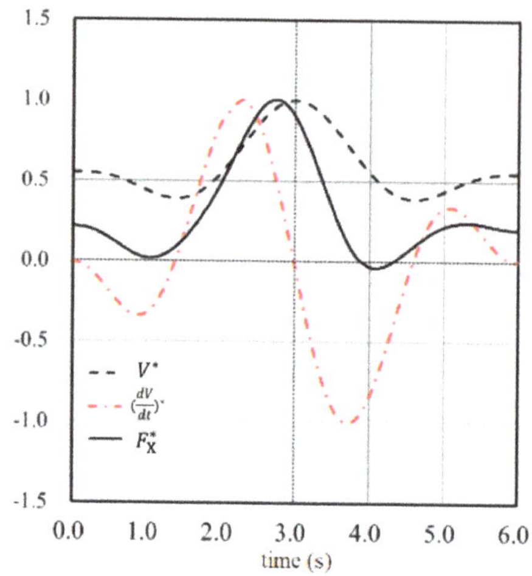

Figure 17. Time history of amplification factor for F_x.

Table 7. Dynamic amplification factor for the wind turbine with a blade configuration of 0°–120°–240°.

Load (kN/MN-m)	Gust	Steady Wind	*DAF*
F_x	693.8	451.6	1.54
Q_A	5.31	3.80	1.40
Q_P	0.18	0.167	1.10
Q_x	1.78	5.00	0.36
Q_y	42.04	32.27	1.30
Q_z	0.84	0.62	1.35

5. Conclusions

Taiwan, due to its geographical location, has an excellent wind conditions and high potential for wind power production as well as strong typhoons which have damaged several wind turbines in recent years. The aerodynamic loads during typhoons were investigated by using the extreme wind conditions prescribed in IEC 61400-1 together with a measurement-calibrated gust factor. The extreme wind conditions proposed in the present study were based on meteorological measurements from the Zhangbin area during typhoon invasions from 2007 to 2014. A statistical approach was employed to convert the raw data into a fitting formula to quantitatively describe the extreme wind conditions for typhoons. Our study shows that the 1-min average wind speed was 36 m/s, the maximum 3-s average wind speed was 45.36 m/s, and the maximum instantaneous wind speed was 64.85 m/s, where the maximum instantaneous wind speed in Taiwan was about 20 m/s higher than the values suggested in the IEC standards. Additionally, the gust factor for typhoons was estimated at 1.26 with a gust period of 6 s. This local gust factor was used to calibrate the EOG model constant, such that a transient gust time series was reconstructed for the unsteady simulation. The turbulence intensity during typhoons was found to be higher than those of turbulence types defined in the IEC standards at low wind speeds below 30 m/s, but it declines quickly with the growth of wind speed.

The aerodynamic loads acting on the NREL 5-MW wind turbine in a steady extreme wind condition as well as an unsteady extreme gust were studied by RANS simulations. The simulation results show that the extreme aerodynamic loads acting on the wind turbine are obviously underestimated when the extreme wind condition only adopts a constant wind speed. The amplification factor of aerodynamic loads is predicted as follows: 1.54 for the drag force, 1.4 for the rotor torque, 1.3 for the overturning moment along the y direction,

1.1 for the blade pitch moment, and the yawing moment by 1.35 times. The amplification factor is governed by the nature of aerodynamic loading. It is close to the square of the gust factor in a stagnation-pressure dominant case while it approximates to the gust factor when viscous shear plays a major role. The quantitative amplification factors may not be universal for different wind turbines in different areas because the present study is limited to the Taiwan local wind conditions and this specific wind turbine. With a concern of lacking validation under extreme typhoons in this study, more load measurements of in-situ wind turbines are expected in the future. Lastly, it is suggested to conduct future studies on the dynamic amplification of this reference turbine by adopting the proposed extreme turbulence intensity in EWM, such that the full aerodynamics can be resolved and compared.

Author Contributions: Conceptualization, C.-Y.Y. and J.-S.K.; methodology, C.-Y.Y., T.-Y.L. and S.-W.C.; writing—original draft preparation, C.-Y.Y.; writing—review and editing, T.-Y.L. and S.-W.C.; visualization, C.-Y.Y. and T.-Y.L. All authors have read and agreed to the published version of the manuscript.

Funding: This work was funded by the Ministry of Science and Technology (MOST), Taiwan [grant No. 109-3116-F-006-013-CC1] and [grant No. 109-2221-E-002-090-].

Institutional Review Board Statement: Not applicable.

Informed Consent Statement: Not applicable.

Data Availability Statement: Not applicable.

Acknowledgments: The authors would like to thank Bryan Nelson, a wind energy specialist at CR Classification Society, for English proofreading.

Conflicts of Interest: The authors declare no conflict of interest.

References

1. Liu, J.-H.; Chen, J.-C. Typhoon wind condition analysis—A case study to the 2 MW wind turbine during the typhoon soudelor in 2015. In Proceedings of the 15th World Wind Energy Conference and Exhibition, Tokyo, Japan, 31 October–1 November 2016.
2. *International Standard: Wind Turbines Part 1: Design Requirements*; International Electrotechnical Commission: Geneva, Switzerland, 2005.
3. Ishizaki, H. Wind profiles, turbulence intensities and gust factors for design in typhoon-prone regions. *J. Wind Eng. Ind. Aerodyn.* **1983**, *13*, 55–66. [CrossRef]
4. Cao, S.; Tamura, Y.; Kikuchi, N.; Saito, M.; Nakayama, I.; Matsuzaki, Y. Wind characteristics of a strong typhoon. *J. Wind Eng. Ind. Aerodyn.* **2009**, *97*, 11–21. [CrossRef]
5. Clausen, N.E.; Ott, S.; Tarp-Johnsen, N.J.; Norgard, R.; Guo, X.L. Design of wind turbine in an area with tropical cyclones. In Proceedings of the European Wind Energy Conference and Exhibition, Athens, Greece, 27 February–2 March 2006.
6. Garcino, L.E.O.; Koike, T. New reference wind speed for wind turbine in typhoon-prone areas in the Philippines. *J. Struct. Eng.* **2010**, *136*, 463–467. [CrossRef]
7. Ishihara, T.; Yamaguchi, A.; Takahara, K.; Mekaru, T.; Matsuura, S. An analysis of damaged wind turbines by typhoon Maemi in 2003. In Proceedings of the sixth Asia-Pacific Conference on Wind Engineering, APCWE-VI, Seoul, Korea, 12–14 September 2005.
8. Uchida, T.; Maruyama, T.; Ishikawa, H.; Zako, M.; Deguchi, A. Investigation of cause of wind turbine blade damage at Shiratakiyama wind farm in Japan, a computer simulation based approach. In *Reports of Research Institute for Applied Mechanics*; Kyushu University: Fukuoka, Japan, 2011; pp. 13–25.
9. Länger-Möller, A. Simulation of transient gusts on the NREL 5MW wind turbine using the URANS solver THETA. *Wind Energy Sci.* **2018**, *3*, 461–474. [CrossRef]
10. Jonkman, J.M. *Dynamics Modeling and Loads Analysis of an Offshore Floating Wind Turbine*; Technical Report NREL/TP-500-41958; National Renewable Energy Laboratory: Golden, CO, USA, 2007.
11. Amaechi, C.V.; Wang, F.; Hou, X.N.; Ye, J.Q. Strength of submarine hoses in Chinese-lantern configuration from hydrodynamic loads on CALM buoy. *Ocean Eng.* **2019**, *171*, 429–442. [CrossRef]
12. Haddadin, S.; Aboshosha, H.; El Ansary, A.; El Damatty, A. Sensitivity of wind induced dynamic response of a transmission line to variations in wind speed. In Proceedings of the Conference of Canadian Society of Civil Engineers—Resilience Infrastructure, London, UK, 1–4 June 2016.
13. Pu, Z.R. A Study of Typhoon Wind Speed Profile's Parameters in Taiwan. Master's Thesis, Department of Harbor and River Engineering, National Taiwan Ocean University, Zhongzheng, Taiwan, 2008.

14. Harper, B.A.; Kepert, J.D.; Ginger, J.D. *Guidelines for Converting between Various Wind Averaging Periods in Tropical Cyclone Condition*; World Meteorological Organization: Geneva, Switzerland, 2010.
15. Lin, P.H.; Tu, C.T. The characteristics of gust wind at western coast of Taiwan during typhoon weather condition—An example from typhoon. In Proceedings of the Taiwan Wind Energy Conference, Penghu, Taiwan, 17 December 2010.
16. Jonkman, J.; Butterfield, S.; Musial, W.; Scott, G. *Definition of a 5-MW Reference Wind Turbine for Offshore System Development*; Technical Report NREL/TP-500-38060; National Renewable Energy Laboratory: Golden, CO, USA, 2009.
17. *Ansys®FLUENT*; Release 16—User's Guide; ANSYS, Inc.: Canonsburg, PA, USA, 2015.

MDPI
St. Alban-Anlage 66
4052 Basel
Switzerland
Tel. +41 61 683 77 34
Fax +41 61 302 89 18
www.mdpi.com

Journal of Marine Science and Engineering Editorial Office
E-mail: jmse@mdpi.com
www.mdpi.com/journal/jmse

www.ingramcontent.com/pod-product-compliance
Lightning Source LLC
Chambersburg PA
CBHW061820210926
51850CB00006B/763